ASIAN POLITICAL, ECONOMIC AND SECURITY ISSUES

ASSOCIATION OF SOUTHEAST ASIAN NATIONS (ASEAN) AND U.S. INTERESTS

ASIAN POLITICAL, ECONOMIC AND SECURITY ISSUES

Additional books in this series can be found on Nova's website under the Series tab.

Additional e-books in this series can be found on Nova's website under the e-books tab.

ASIAN POLITICAL, ECONOMIC AND SECURITY ISSUES

ASSOCIATION OF SOUTHEAST ASIAN NATIONS (ASEAN) AND U.S. INTERESTS

MARGARET E. STAMLIN
EDITOR

New York

LIBRARY OF CONGRESS CATALOGING-IN-PUBLICATION DATA

Association of Southeast Asian Nations (ASEAN) and U.S. interests / editor, Margaret E. Stamlin.
p. cm.
Includes index.
ISBN: 978-1-62417-982-2 (softcover)
1. Southeast Asia--Foreign economic relations--United States. 2. United States--Foreign economic relations--Southeast Asia. 3. Southeast Asia--Economic integration. I. Stamlin, Margaret E. II. ASEAN.
HF1591.Z4A87 2011
337.1'59--dc22
2010054076

Published by Nova Science Publishers, Inc. † New York

CONTENTS

PREFACE

The Association of Southeast Asian Nations (ASEAN) is Southeast Asia's primary multilateral organization. Established in 1967, it has grown into one of the world's largest regional fora, representing a strategically important group of 10 nations that spans critical sea lanes and accounts for 5% of U.S. trade. This new book examines U.S. diplomatic, security, trade and aid ties with ASEAN, analyzes major issues affecting Southeast Asian countries and U.S.-ASEAN relations, and explores ASEAN's relations with other regional powers with a focus on mulitlateral diplomacy.

Chapter 1- The Association of Southeast Asian Nations (ASEAN) is Southeast Asia's primary multilateral organization. Established in 1967, it has grown into one of the world's largest regional fora, representing a strategically important group of 10 nations that spans critical sea lanes and accounts for 5% of U.S. trade. This chapter discusses U.S. diplomatic, security, trade, and aid ties with ASEAN, analyzes major issues affecting Southeast Asian countries and U.S.-ASEAN relations, and examines ASEAN's relations with other regional powers. Much U.S. engagement with the region occurs at the bilateral level, but this chapter focuses on multilateral diplomacy.

Chapter 2- On July 22, 2009, during Secretary of State Hillary Rodham Clinton's visit to Southeast Asia, the United States acceded to the Association of Southeast Asian Nations' (ASEAN) Treaty of Amity and Cooperation (TAC), one of the 10-nation organization's core documents, as had been amended by the 1987 and 1998 TAC Protocols. The move came less than six months after Secretary of State Clinton announced in Jakarta that the Obama Administration would launch its formal interagency process to pursue accession. This chapter analyzes the legal and diplomatic issues involved with accession to the TAC.

Chapter 3- This chapter describes trends in regional integration, export competitiveness, and inbound investment for six industries within the Association of Southeast Asian Nations (ASEAN): computer components, cotton woven apparel, hardwood plywood and flooring, healthcare services, motor vehicle parts, and palm oil. The six profiled industries are a subset of 12 priority sectors that ASEAN members identified in 2004 in order to promote regional integration. The members created a regional "Roadmap for Integration" (Roadmap) for each priority sector, and while these Roadmaps have promoted tariff reductions and streamlined certain administrative procedures, their success in promoting regional integration has been mixed

In general, economic factors and national government policies have had more influence than the Roadmaps over regional industrial structures. ASEAN members tend to view each other as competitors for inbound investment and jobs, and the members have no legally binding means to enforce compliance with the objectives of the Roadmaps.

In: Association of Southeast Asian Nations (ASEAN)... ISBN: 978-1-62417-982-2
Editor: Margaret E. Stamlin

Chapter 1

UNITED STATES RELATIONS WITH THE ASSOCIATION OF SOUTHEAST ASIAN NATIONS (ASEAN)

Thomas Lum, Ben Dolven, Mark E. Manyin[1], Michael F. Martin and Bruce Vaughn

SUMMARY

The Association of Southeast Asian Nations (ASEAN) is Southeast Asia's primary multilateral organization. Established in 1967, it has grown into one of the world's largest regional fora, representing a strategically important group of 10 nations that spans critical sea lanes and accounts for 5% of U.S. trade. This chapter discusses U.S. diplomatic, security, trade, and aid ties with ASEAN, analyzes major issues affecting Southeast Asian countries and U.S.-ASEAN relations, and examines ASEAN's relations with other regional powers. Much U.S. engagement with the region occurs at the bilateral level, but this chapter focuses on multilateral diplomacy.

The United States has deep-seated ties in Southeast Asia, and it has viewed ASEAN as a useful organization since its inception during the Cold War. Today, U.S. policy towards ASEAN and Southeast Asia is cast against the backdrop of great power rivalry in East Asia, and particularly China's emergence as an active diplomatic actor in its geographic backyard. Some worry that the United States, preoccupied with other priorities, has been neglectful of ASEAN and of Asian multilateral diplomacy in recent years. The Obama Administration has expressed an intent to work more closely with multilateral organizations, particularly ASEAN. A number of steps in this direction include Secretary of State Hillary Clinton's visit to the ASEAN Secretariat in Jakarta in February 2009, the U.S. accession to ASEAN's Treaty of Amity and Cooperation (TAC) in July 2009, and President Obama's attendance at an ASEAN leaders meeting in November 2009.

Congress has frequently played an important role in shaping U.S. diplomatic, security, and economic relations with Southeast Asia and ASEAN. Major U.S. and congressional interests in Southeast Asia include maritime security, the promotion of democracy and human rights, the encouragement of liberal trade and investment regimes, counterterrorism, combating narcotics trafficking, environmental preservation, and many others. In October 2009, Senator Richard Lugar introduced S.Res. 311, calling for the start of discussions on a free trade agreement with ASEAN. In August 2009, Senator Jim Webb visited five countries in mainland Southeast Asia and was the first Member of Congress in ten years to visit Burma.

The United States exerts a strong military and economic presence in Southeast Asia, and through diplomacy it seeks to remain a major power—perhaps the major power—in the region. ASEAN, however, has been active in recent years in exploring a variety of diplomatic architectures for East Asia and the Pacific. ASEAN is at the center of several broader security- and trade-related groupings in the Asia-Pacific region, through which it has aimed to maintain regional multi- polarity or a balance of powers among itself and other states including the United States, China, and Japan. ASEAN is also the nexus for discussion of regional economic integration. ASEAN has launched an internal free trade accord, the ASEAN Free Trade Agreement (AFTA), which will go into full effect in 2015. ASEAN has also concluded FTAs with many external trade partners, though not with the United States. ASEAN has also been exploring ways to advance the ultimate creation of a broader European Union-like East Asia Community. Some within the group—but not all—support the inclusion of the United States in such a community.

Human rights conditions, particularly in some ASEAN members such as Burma, have long been a source of friction between the organization and the United States. ASEAN's new Charter, enacted in 2007, attempts to bring more pressure to bear upon recalcitrant member states. However, ASEAN still operates on principles of consensus and non-interference in the internal affairs of its members, so it remains unclear how active an actor it will be in this area.

Overview: U.S. Interests towards ASEAN

The Association of Southeast Asian Nations (ASEAN) is Southeast Asia's primary multilateral organization, a 10-member grouping of nations with a combined population of 580 million and an annual gross domestic product (GDP) of around $1.5 trillion. Established in 1967 to foster regional dialogue during the turbulent post-colonial, Cold War period, it has grown into one of the world's largest regional fora, representing a strategically and economically important region that spans some of the world's most critical sea lanes and accounted for around 5% of the United States' total trade in 2008.[1]

The Obama Administration is pursuing a policy of expanding and upgrading U.S. relations with Southeast Asia, and with ASEAN itself. Although the Bush Administration took steps to develop ties with the region, it was widely perceived among members of ASEAN as narrowly focused on terrorism, neglectful of other issues, and not sufficiently committed to multilateral dialogue. By contrast, the Obama Administration has explicitly expressed an intent to pay greater attention to Southeast Asia, listen more carefully to regional concerns, and work with multilateral organizations, particularly ASEAN, to cooperate on issues of mutual interest.

The United States has deep-seated interests in Southeast Asia, such as maritime security, the promotion of democracy and human rights, the encouragement of liberal trade and investment regimes, counterterrorism, the combating of illegal trafficking of narcotics and human trafficking, and many others. As China has deepened its economic and cultural ties in

Southeast Asia, and even taken some steps to build security ties, some analysts believe the region has also become an important site of "soft power" rivalry, in which the longstanding leadership role of the United States could be challenged by a rising China. Other external powers also have shown renewed or greater interest in the region, including Japan, the EU, and India.

Engagement with ASEAN has presented the United States with an important foreign policy dilemma. Despite considerable U.S. security, economic, and foreign assistance initiatives in the region, particularly at the bilateral level, in recent years a perception has developed among Southeast Asian elites that the United States has placed relatively little priority on ASEAN itself and has, thereby, demonstrated a lack of commitment to Southeast Asia as a whole. Southeast Asian diplomats frequently note that other nations, including China and Japan, have given ASEAN meetings a considerably higher diplomatic commitment than has the United States. Indeed, in some ASEAN countries, one of the largest irritants to *bilateral* relationship with the United States is the fact that it is perceived as insufficiently engaged with the multilateral body of ASEAN.[2]

The United States has long had close bilateral relations with many of Southeast Asia's nations. Two ASEAN members, Thailand and the Philippines, are U.S. treaty allies, and a third, Singapore, is a close security partner. Indonesia and Malaysia have long had strong ties with Washington, and both are seen as important models of progressive governance and economic development in majority Muslim nations. In recent years, Vietnam has also become an increasingly important voice in regional affairs, and the United States has moved to normalize and deepen ties with its one-time adversary.

Some feel that these strong sets of bilateral ties are sufficient to anchor the U.S. role in the region, arguing moreover that ASEAN's consensus-based decision-making makes it difficult for the organization to accomplish much, given its broad membership, which includes highly developed financial centers, vibrant developing-nation democracies, and impoverished military dictatorships.

Still, symbolic commitment is particularly important in a region that places a heavy emphasis on process and informal networking. Many observers argue that the United States needs to "show up" more frequently and at higher official levels, lest it lose influence in the region and risk being cut out of emerging Asian diplomatic and economic architectures. Recent actions by the Obama Administration suggest that it accepts this argument, at least on a symbolic level.

U.S. Policy Developments toward the Region

The United States has been steadily expanding and deepening its relations with ASEAN since the middle of the decade. A common goal of both the Bush and the Obama Administrations appears to be to increase the multilateral dimension of U.S. policy in Southeast Asia, which traditionally has been organized along bilateral lines. However, many of the Bush Administration's initiatives—which included becoming the first country to appoint an ambassador to ASEAN, providing assistance to the ASEAN Secretariat to upgrade its capabilities, and launching the US- ASEAN Trade and Investment Framework Agreement (TIFA)—were undermined by a belief among Southeast Asian elites that the United States lacked a strong commitment to ASEAN and Southeast Asia.[3] The piece of evidence cited most often by critics was former Secretary of State Condoleezza Rice's decision to not attend two of the four ASEAN Regional Forum (ARF) Foreign Ministerial meetings during her tenure. Considerable attention was focused on President Bush's decision to cancel the scheduled US-ASEAN Summit in September 2007 to focus on the security situation in Iraq. A number of countries have regular summits with ASEAN leaders, including China, Japan, South Korea, and India.

Thus far in 2009, the Obama Administration has taken steps with ASEAN that some see as explicitly designed with symbolic diplomacy in mind. First, in February, Secretary of State Hillary Clinton visited the ASEAN Secretariat in Jakarta, a first for a U.S. Secretary of State. Second, in July, during Clinton's second visit to Southeast Asia to participate in the ARF Foreign Ministerial in Thailand, the United States acceded to ASEAN's Treaty of Amity and Cooperation (TAC), which promotes the settlement of regional differences or disputes by peaceful means and is one of the organization's core documents.[4] Third, President Obama attended a first-ever U.S.- ASEAN leaders meeting on the sidelines of the November 2009 Asia-Pacific Economic Cooperation (APEC) forum summit in Singapore.[5] In a joint statement, the leaders pledged continued or enhanced dialogue and cooperation in many areas, including engagement with the government of Burma (Myanmar), human rights, trade, regional security, nuclear nonproliferation and disarmament, counterterrorism, energy, climate change, educational exchanges, and support for the Lower Mekong Basin countries (Cambodia, Laos, and Vietnam).[6]

The Obama Administration has taken other potentially noteworthy steps. Divergent U.S. and ASEAN approaches to Burma have also been an irritant to U.S.-ASEAN relations since Burma became a member of the organization in 1997. The United States has pursued a policy of diplomatically shunning the Burmese regime and imposing stringent economic sanctions against the country—creating difficulties in engaging both politically and economically with a grouping that includes it. In the fall of 2009, the State Department announced a new Burma policy, in which the United States would hold dialogues with the Burmese leadership while still maintaining U.S. sanctions. This move, which brings Washington closer to ASEAN policy, could help to improve U.S.-ASEAN ties.

Additionally, on the sidelines of the July 2009 ARF meeting, Secretary Clinton met with the foreign ministers of the lower Mekong countries, excluding Burma (i.e. Vietnam, Cambodia, Laos, and Thailand), in the first-ever U.S.-Lower Mekong Ministerial Meeting. The Ministers issued a joint statement outlining the wide-ranging areas of discussion, which included responses to climate change, fighting infectious disease, and education policy.[7]

Taken together, the message of the Administration's symbolic and substantive moves appears to be that the United States intends to engage with ASEAN and Southeast Asian countries at a higher level, and do so more persistently. There remain questions about how far this change in approach will persist, particularly as it raises expectations in Southeast Asia. For instance, will the U.S.-ASEAN leaders' meeting be regularized, as many Southeast Asian leaders hope? On the other hand, by raising the profile of U.S.-ASEAN ties, the United States likely will place new pressures on ASEAN to increase its own utility in resolving regional crises, lest a more activist United States eventually bypass it.

ASEAN: FORMATION AND INSTITUTIONS

The Association of Southeast Asian Nations was founded on August 8, 1967, with the adoption of the ASEAN Declaration in Bangkok, Thailand. Originally, the association had five members— Indonesia, Malaysia, Philippines, Singapore, and Thailand—and expanded to its current 10 members during the 1980s and 1990s with the addition of Brunei Darussalam, Vietnam, Laos, Burma, and Cambodia. Colonial experiences led to a strong desire by the original members to prevent the domination of the region by any single power. Furthermore the formation of the organization reflected an attempt to forge independent foreign policies in the context of Cold War pressures. As stated in the ASEAN Declaration, the association was created to achieve joint goals including those related to economic growth, regional peace and security, collaboration and mutual assistance in a number of development areas, trade promotion, and linkages with other regional organizations.

On February 24, 1976, ASEAN created the ASEAN Secretariat, located in Jakarta, Indonesia, an administrative body consisting of representatives of each ASEAN member nation. The Secretariat is headed by a Secretary-General, who serves a term of five years. Since its creation, the structure and the duties of the Secretariat have been changed on several occasions. As of 2009, the Secretary-General's main responsibilities are to: organize the annual foreign ministers' meeting; initiate, advise, coordinate, and implement ASEAN activities; serve as spokesman and representative of ASEAN on all matters; and oversee the operations of the ASEAN Secretariat.

ASEAN remains, to a large degree, an informal organization. The ASEAN Secretariat is lightly staffed, without the deep administrative resources and responsibilities of some multilateral organizations such as the European Union. Its current secretary-general is Surin Pitsuwan, a former Thai Foreign Minister, who has sought to institutionalize many of ASEAN's practices and has pushed the introduction of the ASEAN Charter.

Still, much of the diplomatic activity that occurs at meetings of ASEAN leaders and senior officials occurs on the sidelines rather than at the formal level. ASEAN has traditionally operated on principles of consensus and non-interference in the internal affairs of members, which has led to considerable difficulty in the group operating in formal concert. Many analysts note that ASEAN's expansion to include underdeveloped nations such as Laos, Cambodia, and Burma has created a wide range of interests within the group that make formal security and economic moves difficult to agree upon. Although ASEAN is starting to play a more active role in dealing with its members' differences—most notably over Burma's human rights record—much of what the group does is still done through informal channels.

ASEAN Charter

A new ASEAN Charter went into effect on December 15, 2007, superseding the ASEAN Declaration as the organizing document for the organization. The Charter is effectively a constitution for ASEAN, committing the member nations to the formation of an "ASEAN Community in furtherance of peace, progress and prosperity of its peoples." Some aspects of the Charter that may signal a greater willingness to discuss and comment on the internal affairs of the organization's members. Such a potential institutional development may help the organization to deal with members such as Burma which have caused troublesome policy issues both within the region and with ASEAN's relations with outside states.

The new charter establishes a number of goals for ASEAN, including:

- Maintenance of peace, stability, and security in the region;
- Promotion of greater political, security, economic; and socio-cultural cooperation;
- Preservation of Southeast Asia as an area free of weapons of mass destruction, including nuclear weapons;
- Creation of a just, democratic, and harmonious environment in the region;
- Formation of a single market and production base in which there is free flow of goods, services, and investment, as well as facilitated movement of business persons, professionals, talents and labor, and the freer flow of capital;
- Alleviation of poverty and the narrowing of the development gap in the region; and
- Promotion of sustainable development so as to ensure the protection of the region's environment.

According to the new charter, there are to be two ASEAN Summits each year, attended by the members' heads of state or their designated representatives. In addition, the foreign ministers of the ASEAN members are to meet at least twice a year. The ASEAN Charter also creates three Community Councils, dealing with political and security, economic, and socio-cultural issues, respectively, plus preserves the institutions of the ASEAN Secretary-General and the ASEAN Secretariat as the administrative bodies for the association. Article 14 of the Charter calls for the establishment of an ASEAN human rights body, a new development for ASEAN which has traditionally refrained from commenting on the human rights situation in member nations. The first meeting of the ASEAN human rights body—formally called the ASEAN Intergovernmental Commission on Human Rights (AICHR)—took place on October 23, 2009, in Cham-Am, Thailand, following an ASEAN Summit.

ASEAN's Regional Significance

ASEAN is at the center of several other security- and trade-related groupings in the Asia-Pacific region. The ASEAN Regional Forum (ARF), established in 1994 with 26 Asian and Pacific states plus the European Union, was formed to facilitate dialogue on political and security matters in the region. The ASEAN + 3 (China, Japan, and South Korea) was created in 1997, partly as a response to the Asian financial crisis, and partly as a way to balance the northeast Asian powers in the security dialogue process with ASEAN. Created in 2005, the East Asia Summit (EAS) which, in addition to the ASEAN + 3 members, includes Australia, New Zealand, and India,[8] represented an effort by some countries in the region, particularly Japan, to balance China's influence in the region through the inclusion of additional, non-East Asian powers.

More recently, the geopolitical discussion in Asia has turned to the issue of the formation of a EU-style association of Asian nations. While this discussion is in its early stages, there are already advocates for the creation of a pan-Asian entity—the East Asian Community

(EAC)— that would include closer economic and trade relations among its members, possibly even the creation of a single Asian currency.

At the same time, there has been a separate ongoing discussion about greater regional economic and trade integration in Asia taking place in various fora. The Asia-Pacific Economic Cooperation (APEC) forum was formed in 1989 with the express mission of accelerating regional economic integration and fostering greater trade and investment liberalization through a process known as "open regionalism."[9] ASEAN has also formed the core at periodic meetings of ASEAN + 3 and the EAS to consider ways and means of fostering closer economic and trade ties. For its own part, ASEAN has been pursuing ways to expedite closer economic ties amongst its 10 member nations with the goal of creating an ASEAN Economic Community.

POLICY ISSUES

Security

While security concerns were downplayed in the original ASEAN Declaration, the importance of regional peace and security was a major purpose behind ASEAN's formation.[10] ASEAN has sought to maximize its security interests by developing a set of norms for its members and beyond that has increasingly relied on consensus building and discussion as the preferred means of conflict resolution. That said, all is not tranquil among ASEAN members or between ASEAN states and external powers. There continue to be bilateral tensions among ASEAN states as recently demonstrated by the border clashes between Cambodia and Thailand near the 11th century Preah Vihear Temple, and maritime disputes between Indonesia and Malaysia over the energy-rich Ambalat sea bloc in the Sulawesi Sea. Nevertheless, it does appear that ASEAN has played a key role in promoting a normative order that has minimized inter-state conflict in Southeast Asia since the group's formation during the Cold War.

Geopolitical Importance

ASEAN's key strategic value emanates from its geographic position as well as its economic development. ASEAN is situated astride the key sea lanes that link the energy-rich Persian Gulf and the economic power centers of East Asia. Maintaining the free flow of goods and energy through the strategically vital Malacca, Sunda, and Lombok Straits is a key geostrategic interest for ASEAN members, as well as the United States, Japan, China, and South Korea. Energy reserves in and around the South China Sea, Indonesia, and Burma also give the region added strategic importance.[11]

While ASEAN has been a key player in the creation of emerging economic and strategic architectures in Asia, such as the ASEAN + 3 and the East Asia Summit, it faces the increasingly challenging task of maintaining strategic balance and its pivotal role in this process. The emergence of China and India as great powers in an increasingly multi-polar world, and the continued engagement of the United States, present diplomatic challenges for ASEAN as it seeks to shape an international order that will promote peace and stability for the region.

The United States and ASEAN share a mutual interest in preventing conflict and maintaining the independence of regional states. ASEAN as an organization will likely seek to balance external actors in the region while seeking to avoid antagonizing great powers. America's military posture in Asia supports ASEAN's goal of ensuring that no hegemon can arise that could dominate the region.[12] As such, America is generally a valued offshore balancer relative to the perceived rising influence of China, though some ASEAN members—Laos, Cambodia, and Burma, in particular—are relatively closer to China than others. China also acts as a balancer to American presence in the region.

While securing sea lanes of communication and trade that transit maritime Southeast Asia is of mutual importance among all interested states, there is the potential that increasingly intense competition for energy resources could lead to increased tensions. This could be the case should Chinese efforts to secure energy resources and routes entangle China and India in a security dilemma where "defensive" moves by one party are viewed as "offensive," or threatening, by the other. This could also be the case should Chinese activity in Burma intensify. China is interested in developing an energy and trade corridor from Sittwe, Burma, to Kunming, China, which could be viewed as a means of lessening China's strategic vulnerability at the Strait of Malacca. Some in India are increasingly concerned that this move by China in Southeast Asia could be part of a larger strategy to encircle India.

South China Sea Disputes

Territorial disputes in the South China Sea have been at the center of some of ASEAN's most active security-related diplomacy in recent decades—but also serve as an illustration of the difficulty of marshaling the group's diverse membership to act in concert. For decades, the Paracel and Spratly Islands have been the site of regional competition for control of the South China Sea among ASEAN members and China, and between individual ASEAN members themselves. The source of competition over this region is the desire to extend sovereignty over sea beds by establishing claims to the islands and thereby control important fishing areas and what are thought to be rich energy reserves beneath the sea.

In the late 1990s and early 2000s, ASEAN's push for a Code of Conduct on the South China Sea to promote the norms of peaceful resolution of conflict—which resulted in the Declaration on the Conduct of Parties in the South China Sea signed by ASEAN's members and China in November 2002—can be viewed as one of the group's successes in acting in concert to promote common security interests of organization members. In 1992, following a series of incidents including China's sinking of three Vietnamese vessels near Fiery Cross Reef in the Paracels in 1988, ASEAN issued a declaration on the South China Sea, calling for a mutual code of conduct for nations navigating in the waters. This led to a decade of active diplomacy in which the organization's members largely held together to promote multilateral security in the area. By acting as a group, ASEAN states arguably have collectively more weight when dealing with any outside actor than they do when acting individually.

However, the continuation of flare-ups in these waters is also an illustration of the limits of ASEAN's ability or willingness to act in concert to deal with external powers. In recent years, continued Chinese disputes with Vietnam and the Philippines have been kept largely bilateral, with ASEAN as a grouping opting not to lend formal support its members in their disputes with China. This has had an effect on U.S. interests. In 2008, for instance, China warned international oil firms, including ExxonMobil, against exploring for energy resources in blocks leased by the Vietnamese government.

Historical Context

Of ASEAN states, only Thailand was able to maintain a fair degree of political autonomy throughout the colonial period in Southeast Asia. The later colonial period witnessed the domination of Indo-China by France; Burma, Malaya, and Singapore by the United Kingdom; Indonesia by the Netherlands; and the Philippines by Spain and the United States. During WWII, the region came under the control of imperial Japan. These experiences led to a strong desire of ASEAN members to prevent their newly independent states from being dominated by any single power, as Japan did during WWII, and to preserve and expand their independence of action from external great powers.

ASEAN was formed through the Bangkok Declaration of 1967 at the height of the Cold War when external powers were directly or indirectly militarily engaged in the region. ASEAN was created largely as a reaction to Cold War pressures on the region.[13] At the time, the United States was deeply engaged in the war in Vietnam and the ongoing global struggle between the West and the Soviet bloc was intense. Small Southeast Asian states also sought in part to bring Indonesia into a regional grouping as a way of curbing its previously demonstrated ability to threaten regional neighbors, as it did with Malaya under its policy of *Konfrontasi,* which included a guerilla war on Borneo from 1963 to 1966 against British, Australian, New Zealand, and Malay security forces.

While the Cold War is now history, ASEAN continues to be faced with the diplomatic, strategic and foreign policy challenges of how to deal with external great power actors in its region. Today, Soviet influence has faded and Chinese influence has expanded while the United States has sought to remain engaged in the region. ASEAN-China relations have become deeper, as China has engaged in a "charm offensive" since the late 1990s, seeking better diplomatic and trade relations with Southeast Asian states. The potential for larger Indian engagement with the region is also developing as demonstrated by India's inclusion in the East Asia Summit.

External Security Ties

Some regional states continue to have outside bilateral or multilateral defense ties, some of which can be viewed as legacies of the colonial, post-WWII, and Cold War periods. These security relationships include the Five Power Defense Agreement between the United Kingdom, Malaysia, Singapore, Australia, and New Zealand and the U.S. alliances with the Philippines and Thailand that were originally part of the San Francisco system formed in the early 1950s. In addition, Indonesia has moved on somewhat from its Non-Aligned position by developing bilateral security ties with Australia.

While there are some relatively low level security concerns either between ASEAN states or at the sub-state level, as is the case with insurgents in Southern Thailand and the Southern Philippines, the largest threats to stability in the region as a whole emanate largely from outside the region and relate to the evolving correlates of power in Asia as a whole. It is for this reason that much of ASEAN's diplomatic activity and initiative has been focused at establishing a new Asian or trans-Pacific economic and strategic group that can seek to prevent or ameliorate conflict between the extra-regional powers that are active in the region, including the United States, China, Japan, South Korea, and India. For example, a conflict between China and the United States over Taiwan would likely have a devastating impact on regional trade and would place unwanted pressure on ASEAN states to pick sides.

The Emerging Security Agenda in Southeast Asia

The general trend in recent decades of re-conceptualizing security as more than simply the realm of cross border conflict between the armed forces of sovereign nation states, or internal counterinsurgency operations, is clearly evident in Southeast Asia. This is evident from the negative impact of terrorist groups active in the region, such as Jemaah Islamiya and Abu Sayyaf, as well as from the relatively high incidence of piracy in maritime Southeast Asia.

Jemaah Islamiya in Indonesia and Abu Sayyaf in the Philippines are two key terrorist groups that are a threat to Americans and Western interests in the region. Counterterrorism efforts by ASEAN states working with the United States and Australia have done much to hunt down regionally based terrorists. While the cultural heart of Islam is Mecca, the demographic heart of Islam is closer to Southeast Asia, as Indonesia has the world's largest Muslim population. Indonesia and Malaysia generally recognize a more tolerant and less fundamentalist form of Islam, which some argue could be a good starting point for increased engagement by the United States in the Muslim world.[14]

Contemporary security interests also encompass other sub-national and trans-regional levels of conflict in addition to interstate conflict. The conflicts that Indonesia has had in East Timor, Aceh, the Moluccas, and Papua, some of them still festering; ongoing insurgencies in Muslim areas of Thailand and the Philippines; and Burma's restive minority groups can be viewed in this context. ASEAN's reluctance to become involved in the internal affairs of its members has largely kept such issues from becoming the business of the group as a whole.

Concepts of human security have also brought many analysts of security dynamics in the region to increasingly focus on the negative impacts of environmental degradation and the impact that climate change may have on the region. The "haze" generated by the burning of forests after logging operations brought this to the attention of regional governments concerned over public health risks in 1997. The damming of the upper reaches of the Mekong in China have also raised concerns over the long term viability of that river system as a source of food for the region. Increased temperatures associated with climate change may undermine regional food production and cause sea level rise that would negatively impact low-lying coastal areas where many in the region live.

Piracy in Southeast Asia has been a relatively large problem as compared with other areas of the world with the exception of the Gulf of Aden and the Arabian Sea in recent years. Human and narcotics trafficking and the plight of refugees in the region are other human security issues worthy of attention.

Trade and Trade Relations

ASEAN's Role in Global Trade

The ASEAN economies have become a major regional hub for globalized manufacturing. According to official ASEAN statistics, ASEAN's total merchandise trade exceeded $1.7 trillion (see **Table 1**). A little more than a quarter of its trade was between ASEAN members. Another third was with the European Union (EU-25), Japan, and the United States. Trade with China claimed about one-tenth of the association's merchandise trade. The rest of ASEAN's trade was distributed around the world.

In terms of the types of goods and commodities traded by ASEAN in 2008, three different groups far surpassed all other categories—electrical machinery, mineral fuels and oils, and mechanical appliances (see **Table 2**). Taken together, these items account for nearly 60% of ASEAN's exports and almost two-thirds of its imports.

Table 1. ASEAN's Trade with Selected Partners, 2008 (in US$ billions)

Partner	Value			Share		
	Exports	Imports	Total Trade	Exports	Imports	Total Trade
ASEAN	242.498	215.617	458.114	27.6%	25.9%	26.8%
Japan	104.862	107.054	211.916	11.9%	12.9%	12.4%
EU-25	112.887	89.472	202.358	12.8%	10.8%	11.8%
China	85.558	107.114	192.672	9.7%	12.9%	11.3%
USA	101.129	79.911	181.039	11.5%	9.6%	10.6%
Rest of World	232.318	232.002	464.323	26.5%	27.9%	27.1%
Total	879.252	831.170	1,710.422	100.0%	100.0%	100.0%

Source. ASEAN's official database.

Note. Differences in row and column sums due to rounding.

Table 2. ASEAN's Top Three Trade Commodity Groups, 2008 (in US$ billions)

		Commodity Group		
		Electrical Machinery	Mineral Fuels and Oils	Mechanical Appliances
Value	Exports	175.493	150.380	121.640
	Imports	166.070	146.519	118.461
	Total Trade	341.563	296.899	240.101
Share of Total ASEAN Trade	Exports	23.4%	20.0%	16.2%
	Imports	25.4%	22.4%	18.1%
	Total Trade	24.3%	21.1%	17.1%

Source. ASEAN's official database.

Note. Electrical Machinery includes all products under HS Chapter 85; Mineral Fuels and Oils includes HS Chapter 27; and Mechanical Appliances includes HS Chapter 84.

This pattern can be partially explained by ASEAN's role in the globalized manufacturing of electrical machinery and mechanical appliances. As described in a number of studies, the production of home appliances, computers, telecommunications equipment and other products that fall into these two categories has become a multi-country process, with components and parts being shipped between nations for final assembly in multiple competing countries. Much of ASEAN's intra-regional trade in intermediate goods end up as components used in final assembly work done in China.[15] While this multi-country assembly process is comparatively mobile and fluid, in recent years, the ASEAN nations—along with China—have become regionally integrated manufacturing hub for selected products.

According to official U.S. trade statistics, ASEAN's trade with the United States—like with the rest of the world—is dominated by electrical machinery (HTS 85) and mechanical appliances (HTS 84) (see **Table 3**). Over 40% of U.S. imports from ASEAN and nearly half of U.S. exports to ASEAN are in these two categories. Knit and non-knit clothing (HTS 61 and 62), plus rubber and articles made of rubber (HTS 40) are also major products imported from ASEAN. Other top five U.S. exports to ASEAN are aircraft (HTS 88); optical and scientific equipment (HTS 90); and mineral fuels and oils (HTS 27). U.S. trade statistics show a larger U.S. trade deficit with ASEAN than ASEAN's statistics.

ASEAN's Trade Relationships

Since the early 1990s, the ASEAN members have been gradually moving towards the creation of a free trade area encompassing the 10 members of the association. The ASEAN Free Trade Area (AFTA) is to be fully implemented in 2010 by six ASEAN countries and 2015 for the remaining signatories.[16] Under AFTA's Common Effective Preferential Tariff (CEPT) Scheme, more than 99% of the product categories will have their intra-ASEAN tariff rates reduced to below 5%.[17]

In addition, the 10 ASEAN members have agreed to the goal of creating an ASEAN Economic Community (AEC) by 2015. During the ASEAN Summit held in Cha-Am, Thailand on October 23-25, 2009, there was a recommitment to the 2015 goal for the creation of the AEC, as well as discussion of alternative ways of forming closer economic and trade ties with several Asian nations, including China, India, Japan, and South Korea.

Table 3. Top Five U.S. Imports and Exports in ASEAN Trade (in US$ billions)

	2008 Imports (CIF)			2008 Exports (FAS)	
HTS Number	**Value**	**Share**	**HTS Number**	**Value**	**Share**
Electrical machinery (HTS 85)	25.406	22.1%	Electrical machinery (HTS 85)	20.439	30.0%
Mechanical appliances (HTS 84)	24.920	21.7%	Mechanical appliances (HTS 84)	12.869	18.9%
Knit clothing (HTS 61)	9.217	8.0%	Aircraft (HTS 88)	4.542	6.7%
Non-knit clothing (HTS 62)	7.055	6.2%	Optical and scientific equipment (HTS 90)	3.420	5.0%
Rubber (HTS 40)	5,051	4.4%	Mineral fuels and oils (HTS 27)	2.677	3.9%
Total	114.715		Total	68.1501	

Source. U.S. International Trade Commission.

ASEAN's efforts to create the AEC has been complemented by its interest in negotiating trade agreements with key Asian nations. The United States is the only major power in the region that has not agreed to some form of formal free trade agreement (FTA) with ASEAN. As of October 2009, ASEAN had concluded trade agreements with the following countries:

Australia and New Zealand—On February 29, 2009, ASEAN, Australia, and New Zealand signed the ASEAN-Australia-New Zealand Free Trade Area (AANZTA) Agreement. The agreement commits the parties to the progressive reduction of tariff and non-tariff trade barriers.

China—On November 5, 2002, ASEAN and China lowered some of their trade barriers and set the goal of establishing an ASEAN-China Free Trade Area (ACFTA) within 10 years. On August 15, 2009, ASEAN and China expanded the ACFTA to include provisions pertaining to investment.

India—On August 13, 2009, ASEAN and India concluded an agreement on trade in goods that provides for the gradual reduction of tariff and non-tariff trade barriers, plus commits the parties to the establishment of an ASEAN-India Free Trade Area (AIFTA).

Japan—In April 2008, ASEAN and Japan concluded negotiations for the creation of an ASEAN- Japan Free Trade Area (AJFTA).

South Korea—On August 24, 2006, ASEAN and the Republic of Korea concluded the ASEAN- Korea Free Trade Agreement (AKFTA). The original document covered trade in goods. Since then ASEAN and South Korea have extended their trade arrangement to cover investment as well.

ASEAN has also held talks with the European Union (EU) about a possible free trade agreement, but progress has been slow and prospects arc unclear.

Beyond their efforts to negotiate bilateral trade agreements with selected countries, ASEAN has also been actively promoting the creation of a larger, Asia-based free trade area. This possible regional economic association has been referred to by different names at different times, including the more recent East Asian Community (EAC). In some cases, the discussants have been limited to ASEAN + 3. In other cases, the group of nations has been expanded include the EAS (ASEAN + 6).

During the East Asia Summit held in Hua-Hin, Thailand, on October 25, 2009, there was discussion about the nature of a possible EAC as well as which nations ought to be members. While there appeared to be some consensus to create a regional free trade area by 2020, there was no agreement on which nations should he part of such an arrangement. In particular, there were apparently sharp differences of opinion over the inclusion of the United States in the free trade area. Similarly, although Russia has applied for membership in the East Asia Summit, it is unclear if Russia is being considered for inclusion in the EAC.

While some have suggested the possibility of an ASEAN-U.S. Free Trade Agreement, there are several structural problems to negotiating such an agreement. First, the United States would probably require that the trade agreement comply with the U.S. model FTA, a condition that ASEAN may not find acceptable. Second, the United States has a comprehensive ban on direct trade with Burma. Third, the ASEAN economies vary in their level of economic and legal development, which would make the FTA's compliance requirements difficult to specify.

U.S. Burma Policy and ASEAN

The Obama Administration's revision of U.S. policy towards Burma has coincided with a similar review by ASEAN of its stance on relations with the military junta. While the new U.S. policy may be viewed as a tacit admission that sanctions alone were not sufficient to effect change in Burma, recent statements and actions by ASEAN may indicate that their past policy of "constructive engagement" had proven equally ineffective. As a result, there may be an opportunity for ASEAN and the United States to confer and coordinate their policies towards Burma's military government.

Although there was interest in including Burma as an original member of ASEAN in 1967, it did not join the association until 1997. From the start, ASEAN as an organization adopted a policy of "constructive engagement" towards Burma, refraining from public comments in its "internal affairs," while some members sought closer economic, trade, and investment relations with Burma.

Some of the strongest supporters of ASEAN's policy of "constructive engagement" towards Burma have been the governments of Indonesia, Malaysia, Singapore, and Thailand for slightly different reasons. Thailand has had an ambivalent view of Burma, as its domestic unrest has had a more immediate adverse impact on Thailand. In addition, at the time of Burma's admission to ASEAN, Thailand was emerging from a period of military coups. However, the new civilian government in Thailand decided to support Burma's admission and follow a policy of engagement. Under the Suharto regime, the Indonesian government shared some ideological views with Burma's military government that led to its support of Burma's ASEAN membership and closer relations, although Jakarta has taken a harder line as the country has democratized. Malaysia at the time was concerned about both Chinese and U.S. influence in the region, and found similar views among Burma's military rulers. The Singaporean government saw economic opportunity in closer relations with Burma, and for a time was a major supplier of equipment and arms for the Burmese military.

The adoption of a new ASEAN Charter in 2007 may signal a greater willingness to address issues such as human rights and democracy. As previously mentioned, the new charter states that among ASEAN's purposes are strengthening democracy and protecting human rights, and mandated the establishment of an "ASEAN human rights body." However, among ASEAN's founding principles is a commitment to "non-interference in the internal affairs of ASEAN Member States."

In practice, under the new charter, ASEAN has shown a greater willingness to express its opinion about the situation in Burma. In response to the conviction of Aung San Suu Kyi in August 2009, ASEAN's Chairman issued a statement expressing ASEAN's "deep disappointment" at the verdict, calling for the immediate release of Aung San Suu Kyi and other political prisoners, and asserting that "such actions will contribute to national reconciliation among the people of Myanmar, meaningful dialogue and facilitate the democratization of Myanmar."[18] In addition, ASEAN has indicated that the Burmese military's treatment of opposition groups and ethnic minorities will affect how the election results will be perceived by the Association.

Although ASEAN appears to be more willing to publicly criticize Burma's military government, it has not shown a greater willingness to impose economic sanction on the country. Malaysia, Singapore, and Thailand are major trading partners with Burma, and may be reluctant to forswear the economic benefits of bilateral trade and investment. Indonesia's civilian government may be more willing to consider economic pressure on Burma, in part because of its history of military rule and in part because of its concern about Burma's Muslim minority.

The outcome of Burma's 2010 parliamentary elections may prove critical to ASEAN's future relationship with Burma. While few expect a free and fair election, if the results provide some space for opposition views in the government and indicate a possible shift in power to civilian rule, then ASEAN will likely continue its policy of modified "constructive engagement." If, however, the election results provide only a veil of cover to the continuation of military rule, then ASEAN may be willing to consider adopting a policy closer to that of the Obama Administration.

The actions of Burma's military in response to Burma's last two national plebiscites—staging a military coup in 1990 and producing fraudulent results in 2008—may be indicative of how the elections of 2010 will be handled.

U.S. Assistance to ASEAN

U.S. assistance for Southeast Asian multilateral efforts focus on trade facilitation, counterterrorism, security sector reform, and the environment. Other program areas include good governance, combating transnational crime, and education. U.S. funding for East Asia Pacific Regional programs, a large portion of which support ASEAN, ARF, and APEC objectives, totaled an estimated $20 million in 2009. In the area of security, U.S. foreign assistance supports the Counter-terrorism Regional Strategy Initiative which focuses on transnational aspects of terrorism and regional responses. U.S. assistance to ARF includes funding for regional programs in counter-terrorism, combating transnational crime, disaster preparedness, and non-proliferation. USAID 's Regional Development Mission Asia (RDMA) supports efforts to strengthen the capacity of the ASEAN Secretariat, develop regional economic institutions, and enhance ASEAN's Food Security Information System. RDMA also provides trade-related technical assistance and supports U.S. commitments under the ASEAN-US Enhanced Partnership.

In terms of bilateral assistance, the United States provided an estimated $526 million in FY2009 to nine ASEAN countries (Brunei Darussalam does not receive U.S. assistance). Since 2001, the Philippines and Indonesia have received large increases in U.S. assistance, largely for counterterrorism programs. Vietnam also has received large growth in U.S. aid, reflecting significant funding for HIV/AIDS programs. Among providers of bilateral official development assistance (ODA) as measured by the Organization for Economic Cooperation and Development (OECD), Japan is by far the largest donor in the region, followed by the United States, although Japanese ODA includes a relatively large loan component. France, Germany, the United Kingdom, and Australia also provide significant ODA in the region. China has become a key source of financing and assistance for infrastructure, energy, and industrial development in Southeast Asia.[19]

Regional Powers and Their Relations with ASEAN

ASEAN's most critical external relations continue to be with the United States, the region's primary security guarantor; Japan, the major provider of development assistance; and China, a rising source of aid, trade, and, according to some, strategic influence in the region. Many analysts argue that China's "soft power"—global influence attained through economic, diplomatic, cultural, and other non-coercive means—has grown significantly in the past decade. Furthermore, many observers contend that China's diplomatic outreach, including building links to ASEAN, has surpassed that of the United States during the past several years.[20] Most Southeast Asian leaders and foreign policy experts have welcomed engagement from both the United States and China because of the benefits that strong relations bring; they do not want a single foreign influence to dominate the region, and excluding either power is "not an option." Although Japan is a close development partner in the region, some Southeast Asians would welcome a more robust Japanese diplomatic and security presence. Many analysts view India as an ascendant but still nascent regional power that has an interest in balancing China's rise in the region.

The United States

The United States exerts the most established and forceful military presence the region, including alliances with the Philippines and Thailand (Major Non-NATO Allies), strong security cooperation with Singapore, counterterrorism cooperation with Indonesia and Malaysia, and military education programs in Vietnam, Cambodia, and Laos. The United States is also engaged economically. It is ASEAN's fourth largest trading partner, having been surpassed by China in recent years. The United States is a larger export market than China and the third largest source of FDI from outside the region after the EU and Japan, followed by China (including Hong Kong) and South Korea.[21]

In terms of diplomacy and trade, many in ASEAN considered Washington neglectful of the organization under the Bush Administration, although some foundations were established upon which the Obama Administration has developed its policy of engagement. The United States was the first country to nominate an Ambassador to ASEAN (2008). In 2009, the United States acceded to the Treaty of Amity and Cooperation (TAC), which was seen by many as a symbolic recognition of the value of a multilateral approach to regional security issues.[22] The United States was the last major power in the region to sign the treaty. Although the United States has met the requirements for joining the EAS through its accession to the TAC, the Obama Administration remains undecided about its intent to do so.

In 2005, the United States created a framework for U.S. assistance to ASEAN—the ASEAN-U.S. Enhanced Partnership—encompassing cooperation on political, security, economic, and development issues. This initiative was followed in 2007 by the ASEAN Development Mission Vision to Advance National Cooperation and Economic Integration (ADVANCE). Among the goals of the mission are to help ASEAN and its members work towards an ASEAN community, support the Enhanced Partnership, and promote the U.S.-ASEAN Trade and Investment Framework Agreement (TIFA), signed in 2006, which could be a precursor to a possible FTA with ASEAN.

China

China's ties with ASEAN have reflected attempts to defuse security tensions in the South China Sea, promote economic integration, support infrastructure development, and cultivate diplomatic influence. Some experts argue that China's power projection in the region amounts to a coordinated attempt to dominate the region economically and ultimately militarily. Others contend that although China's influence is growing, in part due to declining American engagement, Bejing has neither the will nor the capacity to aggressively pursue such a strategy, and is content with the U.S. security role in the region, at least in the medium term.[23] Moreover, many Southeast Asian countries remain wary of China's power and intentions and may seek ways to engage China while hedging against its rise.

In 2002, China and ASEAN agreed to the Declaration on Conduct of Parties in the South China Sea as well as several other agreements on economic and agricultural cooperation and nontraditional security threats. China reportedly has favored the ASEAN + 3 (ASEAN, Japan, China, and South Korea) summit process, inaugurated in 1997, over other forums such as the ASEAN Regional Forum and the East Asia Summit. Nonetheless, China has become more active in ARF, which focuses on security issues and dialogue, exceeding U.S. involvement in recent years, according to some analysts. The formation of the EAS in 2005 represented an effort by some countries in the region, including Japan, to balance China's influence by including powers that generally are more aligned with the United States than China on security matters. However, some analysts perceive the U.S. absence in the grouping as working to China's advantage. In 2003, the PRC became the first country to accede to ASEAN's Treaty of Amity and Cooperation.[24]

China committed relatively early to a free trade agreement with ASEAN, signing a framework agreement in 2002 that set a ten-year deadline. In August 2009, China and ASEAN signed a new Investment Agreement to accompany the FTA. A major provider of bilateral development financing in the region and economic assistance to Laos, Cambodia, and Burma, in particular, Chinese leaders announced in April 2009 a plan to set up a $10 billion China-ASEAN Fund on Investment Cooperation to support new infrastructure. Other assistance promised at the time included $15 billion in loans to ASEAN countries to be allocated over three to five years, nearly $40 million to Cambodia, Laos, and Burma "to meet urgent needs," $5 million for the China- ASEAN Cooperation Fund, and rice for a regional emergency rice reserve.[25]

Japan

Japan has been a close partner to ASEAN and the principal provider of development assistance to Southeast Asia, but its role has been relatively low-profile.[26] In the past few years, Japanese governments have pledged to strengthen ties to the organization and to Indonesia, in part to balance China's rising influence. In November 2009, Japanese Prime Minister Yukio Hatoyama pledged $5.5 billion in assistance to the Mekong delta region, in large part to bolster Japan's role in a part of Southeast Asia that is becoming economically integrated with China.[27]

Tokyo has long been actively involved in the three major satellite groupings—ASEAN + 3, ARF, and the EAS. While stressing the importance of Japanese ties with the United States, Japanese governments have supported the formation of an East Asian Community, which may include members of the EAS (ASEAN + 6) as its core (excluding the United States). ASEAN reportedly is divided over whether to include the United States in such as grouping.[28] Japan acceded to the TAC in 2004 and appointed an ambassador to ASEAN in 2008. In 2005, the Japanese government reportedly pledged $70 million for ASEAN regional integration projects. Cooperation and aid activities with ASEAN have included counterterrorism, environmental protection, and preventing the spread of infectious diseases.

In addition to the Japan-ASEAN FTA, Tokyo has signed Economic Partnership Agreements with Singapore, Thailand, Indonesia, Malaysia, and the Philippines, which involve not only trade liberalization but also the areas of labor movement, investment, intellectual property rights, and cultural and educational cooperation.

ISSUES FOR CONGRESS

Much of the Congressional activity concerning Southeast Asia deals with bilateral relations and issues with individual Southeast Asian nations. In recent years, however, Congress has also sometimes played a leadership role in initiatives towards ASEAN. In 2006, Senator Richard Lugar introduced the U.S. Ambassador for ASEAN Affairs Act (S. 2697), urging the Bush Administration to name an Ambassador to the grouping. Its passage helped lead to the naming of Scot Marciel as the first U.S. Ambassador to ASEAN and the first ambassador to the organization from outside the region.

There are several ways in which shifts in the U.S. approach towards ASEAN could be of importance to Congress. Congress may also seek to provide further assistance to support ASEAN's Secretariat and organizational capacity building. In trade policy, Congress may consider, on the one hand, pushing for further economic engagement and the passage of FTAs or other agreements with ASEAN and/or its member countries. In October 2009, Senator Richard Lugar introduced S.Res. 311, calling for the start of discussions on a free trade agreement with ASEAN. Stalled FTA discussions with Malaysia and Thailand could potentially be considered by Congress, although this does not appear to be on the near-term agenda.[29] On the other hand, Congress could prevent further FTA negotiations with Southeast Asian countries or ensure that labor and environmental concerns are addressed in such negotiations.

Shifts in U.S. policy toward Burma and the implications for relations with ASEAN have been a major focus in 2009 and will likely continue to be of congressional interest. Senator Jim Webb, Chair of the Senate East Asia and Pacific Affairs Subcommittee, in August 2009 became the first Member of Congress in ten years to visit Burma. Senator Webb also traveled to Thailand, Cambodia, Laos and Vietnam, where he reportedly told leaders that ASEAN should call for the release of Aung San Suu Kyi.[30]

Over recent years, Congress has been a leader of the U.S. sanctions policy towards the Burmese regime through legislation such as the Burmese Freedom and Democracy Act of 2003 and the Tom Lantos Block Burmese JADE (Junta's Anti-Democratic Efforts) Act of 2008. The outcome of parliamentary elections set to be held in Burma in 2010 may play a significant role in how the Obama Administration implements its new Burma policy, and it its relations with ASEAN vis-a- vis Burma. Congress may seek to play an active role in the development of U.S. policy towards Burma and ASEAN, both before and after Burma's elections.

The development of ASEAN's human rights body may also merit attention. Congress has frequently considered legislation and resolutions concerning human rights conditions in Southeast Asia, and ASEAN's emerging human rights approaches may be of interest in future consideration of how to promote human rights in the region.

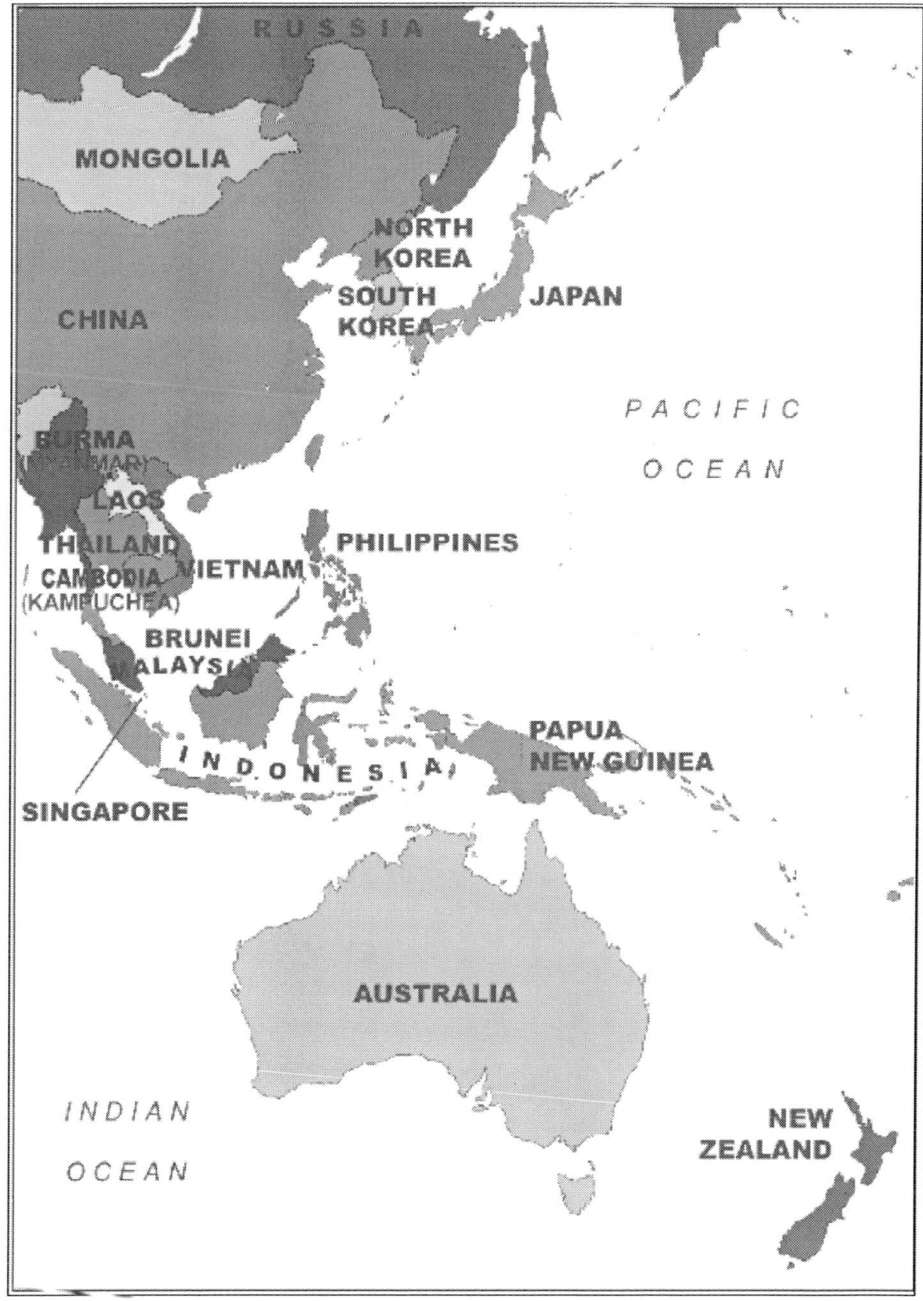

Source: Congressional Research Service.

Figure 1. Southeast Asia and Surrounding Regions.

Table 4. Selected CRS Reports on Southeast Asia

- CRS Report R40495, *Asia Pacific Economic Cooperation (APEC) and the 2008 Meetings in Lima, Peru*, by Michael F.Martin
- CRS Report RS22737, *Burma: Economic Sanctions*, by Larry A. Niksch and Martin A. Weiss
- CRS Report RL34225, *Burma and Transnational Crime*, by Liana Sun Wyler
- CRS Report RL32986, *Cambodia: Background and U.S. Relations*, by Thomas Lum
- CRS Report R40361, *China's Foreign Aid Activities in Africa, Latin America, and Southeast Asia*, by Thomas Lum et al.
- CRS Report RL34620, *Comparing Global Influence: China's and U.S. Diplomacy, Foreign Aid, Trade, and Investment in the Developing World*, coordinated by Thomas Lum
- CRS Report RL33653, *East Asian Regional Architecture: New Economic and Security Arrangements and U.S. Policy*, by Dick K. Nanto
- CRS Report RL32394, *Indonesia: Domestic Politics, Strategic Dynamics, and American Interests*, by Bruce Vaughn
- CRS Report RL34320, *Laos: Background and U.S. Relations*, by Thomas Lum
- CRS Report RL33445, *The Proposed U.S.-Malaysia Free Trade Agreement*, by Michael F. Martin
- CRS Report RL33233, *The Republic of the Philippines: Background and U.S. Relations*, by Thomas Lum and Larry A. Niksch
- CRS Report RS20490, *Singapore: Background and U.S. Relations*, by Emma Chanlett-Avery
- CRS Report RL34194, *Terrorism in Southeast Asia*, coordinated by Bruce Vaughn
- CRS Report RL32593, *Thailand: Background and U.S. Relations*, by Emma Chanlett-Avery
- CRS Report R40583, *U.S. Accession to the Association of Southeast Asian Nations' Treaty of Amity and Cooperation (TAC)*, by Mark E. Manyin, Michael John Garcia, and Wayne M. Morrison
- CRS Report RL31362, *U.S. Foreign Aid to East and South Asia: Selected Recipients*, by Thomas Lum
- CRS Report R40208, *U.S.-Vietnam Relations in 2009: Current Issues and Implications for U.S. Policy*, by Mark E. Manyin
- CRS Report R40755, *U.S.-Vietnam Economic and Trade Relations: Issues for the 111th Congress*, by Michael F. Martin

End Notes

[1] ASEAN's 10 members are: Brunei Darussalem, Burma (Myanmar), Cambodia, Indonesia, Laos, Malaysia, the Philippines, Singapore, Thailand, and Vietnam.

[2] U.S. Alliances and Emerging Partnerships in Southeast Asia: Out of the Shadows. Center for Strategic & International Studies. July 2009.

[3] The 2008 appointment of a U.S. ambassador to ASEAN was in part prompted by congressional legislation. In the 109th Congress, the Senate passed by unanimous consent S. 2697 (Lugar), the United States Ambassador for ASEAN Affairs Act. More than twenty-five countries have since appointed an ambassador to ASEAN.

[4] See CRS Report R40583, *U.S. Accession to the Association of Southeast Asian Nations' Treaty of Amity and Cooperation (TAC)*, by Mark E. Manyin, Michael John Garcia, and Wayne M. Morrison.

[5] In the last years of the Bush Administration, there were plans to follow up on the meeting with a first-ever ASEAN-U.S. summit. However, these plans did not come to fruition.

[6] "Joint Statement—1st ASEAN-U.S. Leaders' Meeting," Singapore, November 15, 2009, http://www.aseansec.org/24020.htm.

[7] State Department Press Release, "Joint Press Statement of the U.S.-Lower Mekong Ministerial Meeting," July 23, 2009.

[8] EAS is also referred to as "ASEAN + 6".

[9] For more about APEC's history and objectives, see CRS Report R40495, *Asia Pacific Economic Cooperation (APEC) and the 2008 Meetings in Lima, Peru,* by Michael F. Martin.

[10] Rodolfo Severino, "ASEAN," Southeast Asia Background Series No. 10, *Institute of Southeast Asian Studies,* Singapore, 2008.

[11] As a region, ASEAN has an estimated 22 billion barrels of oil, 227 trillion cubic feet of natural gas, 46 billion tons of coal, 234 gigawatts of hydropower and 20 gigawatts of geothermal capacity "ASEAN Plan of Action for Energy Cooperation 2004-2009," ASEAN Energy Ministers Meeting, June 9, 2004, Manila.

[12] Sheldon Simon, "ASEAN and the New Regional Multilateralism," in David Shambaugh and Michael Yahuda, *International Relations of Asia, (*New York: Rowman and Littlefield Publishers, 2008), p. 210-11.

[13] Donald Weatherbee, *International Relations in Southeast Asia: The Struggle for Autonomy* (New York: Rowman and Littlefield Publishers, 2009).

[14] Walter Lohman, "Guidelines for U.S. Policy in Southeast Asia," *Heritage Foundation Backgrounder #2017,* March 20, 2007.

[15] Intra-ASEAN trade contributes indirectly to the U.S. bilateral trade deficit with China. As a result, ASEAN members are concerned about China-U.S. trade relations for their own economic reasons. In addition, any U.S. actions designed to reduce imports from China are likely to increase U.S. imports from ASEAN nations as manufacturers shift final assembly out of China and into Southeast Asia.

[16] The six members who are to be party of AFTA in 2010 are: Brunei Darussalam, Indonesia, Malaysia, the Philippines, Singapore and Thailand; Burma, Cambodia, Laos, and Vietnam are to become parties in 2015.

[17] Under AFTA's provisions, parties to the agreement can designate products for inclusion on the Highly Sensitive List (i.e. rice) and the General Exception List, and are excluded from the tariff reduction requirements.

[18] "ASEAN Chairman's Statement on Myanmar," ASEAN Secretariat, 11 August 2009, available online at http://www.aseansec.org/PR-090812-1.pdf.

[19] For further information, see CRS Report RL3 1362, *U.S. Foreign Aid to East and South Asia: Selected Recipients,* by Thomas Lum; Organization for Economic Cooperation and Development (OECD), http://www.OECD.org, *Development Database on Aid from DAC Members.*

[20] For further information, see CRS Report RL34620, *Comparing Global Influence: China's and U.S. Diplomacy, Foreign Aid, Trade, and Investment in the Developing World,* coordinated by Thomas Lum.

[21] ASEAN data

[22] Stanley Foundation, "New Power Dynamics in Southeast Asia: Changing Security Cooperation and Competition," *Policy Dialogue Brief,* October 2007.

[23] For divergent views on China's potential impact on the region, see Dan Blumenthal, Testimony before the Senate Foreign Relations Committee, Subcommittee on East Asian and Pacific Affairs, Hearing on Maritime Territorial Disputes and Sovereignty Issues in East Asia, July 15, 2009; and Robert G. Sutter, Testimony before the U.S.-China Economic and Security Review Commission, "China's Role in the World: Is China a Responsible Stakeholder?" August 3, 2006.

[24] For further information, see CRS Report RL33653, *East Asian Regional Architecture: New Economic and Security Arrangements and U.S. Policy*, by Dick K. Nanto. See also Frank Frost, "ASEAN's Regional Cooperation and Multilateral Relations: Recent Developments and Australia's Interests," Research Paper, Parliament of Australia, October 9, 2008.

[25] "China's $10-B ASEAN Dev't Fund Starts with $1 B," *Manila Bulletin*, October 23, 2009; "China Rolls out Aid Package for ASEAN," *China Daily*, April 12, 2009.

[26] Richard Halloran, "Rudderless Region at Sea; No Country Ready or Able to Take Charge," *Washington Times,* August 21, 2009.

[27] "Japan Vows $5.5 Bln Aid to Mekong Region at Summit," *Reuters*, November 6, 2009.

[28] "ASEAN Said Divided over Inclusion of US in East Asia Community," *Kyodo News Service*, October 25, 2009.

[29] For further information about the U.S.-Malaysia FTA, see CRS Report RL3 3445, *The Proposed U.S.-Malaysia Free Trade Agreement*, by Michael F. Martin.

[30] "ASEAN Could Call for Suu Kyi Release: U.S. Senator," *Agence France-Presse*, August 18, 2009.

In: Association of Southeast Asian Nations (ASEAN)... ISBN: 978-1-62417-982-2
Editor: Margaret E. Stamlin

Chapter 2

U.S. Accession to the Association of Southeast Asian Nations' Treaty of Amity and Cooperation (TAC)

Mark E. Manyin, Michael John Garcia and Wayne M. Morrison

Summary

On July 22, 2009, during Secretary of State Hillary Rodham Clinton's visit to Southeast Asia, the United States acceded to the Association of Southeast Asian Nations' (ASEAN) Treaty of Amity and Cooperation (TAC), one of the 10-nation organization's core documents, as had been amended by the 1987 and 1998 TAC Protocols. The move came less than six months after Secretary of State Clinton announced in Jakarta that the Obama Administration would launch its formal interagency process to pursue accession. This chapter analyzes the legal and diplomatic issues involved with accession to the TAC.

ASEAN is Southeast Asia's primary multilateral organization. Its 10 member-nations include over 500 million people. Collectively, ASEAN is one of the United States' largest trading partners, constituting about 5%-6% of total U.S. trade. Geographically, Southeast Asia includes some of the world's most critical sea lanes, including the Straits of Malacca, through which pass a large percentage of the world's trade. The TAC was first negotiated in 1976 and subsequently amended to allow non-regional countries to accede. Fifteen countries have done so, including U.S. allies Japan, South Korea, and Australia, as well as China, Russia, and India.

Within ASEAN, accession to the TAC by non-members often is seen as a symbol of commitment to engagement in Southeast Asia, and to the organization's emphasis on multilateral processes. The United States is the last major Pacific power to have acceded. The fact that the United States was not a party to the TAC had been one of many pieces of evidence that Southeast Asian leaders cited in arguing that the United States neglected Southeast Asia generally, and ASEAN specifically. Southeast Asian leaders generally have welcomed the Obama Administration's move, which seems to be designed to boost the United States' standing in Southeast Asia by expanding the multilateral component of U.S.

policy in the region. Some U.S. and Southeast Asian officials and analysts say that expanding U.S. engagement with ASEAN will help boost Southeast Asia's political stature, particularly as China seeks to continue expanding its influence in the region.

The major concern with accession is whether the TAC's emphasis on non-interference in other countries' domestic affairs will constrain U.S. freedom of action, particularly its ability to maintain or expand sanctions on Burma. Proponents of accession often note that Australia has imposed and expanded financial and travel restrictions on Burma since it acceded in 2005. Canberra's restrictions are far less extensive than the sanctions the United States maintains on Burma. The Administration and ASEAN negotiated and exchanged side letters designed to alleviate these concerns. Other objections to accession included arguments that it will accord greater legitimacy to the ruling Burmese junta; a view that ASEAN is insufficiently "action- oriented"; and a belief that the TAC is an untested, arguably ineffectual agreement.

One issue for U.S. policymakers was whether accession to the TAC should take the form of a treaty, subject to the advice and consent of the Senate, or whether the President already has sufficient authority to enter the TAC without further legislative action being necessary. Ultimately, after consulting with selective offices in the Senate, the Administration decided that accession would take the form of an executive agreement, which does not require Senate approval.

INTRODUCTION

On July 22, 2009, in Phuket, Thailand, Secretary of State Hillary Rodham Clinton and representatives from the 10-member Association of Southeast Asian Nations' (ASEAN) signed the Instrument of Extension and the Instrument of Accession to ASEAN's Treaty of Amity and Cooperation (TAC). Secretary Clinton was in Thailand to attend the annual ASEAN Regional Forum Foreign Ministerial. The move came less than six months after she had announced, during a visit to Jakarta, that the Obama Administration would launch its formal interagency process to pursue accession.[1] One of ASEAN's pillars, the TAC was first negotiated in 1976 and subsequently amended to allow non-regional countries to accede.[2] The Administration's move is designed to symbolically boost the United States' standing in Southeast Asia by expanding the multilateral component of U.S. policy in the region. Following consultations with selective offices in the Senate, the Administration decided that accession would take the form of an executive agreement, which does not require Senate approval.

The debate over whether the United States should accede to the TAC raised at least three issues for the Obama Administration and the Congress:

1. how would accession to the TAC advance U.S. interests in Southeast Asia?
2. would accession to the TAC constrain U.S. policy in Southeast Asia, particularly with respect to Burma?
3. should the Administration send the TAC to the Senate for ratification?

Periodically, Congressional measures have called attention to ASEAN and/or called for upgrading U.S. engagement with ASEAN. In the 109th Congress, the Senate passed by unanimous consent S. 2697 (Lugar), the United States Ambassador for ASEAN Affairs Act, which mandated the naming of an Ambassador to the organization. None of the congressional measures dealing with U.S. engagement with ASEAN mentioned U.S. accession to the TAC.

Overview of U.S. Interests in the TAC, ASEAN, and Southeast Asia

Motivations for and Reservations against Acceding to the TAC

The Obama Administration's primary motivation for acceding to the TAC appears to have been to send a signal that the United States seeks to upgrade its presence in Southeast Asia.[3] Many leaders in the region have felt neglected by the United States in recent years.[4] ASEAN leaders have long viewed the TAC not only as a constitutional document for the organization, but also as establishing guiding principles that have built confidence among members, thereby contributing to maintaining regional peace and stability.[5] Accession to the TAC by non-members often is seen as a symbol of their commitment to engagement in Southeast Asia and the organization's emphasis on multilateral processes. As shown in **Table 1**, prior to July 2009, the United States was the only major Pacific power that had not joined the TAC; traditionally, the U.S. presence in Southeast Asia has been organized primarily along bilateral lines.

Additionally, acceding to the TAC is also one of the three requirements for joining the East Asia Summit (EAS), a four-year old forum that features an annual meeting among the heads-of-state of the ASEAN members, China, Japan, South Korea, India, Australia, and New Zealand. The other two requirements are dialogue partnership and significant economic relations with ASEAN, both of which the United States already meets. It is unclear whether the Obama Administration plans to join the EAS, or to what extent U.S. participation would be resisted by EAS members, particularly China and Malaysia, which in the past have voiced reservations with U.S. participation. Australia's accession to the TAC in 2005, which reversed years of official policy, was primarily motivated by Canberra's desire to be a founding member of the EAS.

Joining the TAC had been proposed by many in the Asia policy community for several years, and the idea was debated in the George W. Bush Administration. Objections to joining the TAC included arguments that the TAC's emphasis on non-interference in domestic affairs (particularly in Articles 2, 10, and 13) would constrain U.S. freedom of action, particularly its ability to penalize Burma; a concern that the treaty would undermine U.S. security agreements with Asian allies, notably Japan, South Korea, and Australia; a belief that acceding would accord greater legitimacy to the ruling Burmese junta; a view that ASEAN is insufficiently "action-oriented"; and a belief that the TAC is an ineffectual, largely symbolic agreement.

Table 1. The 16 Non-ASEAN Countries that Have Acceded to the TAC

Papua New Guinea (5 July 1989)	Pakistan (2 July 2004)	New Zealand (28 July 2005)	Sri Lanka (30 July 2007)
China (8 October 2003)	Republic of Korea (27 November 2004)	Australia (10 December 2005)	Bangladesh (30 July 2007)
India (8 October 2003)	Russian Federation (29 November 2004)	France (13 January 2007)	Democratic People's Republic of Korea (24 July 2008)
Japan (2 July 2004)	Mongolia (28 July 2005)	Timor-Leste (13 January 2007)	United States (22 July 2009)

Source. Adapted by CRS from ASEAN Secretariat website, Jakarta, in English February 18, 2009.

Proponents of accession countered that the decisions by U.S. allies Australia, South Korea, and Japan to accede to the TAC should negate concerns that the TAC would constrain U.S. policy and/or undermine U.S. alliances. As discussed in detail below, as part of their accession negotiations, Australia and South Korea both signed side letters with ASEAN that were designed to alleviate similar concerns. (See **Appendix B**.) Along the same lines, Japan reached an understanding with ASEAN prior to its accession to the TAC. Following the Australian and South Korean model, the Obama Administration negotiated and exchanged side letters with the Chairman of ASEAN (who in July 2009 was the Thai Foreign Minister) designed to prevent the TAC from constraining U.S. foreign policy actions.

During the debate over accession, Australia's October 2007 promulgation of targeted financial and travel restrictions on over 400 members of the Burmese regime in the aftermath of the regime's September 2007 crackdown against peaceful protesters was cited as evidence that the TAC would not necessarily constrain U.S. policy. Australia, like Japan, has generally followed a policy of quiet engagement of Burma, seasoned with occasional public criticisms and targeted penalties. Canberra's restrictions against the Burmese regime are not nearly as expansive as U.S. sanctions.[6] During its first weeks in office, the Obama Administration announced it would initiate a review of U.S. policy toward Burma. During her February 2009 visit to Asia, Secretary Clinton said that neither sanctions nor the engagement strategies pursued by ASEAN members were working.[7]

U.S. Interests in Southeast Asia

One of the world's largest regional groupings, ASEAN is Southeast Asia's primary multilateral organization. Its 10 member-nations include over 500 million people. Geographically, Southeast Asia includes some of the world's most critical sea lanes, including the Straits of Malacca, through which pass a large percentage of the world's trade. The straits also are important routes for U.S. naval deployments around the globe, including the Middle East and South Asia. Southeast Asia has served as a center and a base for terrorist operations by radical Islamist groups, including Al Qaeda, though the threat posed by such indigenous groups appears to have been significantly reduced since the middle of the decade. The region is a key source and transmission point for many of the world's "human security" problems, including smuggling, narcotics trafficking, piracy, human trafficking, and the spread of contagious diseases such as avian influenza.

Southeast Asia is home to Indonesia, the world's most populous Muslim nation, which is important to the United States for its size, its democratic example for other majority-Muslim countries, and its status as one of the world's largest carbon emitters, largely by virtue of the rapid pace of deforestation. U.S. relations with Malaysia, another core majority-Muslim ASEAN member, also have global and regional importance because of Malaysia's democratic and economic example (it is a middle income country) and because of its attempts to mediate long- running conflicts between Christian and Muslim factions in the southern Philippines.[8] The region also includes two formal U.S. treaty allies with functioning, although sometimes troubled, democracies—Thailand and the Philippines—as well as another close U.S. security partner, Singapore.

Furthermore, diplomatically and strategically, Southeast Asia is the site of a contest for influence among China, the United States, and to a lesser extent Japan. China in particular has expanded its presence and influence in Southeast Asia since the early 2000s. Some commentators have argued that Beijing's increased presence has jeopardized U.S. influence. Others contest this assertion, arguing that the U.S. and China are not locked in a "zero sum" situation in Southeast Asia, that some of China's actions since 2007 have made some Southeast Asians wary of Beijing's actions, and/or that Chinese diplomacy in Southeast Asia is perceived as successful because China has tended to prioritize areas of mutual agreement while putting off issues that are more difficult to resolve.[9] Regardless of whether U.S. interests are materially threatened, China's increased presence in Southeast Asia has made many Southeast Asian leaders eager for a strong U.S. presence in the region.[10] Indeed, one factor motivating the United States' increased engagement with ASEAN in the 2000s has been the desire to support Southeast Asia's political stature as China expands its influence in the region.

U.S.-ASEAN Economic Relations

Collectively, ASEAN is a major U.S. trading partner. Total trade between the United States and the 10 ASEAN countries in 2008 totaled $178 billion. If ASEAN were treated as a single trading partner, it would rank as the fourth largest U.S. export market (at $68.2 billion) and the fifth largest source of U.S. imports (at $110.2 billion) in 2008.[11] Since 2005, between 5%-6% of total U.S. exports by value have been shipped to the ASEAN market, slightly more than exports to Japan. Over the same period, ASEAN has been the source for about 5%-6% of total U.S. imports. ASEAN's share of U.S. trade has fallen since 1995, when it was the destination for nearly 7% of U.S. exports and the source of 8.5% of U.S. imports, by value. [12]

Many analysts who see China as a growing power in East Asia point to the surge in its trade with ASEAN countries vis-a-vis that with the United States. **Table 2** compares Chinese and U.S. trade with ASEAN for 1995, 2005, and 2008.[13] Over this period, China's trade with ASEAN has expanded sharply in terms of trade volume, percentage increase, and size relative to U.S. trade levels.[14]

Table 3 provides trade data on the importance of the United States, as well as of China, from ASEAN's perspective (i.e., using ASEAN trade data). These data indicate that:

- From 1995 to 2007, the share of ASEAN's imports that came from China increased from 2.2% to 12.7%, while the share that came from the United States dropped from 14.6% to 9.6%.

- From 1995 to 2008, the share of ASEAN exports that went to China rose from 2.1% to 9.7%, while the share of ASEAN's exports that went to the United States fell from 18.5% to 11.5%.

According to ASEAN official data, in 2008 (the latest year in which comprehensive ASEAN trade data are available), its top trading partners (excluding intra-ASEAN trade) were Japan (11.4% of total), the European Union (11.8%), China (11.3%), and the United States (10.6%). The United States was ASEAN's third largest export market and its fourth largest source of imports, while China was ASEAN's fourth largest export market and its largest source of imports.

Table 2. U.S. and Chinese Trade with ASEAN: Selected Years

	1995	**2005**	**2008**	**2008/1995 % change**
China's Exports to ASEAN ($millions)	10,474	55,459	114,139	989.7
U.S. Exports to ASEAN ($millions)	39,676	49,637	68,151	71.8
China's Exports to ASEAN as a Percent of Total Exports (%)	7.0	7.3	8.0	—
U.S. Exports to ASEAN as a Percent of Total Exports(%)	6.8	5.5	5.2	—
China's Imports From ASEAN ($millions)	9,901	75,017	116,933	1,081.0
U.S. Imports From ASEAN ($millions)	62,176	98,915	110,157	77.2
China's Imports From ASEAN as a Percent of Total (%)	7.5	11.4	10.3	—
U.S. Imports From ASEAN as a Percent of Total (%)	8.4	5.9	5.3	—
China's Total Trade With ASEAN ($millions)	20,375	130,476	231,072	1,034.1
U.S. Total Trade With ASEAN ($millions)	101,852	148,522	178,308	75.1
China's Total Trade With ASEAN as a % of its Total Trade (%)	7.3	9.2	9.0	—
U.S. Total Trade With ASEAN as a % of its Total Trade (%)	7.7	5.8	5.2	—

Source. World Trade Atlas.

Note. Based on official Chinese (PRC) and U.S. trade data. Current dollars.

Table 3. ASEAN Trade with the United States and China, Selected Years, as a Percent of Total ASEAN Trade

ASEAN Imports (% of total)					
	1995	**2000**	**2006**	**2007**	**2008**
United States	14.6	14.0	9.8	9.6	9.6
China	2.2	5.2	11.5	12.7	12.9
ASEAN Exports (% of total)					
	1995	**2000**	**2006**	**2007**	**2008**
United States	18.5	18.0	12.9	12.3	11.5
China	2.1	3.5	8.7	9.1	9.7

Sources. ASEAN Secretariat, World Trade Atlas, Asian Development Bank, and International Monetary Fund.

The United States is a bigger source of ASEAN's foreign direct investment (FDI) than China (although the European Union and Japan are the two largest investors). During 2006-2008, cumulative U.S. FDI flows to ASEAN were $12.8 billion (or 8.2% of the total), making the United States ASEAN's third largest source of FDI (excluding FDI flows from other ASEAN countries). Over this period, China's FDI flows to ASEAN totaled $3.5 billion or 1.9% of the total, making China the ninth largest source of ASEAN's FDI. In 2008, U.S. FDI flows to ASEAN totaled $3.0 billion versus $619 million from China. (See **Table 4**.)

ASEAN's History and Evolution

Established in 1967 with five original members, ASEAN has evolved from its original Cold War- era goal of containing Chinese and Vietnamese communism.[15] Increasingly, ASEAN is a vehicle for Southeast Asian nations to resolve problems through the "ASEAN way" of informal, consensus-based, and confidence-building efforts rather than through binding commitments or agreements. Since the early 1990s, ASEAN also has been playing a leading role in moving the countries of East Asia toward organizing into cooperative multilateral arrangements. ASEAN often takes the lead in building multilateral institutions because it is viewed as less threatening than China or Japan.[16] Some analysts speculate that this role of neutral convener may be losing some of its utility, as evidenced by the first-ever standalone China-Japan-South Korea tripartite summit in December 2008. Follow-on summits are expected. Previously, the leaders of the three countries had met only on the sidelines of the annual ASEAN "Plus Three" gathering.

ASEAN's consensus-based decision-making and policy of non-interference in members' affairs have led some commentators, particularly from outside the region, to dismiss the organization as a mere "talk shop." They cite ASEAN's ineffectiveness in dealing with transnational issues like drug trafficking, human trafficking, wildlife trafficking, and illegal logging. ASEAN also has not appeared to play a role in some conflicts among members, such as the 2008 and 2009 border skirmishes between Thailand and Cambodia.[17] Indeed, frustrations with ASEAN's internal procedures, continued difficulties with Burma, and the expansion of non-ASEAN regional groupings in Asia have led some prominent Southeast Asians to publicly call attention to ASEAN's limitations.[18]

Table 4. FDI Flows to ASEAN From the EU, Japan, the United States, and China: 2006-2008

	2008		2006-2008 (Cumulative)	
	Value	Percent of Total[a]	Value	Percent of Total[a]
European Union	13,125	27.0	42,180	27.0
Japan	7,157	14.7	25,768	16.5
United States	3,013	6.2	12,777	8.2
China	619	1.3	3,520	1.9
Total outside FDI flows to ASEAN[a]	48,619	—	156,076	—

Source. ASEAN Secretariat.

[a.] Data excludes intra-ASEAN FDI flows.

However, many Southeast Asians contend that ASEAN has been critical to fostering stability, reducing conflict, and promoting trade and economic growth. In the 2000s, some ASEAN members—particularly Indonesia and the Philippines—have pushed to expand the organization's powers. These moves often have been resisted by other members, particularly Vietnam, Cambodia, and Burma. For instance, in 2008, ASEAN adopted a new charter, early drafts of which included provisions for sanctions and a system of compliance monitoring for ASEAN agreements. However, these items eventually were stripped from the charter.[19] In July 2009, ASEAN's Foreign Ministers approved the creation of the organization's first-ever human rights body. The ASEAN Intergovernmental Commission on Human Rights is designed to focus on promoting human rights. The body, launched in October 2009, will not have the power to penalize human rights violations and/or violators.

Overview of TAC Provisions

The TAC establishes general principles governing the relations between State parties, with the intention of promoting "perpetual peace, everlasting amity and cooperation" within Southeast Asia.[20] Towards this end, it provides a mechanism for the pacific settlement of regional disputes between TAC parties. As drafted in 1976, the TAC was open to ratification by the five original members of ASEAN, and was only open to accession by other Southeast Asian States. The TAC was subsequently amended in 1987 to permit the accession of States outside Southeast Asia with the consent of the five ASEAN members, and to establish rules concerning when States outside Southeast Asia could participate in the agreement's dispute-settling mechanism.[21] The TAC was further amended in 1998 to reflect the expansion of ASEAN to 10 members, and to make accession to the TAC by any additional States outside Southeast Asia contingent upon the approval of all 10 ASEAN members.[22]

Article 1 of the TAC announces that the purpose of the agreement is to promote peace and cooperation among the parties. Article 2 provides that in their relations with one another, parties shall be guided by six principles:

- Mutual respect for the independence, sovereignty, equality, territorial integrity, and national identity of all nations;
- The right of every State to lead its national existence free from external interference, subversion, or coercion;
- Non-interference in the internal affairs of one another;
- Settlement of differences or disputes by peaceful means;
- Renunciation of the threat or use of force; and
- Effective cooperation among themselves.

While TAC Article 2 describes these principles as "fundamental," it does not specify that they are the sole principles that may inform relations between parties.

TAC Article 3 obliges parties to endeavor to develop and strengthen their mutual relations and fulfill their obligations under the agreement in good faith.[23]

TAC Articles 4-9 outline party obligations concerning mutual cooperation. Articles 4 and 5 provide that parties shall promote and strengthen active cooperation in the economic, social, technical, scientific and administrative fields on the basis of equality, non-discrimination and mutual benefit. Articles 6 and 7 provide that parties shall collaborate (including through the use of international and regional organizations outside Southeast Asia) to accelerate the region's economic growth, including through promotion of greater use of parties' agriculture and industries, the expansion of trade, and the improvement of economic infrastructure. Article 8 states that parties shall strive to achieve cooperation in the form of training and research facilities in the social, cultural, technical, scientific and administrative fields. Article 9 provides that parties shall retain regular contacts with one another on international and regional matters with a view towards coordinating their policies.

TAC Article 10 provides that no party shall "in any manner or form participate in any activity which shall constitute a threat to the political and economic stability, sovereignty, or territorial integrity of another High Contracting Party." The agreement does not elaborate on the types of activity constituting a "threat" to the political or economic stability, sovereignty, or territorial integrity of another party, or what type of conduct is intended to be barred by the agreement's prohibition on "participat[ion] in any activity" constituting a threat to another party. Presumably, prohibited activity would have to be of a particularly severe nature to constitute a threat to the stability, sovereignty, or integrity of another TAC party.[24]

TAC Articles 11 and 12 provide that parties shall endeavor to promote national and regional resilience.

TAC Articles 13-17 concern the pacific settlement of disputes between parties. Article 13 states that parties shall act in good faith to prevent disputes from arising between them. Parties are obliged to "refrain from the threat or use of force," and are instead called upon to settle disputes "through friendly negotiations." Towards that end, Article 14 establishes a High Council, composed of a ministerial level representative of each State party, to resolve disputes. As amended by the 1987 Protocol, the dispute settlement system established by Article 14 is only applicable to State parties outside Southeast Asia when those States are "directly involved in the dispute to be settled."

TAC Article 15 states that in cases where disputes cannot be settled via direct negotiation between TAC parties, the High Council shall take cognizance of the matter and recommend an appropriate means of settlement, such as good offices, mediation, inquiry, or conciliation. The High Council may also, with the consent of the parties to the dispute, act as a committee for mediation, inquiry, or conciliation. When necessary, the Council shall also recommend appropriate measures to prevent further deterioration of the situation. TAC parties are not legally compelled to abide by the High Council's recommendations.

TAC Article 16 limits application of Article 15 to instances where all parties to the dispute agree to its application. Perhaps for this reason, the High Council has never been convened to resolve a dispute arising under TAC.[25]

TAC Article 17 states that nothing in the agreement precludes parties from seeking recourse pursuant to the modes of peaceful settlement contained in Article 33(1) of the U.N. Charter. Article 33(1) of the Charter provides that U.N. Member States that are parties to a dispute threatening international peace and security shall "seek a solution by negotiation, enquiry, mediation, conciliation, arbitration, judicial settlement, [and may] resort to regional agencies or arrangements, or other peaceful means of their own choice." TAC Article 17 also states that parties are encouraged to resolve disputes through friendly negotiations "before resorting to the other procedures provided for in the Charter of the United Nations."[26] This language appears intended to ensure that TAC's dispute-resolution requirements are not interpreted as violating Article 103 of the U.N. Charter, which provides that Member States' obligations under the U.N. Charter override any conflicting obligations under other international agreements.

TAC Articles 18-20 relate to treaty ratification and accession, entry into force, and the authoritative text of the agreement. As amended by the 1987 and 1998 Protocols, Article 18 provides that accession of any State outside Southeast Asia is subject to the consent of all the States in Southeast Asia, which the agreement expressly lists as the 10 current members of ASEAN. Article 19 describes the procedure by which TAC entered into force. Article 20 notes that the treaty is drawn in the equally authoritative language of all contracting parties. A common English text has also been agreed upon, with any divergent interpretation of the common text to be settled by negotiation.

The TAC does not contain provisions concerning withdrawal from the agreement by a State party, the agreement's relationship to other multilateral or bilateral agreements to which TAC parties may belong,[27] or the remedies available to a party in the event that its rights under the agreement are violated by another party and neither direct negotiation by the parties nor the assistance from the High Council resolves the violation. These matters would presumably be handled in accordance with customary practice, absent evidence of a contrary understanding by TAC parties.[28]

NEGOTIATION OF ACCESSION TO THE TAC

TAC Article 18, as amended, requires the consent of all ASEAN members before candidate States may accede to the agreement. Formal exchanges of correspondence and consultation between ASEAN members and candidates for accession to the TAC are generally made between the candidate and the ASEAN Chairman. In some instances, a candidate will sign a declaration signifying its intent to accede to the TAC contingent upon completion of any necessary domestic procedures.[29] If all ASEAN Members consent to a candidate's proposed accession to the TAC, the Chairman is authorized to sign a preliminary declaration of consent to accession on behalf of ASEAN Members.[30] The accession process is completed once all ASEAN foreign ministers sign an instrument formally consenting to the candidate party's accession to the TAC,[31] and the candidate party signs and submits the instrument of accession.[32] The instrument of accession is typically signed and deposited by the acceding State's foreign minister.

Negotiations regarding accession to the TAC may raise issues related to the interpretation and application of the agreement's provisions. In many cases, a party will attach a reservation, declaration, or understanding to an agreement at the time of accession or ratification when questions or concerns arise regarding an agreement's potential application. The TAC does not contain a provision barring this practice. However, ASEAN members have historically been unwilling to permit an acceding State to make a reservation or declaration upon accession.[33]

Some States seeking to accede to the TAC have instead sought to reach common understandings with ASEAN members regarding the interpretation of certain TAC provisions, and have recorded these shared understandings in an exchange of notes ("side letters") with the ASEAN Chairman prior to acceding to the TAC.[34] Although these side letters are not understood to amend or modify the TAC, they may serve as important interpretative guidance as to the meaning of its provisions. The Vienna Convention on the Law of Treaties, which is recognized as an authoritative guide to treaty law and practice, states that "any instrument which was made by one or more parties in connection with the conclusion of the treaty and accepted by the other parties as an instrument related to the treaty" may be relied upon to assist in interpreting the underlying treaty.[35]

The United States memorialized its understanding of certain TAC requirements during communications with the ASEAN Chairman regarding its proposed accession. This side letter is attached as Appendix A. Similar communications appear to have been made by Australia, South Korea, Japan, and New Zealand when they acceded to the TAC. Most of these communications have not been made publicly available. However, the side letters memorializing understandings reached by Australia with ASEAN members during the TAC accession process are attached as Appendix B.

FORM OF U.S. ACCESSION TO THE TAC

One of the most notable issues that U.S. policymakers were required to consider when contemplating whether to join the TAC was the form that U.S. accession should take. Legally binding international agreements entered into by the United States take the form of either a treaty or an executive agreement.[36] If an agreement is entered into as a treaty, the Senate must provide its advice and consent by a two-thirds majority for the agreement to become "the Law of the Land."[37] The great majority of international agreements that the United States enters into are not treaties but executive agreements—agreements made by the executive branch that are not submitted to the Senate for its advice and consent. Depending upon the circumstances, authority to enter an executive agreement may derive from different sources, including from a statute enacted by Congress which authorizes the Executive to enter the agreement (a congressional- executive agreement), or pursuant to the Executive's constitutional authority in a given area (sole executive agreement).[38] There are a number of provisions in the Constitution that may confer limited authority upon the President to promulgate sole executive agreements, including his Commander-in-Chief authority and power in the area of foreign affairs.[39] Ultimately, the executive branch opted to accede to the TAC as a sole executive agreement, presumably because the agreement is deemed to focus solely upon relations between TAC parties and impose no requirements upon parties' domestic activities.[40]

Arguably, U.S. accession to the TAC could have taken the form of a treaty, with accession being subject to the advice and consent of the Senate, or an executive agreement. Agreements concerning friendly relations, consultation, and cooperation between countries have taken both forms. The United States has concluded numerous agreements which concern amity or friendly relations between parties as treaties.[41] However, these agreements have traditionally focused on different matters than the TAC. Agreements concerning amity and cooperation that have taken the form of treaties generally focus on the rights afforded to each party's nationals in the territory of the other State party.[42] In contrast, the ASEAN TAC appears to focus exclusively on State-to-State relations. It does not appear that the TAC is intended to afford parties' nationals with individually enforceable rights in the territory of other TAC parties, or otherwise modify a State's internal practices.

The United States has concluded several international agreements as executive agreements that address one or more issues covered by the ASEAN TAC—e.g., cooperation on matters involving security, economics, and science and technology.[43] In a few instances, the United States has entered international agreements of similar breadth to the TAC (and in some cases greater breadth) by way of executive agreement. For example, in 1982, the United States concluded an executive agreement with Spain in order to "promote their cooperation in the common defense, as well as ... economic, scientific, and cultural cooperation."[44] Beyond establishing a general framework for relations in these areas, the agreement also contained specific provisions related to basing rights, defense procurement, the status of U.S. forces in Spain, and the establishment of joint committees to promote cooperation on economic, scientific, and cultural matters.[45]

The form that U.S. accession to the TAC would take was the subject of discussion between the executive and legislative branches. One issue that was discussed was the significance of the agreement being titled the *Treaty* of Amity and Cooperation.[46] Although agreements similar to the TAC have often been entered via executive agreements, none of these agreements referred to themselves as "treaties." An examination of official compendiums of international agreements and other sources by CRS found only a single agreement currently in force at the time of U.S. accession to the TAC which, although referring to itself as a "treaty," was concluded by the United States via executive agreement[47]—the Budapest Treaty on the International Recognition of the Deposit of Microorganisms for the Purposes of Patent Procedure.[48]

Although the TAC refers to itself as a "treaty," it is important to distinguish the meaning of this term in the context of international law, in which "treaty" and "international agreement" are synonymous terms for all binding agreements,[49] and "treaty" in the context of domestic American law, in which the term more narrowly refers to a particular subcategory of binding international agreements.[50] In other words, the fact that the TAC uses the word "treaty" in its title does not necessarily mean that it must take the form of a "treaty" for purposes of U.S. law. Perhaps to prevent confusion, U.S. negotiators have generally avoided using the term "treaty" whenever drafting language of an international agreement which is unlikely to be presented to the Senate for its advice and consent. Because the United States did not participate in the drafting of the TAC, it was not involved in the decision to entitle it a "treaty."

Shortly after the United States acceded to the TAC by way of executive agreement, Senator Mitch McConnell submitted into the *Congressional Record* a July 10, 2009, letter to Secretary of State Hillary Clinton that had been signed by Senator McConnell as well as Senators John Kerry and Richard Lugar, the Chairman and Ranking Member of the Senate Foreign Relations Committee, respectively.[51] In the letter, written prior to U.S. accession to the TAC, the Senators expressed their view that "it is consistent with U.S. practice for the United States to accede to the TAC as an executive agreement."[52] However, the Senators also expressed their belief that, while the TAC could properly be entered as a sole executive agreement, other agreements labeled "treaties" should generally be submitted to the Senate for its advice and consent. Specifically, the letter stated:

> We note that the title of the agreement refers to the agreement as a "treaty," and we are unaware of any precedent for the United States acceding to an agreement styled as a "treaty" without the advice and consent of the Senate as provided for in Article II, Section 2 of the Constitution. At the same time, we are mindful that other factors apart from the formal name of the agreement could suggest that it is consistent with U.S. practice for the United States to accede to the TAC as an executive agreement. Of particular importance, the agreement is largely limited to general pledges of diplomatic cooperation and would [generally] not appear to obligate the United States to take (or refrain from taking) any specific action ... We also note that the United States did not take part in the negotiations among ASEAN countries leading up to the conclusion of the TAC in 1976, or in the decision to characterize it as a treaty. In light of these unique considerations, we will not object to the Department's plan to accede to the TAC as an executive agreement. We continue to believe, however, that the use of the term "treaty" in the title of an agreement will generally dictate that Senate advice and consent will be required before the United States may accede to the agreement. In this regard, treatment of the TAC as an executive agreement should not be considered a precedent for treating future agreements entitled "treaties" as sole executive agreements. To ensure our understanding that the process surrounding this agreement is not misinterpreted in the future as a precedent, we will submit this letter into the Congressional Record. We would also request that the State Department include it in the next edition of the Digest of United States Practice in International Law.[53]

If Congress believes that accession to the TAC via sole executive agreement is inappropriate, or disagrees with the manner in which the Executive implements TAC requirements, it has several tools available by which it may address these concerns. For example, Congress could enact new legislation that modifies or repudiates U.S. adherence to or implementation of the agreement. It could require the Executive to submit information to Congress or congressional committees regarding U.S. implementation of its TAC commitments. Congress could also limit or prohibit appropriations necessary for the Executive to implement the provisions of the TAC, or condition such appropriations upon the Executive implementing the agreement in a particular manner.

Potential Implications of TAC Accession for U.S. Law

While the TAC may impose obligations upon parties in their mutual relations as a matter of *international law*, its requirements do not appear to be of the kind that would create binding federal law enforceable by U.S. courts.[54] Many of the agreement's clauses address parties' obligations in non-specific terms—e.g., requiring parties to "endeavor to develop and strengthen ... ties" and "achieve the closest cooperation on the widest scale."[55] Such clauses appear to lack the precision necessary to be considered legally enforceable, though they may nonetheless carry political or moral weight for TAC parties.[56] More broadly, while the agreement establishes guidelines for parties in their relations with one another, it does not expressly require parties to modify any of their existing domestic laws, even if such laws arguably conflict with the principles espoused by the TAC.[57] Instead, the TAC expressly calls on parties to resolve any dispute between them through diplomatic means. Accordingly, it does not appear that U.S. accession to TAC would have the effect of modifying or limiting the enforcement of existing domestic laws, though the United States would have an obligation under customary international law to execute its obligations under the TAC in good faith.[58]

Nonetheless, the TAC may have implications for U.S. policies, at least as a matter of international law. Some States that have acceded to the TAC, including the United States, negotiated side letters that address the issues discussed in the following sections.

Right to Individual and Collective Self-Defense

TAC Article 2 provides that a "fundamental" guiding principle in the relationship between parties is the "renunciation of the threat or use of force." This prohibition is similar to that contained in Article 2(4) of the U.N. Charter, which bars U.N. Member States "from the threat or use of force against the territorial integrity or political independence of any state." Unlike the U.N. Charter, however, the TAC does not provide an express exception to this bar in cases where force is used in self-defense.

Even in the absence of a TAC provision recognizing parties' right to use force in self-defense, there is good reason to believe that the agreement is not intended to abrogate this right. Although TAC Article 2 lists six fundamental guiding principles in relations between TAC parties, including renunciation of the threat or use of force, it does not specify that these are the sole principles that may guide relations between State parties. Article 51 of the U.N. Charter recognizes that Members possess "the inherent right of individual or collective self-defense if an armed attack occurs." A nation's right to defend itself from attack is believed by many to be a peremptory norm (*jus cogens*),[59] and accordingly any provision of an international agreement that derogated from this principle would not be legally binding.[60] As such, it seems reasonable to interpret the TAC in a manner that would not be inconsistent with well-established principles concerning a nation's right to defend itself. Nonetheless, in negotiations concerning accession to the TAC, some States have found it necessary to exchange side letters in which it was made clear that parties to the TAC did not interpret the agreement as modifying parties' rights and obligations under the U.N. Charter, including as they relate to the right to self-defense.

Related considerations might be raised with respect to the TAC's effect upon U.S. security arrangements. The United States is a party to numerous bilateral and multilateral security arrangements, including some which oblige parties to assist in the defense of any party that is attacked.[61] Some TAC parties with security arrangements with the United States acceded to the TAC after reaching an understanding with ASEAN members that the TAC was not intended to affect other agreements to which they were parties.[62] Prior to acceding to the TAC, the United States exchanged side letters reflecting its understanding that nothing in the TAC would affect its existing bilateral or multilateral relationships, its rights and obligations under the U.N. Charter, or its ability to take actions it "considers necessary to address a threat to its national interests."[63]

Relationship between the TAC and Other Agreements Concerning Human Rights, Trade, Terrorism, Transnational Crime, and Other Matters

The TAC might also be interpreted as having implications for U.S. policy on matters occurring in the territory of another party to the agreement, if the policy is interpreted as violating the TAC parties' obligations concerning non-interference. TAC Article 2 provides that relations between parties shall be guided by the principle of "non-interference in the internal affairs of one another." This obligation could be interpreted as being a narrowly circumscribed requirement, prohibiting parties from actively attempting to undermine other parties' sovereignty or territorial integrity.[64] However, it is possible that some countries might argue that the prohibition is more broadly applicable to TAC party activities towards one another, including, for example, criticism of another party's domestic human rights record.[65] It is also possible that some countries may argue that the TAC bars the imposition of economic or other sanctions upon a TAC party on account of that party's domestic activities. This issue may be of particular concern for the United States on account of its stringent economic sanctions regime against Burma, a member of ASEAN whose consent is necessary for U.S. accession.

State practice arguably conflicts with the view that the TAC is intended to deter parties from commenting upon or engaging on matters of international interest that arise within the territory of another TAC party.[66] Indeed, some TAC parties have adopted specific measures, including economic sanctions, to deter human rights violations and other practices occurring in the territory of another party. For example, as mentioned above, both Australia and France, which are each parties to the TAC, have imposed sanctions upon Burma.[67] Nonetheless, it is possible that U.S. sanctions policy may become an issue of contention with some TAC members.

The United States and most other parties to the TAC are also parties to international agreements that obligate or permit members to take action to deter specified activities arising in other countries, including with regard to matters involving unfair trade practices, transnational criminal activity, international terrorism, and gross human rights violations.[68] Further, all TAC parties are members of the United Nations, and may be required to comply with Security Council resolutions imposing economic sanctions upon a particular country (including, potentially, another party to TAC).[69] In addition, Article 1 of the U.N. Charter provides that one of the purposes of the organization is "to achieve international cooperation ... in promoting and encouraging respect for human rights and for fundamental freedoms for all without distinction as to race, sex, language, or religion." This language arguably imposes a right and obligation upon U.N. Members to abide by these principles in their relations with other States.[70] Some States acceding to TAC have exchanged side letters with ASEAN Members to clarify the parties' mutual understanding that the TAC is not intended to prevent parties from exercising rights and obligations under other international agreements, including the U.N. Charter. Side letters exchanged by the United States prior to its accession to the TAC also reflect this understanding regarding the TAC's relationship with other agreements.[71]

Application of the TAC to U.S. Relations with Other TAC Parties outside Southeast Asia

Although the TAC was "originally conceived as a legally-binding code of inter-State conduct among Southeast Asian countries,"[72] it was subsequently amended to permit the accession of States located outside Southeast Asia. Since that time, several non-Southeast Asian States have acceded to the TAC (see **Table 1**).

While some provisions of the TAC clearly focus on parties' obligations with respect to the region of Southeast Asia,[73] other provisions of the agreement could be interpreted as being applicable to relations between parties outside the region as well.[74] In acceding to the TAC, some States outside Southeast Asia, including the United States, exchanged side letters to clarify their mutual understanding with ASEAN Members that the agreement was not intended to govern relations between TAC parties outside Southeast Asia, but only TAC parties' relations with Southeast Asian States.[75]

Participation in the High Council

As previously discussed, when the TAC was amended to permit accession by States outside Southeast Asia, it limited their participation in the High Council to instances where the State was directly involved in the dispute to be settled. In contrast, Southeast Asian States are permitted to participate in all High Council meetings, regardless of whether they are parties to the dispute to be settled. The purpose of this limitation appears to have been to ensure that the focus of the TAC remained on Southeast Asia.[76]

In acceding to the TAC, Australia exchanged side letters with ASEAN members clarifying each side's understanding of the rights of States outside Southeast Asia pursuant to TAC Articles 14 and 16. The letters reflected the shared understanding that the High Council could not resolve a dispute in which a State outside Southeast Asia was directly involved without the State's consent, and that if the State agreed to the High Council being convened, it would be permitted to participate in the Council. The United States did not include a similar clarification in its side letter with the ASEAN Chairman.

Appendix A. Side Letter from United States to ASEAN Concerning U.S Accession to the TAC and Mutual Understandings Concerning the Agreement

DEPARTMENT OF STATE
WASHINGTON
July 15, 2009

Excellency:

I have the honor to refer to the Treaty of Amity and Cooperation in Southeast Asia, done at Denpasar, Bali, Indonesia, on February 24, 1976, as amended, (the Treaty), and am pleased to inform you that the Government of the United States of America has decided to accede to the Treaty in accordance with Article 18.

The Government of the United States greatly appreciates the assistance it has derived from the discussions that U.S. officials have had with ASEAN counterparts regarding the Treaty. The U.S. Government, upon acceding to the Treaty, has the further honor to make the following statements. The Treaty will not apply to, or affect, U.S. relationships with states outside of Southeast Asia. The United States' accession to the Treaty does not affect the United States' rights and obligations under other bilateral

His Excellency
Kasit Piromya,
Minister of Foreign Affairs of the Kingdom of Thailand
and Chairman of the ASEAN Coordinating Council,
Bangkok.

DIPLOMATIC NOTE

or multilateral agreements and, noting Article 10, does not limit actions taken by the United States that it considers necessary to address a threat to its national interests. The Treaty will be interpreted in conformity with the United Nations Charter, and the United States' accession does not affect the United States' rights and obligations arising from the Charter of the United Nations.

I have the further honor to note that the United States' accession to the Treaty, which supports the integrity of the Treaty's purpose to promote perpetual peace, everlasting amity and cooperation among the peoples of the High Contracting parties, will further strengthen our substantive ties with ASEAN.

Accept, Excellency, the renewed assurances of my highest consideration.

Hillary Rodham Clinton

Appendix B. Side Letters between Australia and ASEAN Concerning Australia's Accession to the TAC and Mutual Understandings Concerning the Agreement

THE HON ALEXANDER DOWNER MP

MINISTER FOR FOREIGN AFFAIRS
PARLIAMENT HOUSE
CANBERRA ACT 2600

13 JUL 2005

H.E. Mr Somsavat Lengsavad
Deputy Prime Minister and Minister of Foreign Affairs
Lao People's Democratic Republic

Your Excellency

I have the honour to inform Your Excellency that Australia has decided to accede to the Treaty of Amity and Cooperation in South East Asia (the Treaty), in accordance with Article 18.

The Australian Government is pleased to note that Australia's accession to the Treaty will provide further confirmation of the strong friendship, close ties and extensive common interests and objectives which Australia shares with ASEAN Member countries, both individually and collectively.

The Australian Government's decision to accede to the Treaty has been greatly assisted by the extensive discussions which Australian officials have had with ASEAN counterparts on the Treaty. In that context, the Australian Government, in taking the decision to accede to the Treaty, is pleased to note the following understandings of key provisions of the Treaty, on a non-prejudice basis to ASEAN. First, Australia's accession to the Treaty would not affect Australia's obligations under other bilateral or multilateral agreements. Second, the Treaty is to be interpreted in conformity with the United Nations Charter, and Australia's accession would not affect Australia's rights and obligations arising from the Charter of the United Nations. Further, the Treaty will not apply to, nor affect, Australia's relationships with states outside South-East Asia. Finally, Articles 14 and 16 of the Treaty effectively provide that, when a state outside South-East Asia to the Treaty is directly involved in a dispute, the agreement of that state-party is required before the High Council can be convened. Should the High Council be convened, that state would be entitled to participate in the High Council.

The Australian Government is pleased to note that it will lodge a formal instrument of accession to the Treaty following completion of Australia's domestic treaty process, including the necessary consultation with Parliament.

I thank you for your assistance with this matter, and look forward to receiving your reply.

Yours sincerely
SIGNED
Alexander Downer

LAO PEOPLE'S DEMOCRATIC REPUBLIC
Peace, Independence, Democracy, Unity, Prosperity

Deputy Prime Minister,
Minister for Foreign Affairs

Vientiane, 23 July 2005

Excellency,

I would like to express my thanks for Your Excellency's letter dated 13 July 2005 addressed to me as Chairman of the 38th ASEAN Standing Committee, concerning the intention of Australia to accede to the Treaty of Amity and Cooperation in Southeast Asia.

We believe that Australia's accession to the Treaty of Amity and Cooperation in Southeast Asia would further contribute to the strengthening of cooperation between ASEAN and Australia, particularly in the promotion of peace, security and cooperation in the region.

I look forward to receiving Australia's formal instrument of accession to the Treaty.

Yours sincerely,

SIGNED

Somsavat LENGSAVAD
Chairman of the 38th ASEAN Standing Committee

To: His Excellency Alexander Downer MP
Minister of Foreign Affairs
Parliament House
Canberra ACT 2600
AUSTRALIA

CC: - All ASEAN Foreign Ministers
- Secretary-General of ASEAN

End Notes

1 State Department, "United States Accedes to the Treaty of Amity and Cooperation in Southeast Asia," press release, July 22, 2009 (hereinafter "State Dept. press release"); State Department, "Beginning a New Era of Diplomacy in Asia. Remarks With ASEAN Secretary General Dr. Surin Pitsuwan," press release, February 18, 2009, http://www.state.gov/ secretary/rm/2009a/02/1 19422.htm. ASEAN's 10 members are Brunei Darussalam, Burma (Myanmar), Cambodia, Indonesia, Laos, Malaysia, Philippines, Singapore, Thailand, and Vietnam.

2 Treaty of Amity and Cooperation in Southeast Asia, done on February 24, 1976 in Bali, Indonesia (hereinafter "TAC"), available at http://www.aseansec.org/1217.htm.

3 See State Dept. press release, *supra* footnote 1.

4 For instance, in introducing Secretary of State Clinton during her visit to the ASEAN Secretariat, ASEAN Secretary General Surin Pitsuwan said, "your visit shows the seriousness of the United States to end its diplomatic absenteeism in the region." State Department, "Beginning a New Era of Diplomacy in Asia," press release, February 18, 2009.

5 *See* Michael Bliss, "Amity, Cooperation, and Understanding(s): Negotiating Australia's Entry into the East Asia Summit," 26 Australian Book of Int'l L. 63, 79 (2006).

6 For more on Australia's policy toward Burma, see Australian Department of Foreign Affairs and Trade, *Burma Country Brief—March 2009*; Andrew Selth, "Burma's 'Saffron Revolution' and the Limits of International

Influence," *Australian Journal of International Affairs*, vol. 62, no. 3 (September 2008). For more on U.S. policy toward Burma, see CRS Report RL3 3479, *Burma-U.S. Relations*, by Larry A. Niksch, and CRS Report RS22737, *Burma: Economic Sanctions*, by Larry A. Niksch and Martin A. Weiss.

[7] State Department, "Working Toward Change in Perceptions of U.S. Engagement Around the World; Roundtable with Traveling Press," Seoul, South Korea, February 20, 2009.

[8] For more, see CRS Report RL33 878, *U.S.-Malaysia Relations: Implications of the 2008 Elections*, by Michael F. Martin, and CRS Report RL33233, *The Republic of the Philippines: Background and U.S. Relations*, by Thomas Lum and Larry A. Niksch.

[9] For a summary of this debate, see CRS Report RL32688, *China-Southeast Asia Relations: Trends, Issues, and Implications for the United States*, by Bruce Vaughn and Wayne M. Morrison.

[10] *See* Bates Gill et al., *Strategic Views on Asian Regionalism: Survey Results and Analysis*, Center for Strategic and International Studies, Washington, DC, 2009.

[11] These rankings would fall to 5th and 6th respectively, if the 27 countries that make up the European Union are treated as a single trading entity.

[12] Compiled by CRS from U.S. Dept. of Commerce, Bureau of Census via World Trade Atlas.

[13] The global economic crisis has sharply diminished U.S. and Chinese trade with ASEAN in 2009. From January-August 2009, U.S. and Chinese exports to ASEAN were down 30.6% and 18.3%, respectively, over the same period in 2008. U.S. and Chinese imports from ASEAN were down 21.9% and 24.0%, respectively. During this period, ASEAN accounted for 8.6% of China's exports and 5.0% of U.S. exports; it also accounted for 10.3% of China's imports and 5.9% of U.S. imports. These data indicate that the relative importance of ASEAN to China as a trading partner has continued to grow (in the case of exports) or remain steady (in the case of imports) in 2009, but for the United States, the importance of ASEAN trade (both exports and imports) has declined.

[14] This is in line with China's overall trade trends. Between 1995 and 2008, China's global exports and imports rose by 860% and 757%, respectively. U.S. exports and imports rose by 123% and 182%, respectively. For instance, in 1995, U.S. exports to ASEAN were nearly three times those of China, but in 2008, China's exports exceeded those of the United States by 75%. In 1995, U.S. imports from ASEAN were more than six times those of China, but in 2008, China's imports exceeded those of the United States by 6%. As the table shows, ASEAN's economic importance to China has increased; from 1995 to 2008, ASEAN's share of China's total trade rose from 7.3% to 9.0%. On the other hand, the importance of ASEAN for U.S. trade has declined over this period: from 6.8% to 5.2% for U.S. exports, and from 8.4% to 5.3% for imports.

[15] ASEAN's founders were Indonesia, Malaysia, the Philippines, Singapore, and Thailand. Brunei joined in 1984, Vietnam in 1995, Laos and Burma in 1997, and Cambodia in 1999.

[16] For more, see CRS Report RL33653, *East Asian Regional Architecture: New Economic and Security Arrangements and U.S. Policy*, by Dick K. Nanto.

[17] For more on U.S.-ASEAN relations generally and the debate over TAC accession specifically, see Satu Limaye, "United States-ASEAN Relations on ASEAN's Fortieth Anniversary: A Glass Half Full," *Contemporary Southeast Asia*, Vol. 29, No. 3 (2007); Ellen Frost, "Re-Engaging with Southeast Asia," Japanese Institute of Global Communications Commentary, July 28, 2006, http://www.glocom.org/debates/20060728_frost and Richard Cronin, "The Second Bush Administration and Southeast Asia," July 2007, http://www.stimson.org/pub.

[18] See Jusuf Wanandi, "Remodeling Regional Architecture," PacNet #13, February 18, 2009.

[19] For a brief summary of ASEAN's evolution and the debate over the charter, see Julie Ginsberg, "ASEAN: The Association of Southeast Asian Nations," February 25, 2009, Council on Foreign Relations website, http://www.cfr.org.

[20] TAC, art. 1.

[21] Protocol Amending the Treaty of Amity and Cooperation in Southeast Asia, done on December 15, 1987, in Manila, Philippines, available at http://www.aseansec.org/1218.htm. While States outside Southeast Asia may accede to the TAC, membership in ASEAN is limited to Southeast Asian States.

[22] Second Protocol Amending the Treaty of Amity and Cooperation in Southeast Asia, done at Manila, Philippines, July 25, 1998, available at http://www.aseansec.org/702.htm.

[23] Even in the absence of this express language, customary international law establishes that parties to an agreement must execute their obligations in good faith. See Restatement (Third) of Foreign Relations § 321 (1987) (recognizing that "every international agreement in force is binding upon the parties to it and must be performed by them in good faith"); Vienna Convention on the Law of Treaties, entered into force Jan. 27, 1980, 1155 U.N.T.S. 331 (hereinafter "Vienna Convention"), arts. 26, 31. Although the United States has not ratified the Vienna Convention, it recognizes it as generally signifying customary international law. See, e.g., Fujitsu Ltd. v. Federal Exp. Corp., 247 F.3d 423 (2nd Cir. 2001) ("we rely upon the Vienna Convention here as an authoritative guide to the customary international law of treaties ... [b]ecause the United States recognizes the Vienna Convention as a codification of customary international law ... and [it] acknowledges the Vienna Convention as, in large part, the authoritative guide to current treaty law and practice") (internal citations omitted).

[24] For example, the practice of TAC parties in their mutual relations, and more generally in the context of international State practice, suggests that economic sanctions are not typically viewed as an impermissible threat to the sovereignty or integrity of another State. See generally Note, "Economic Sanctions" 11 U. Miami Int'l & Comp. L. Rev. 115 (2003) (discussing legality of economic sanctions under international law). See also Australian Dept. of Foreign Affairs and Trade, "Australian Autonomous Sanctions: Burma," at http://www.dfat.gov.au/un/unsc_sanctions/burma.html (listing sanctions imposed by Australia, a party to the TAC, against Burma, a fellow TAC party and also a member of ASEAN).

[25] See Bliss, *supra* footnote 5, at 79 (noting that "the High Council has never actually convened to consider a dispute under the Treaty").

[26] For example, the U.N. Charter established the International Court of Justice (ICJ), which serves as the "principle judicial organ of the United Nations." U.N. Charter, art. 92. The ICJ may settle international legal disputes between States, and also provide advisory opinions on legal matters referred to it by the U.N. Security Council, General Assembly, or other authorized U.N. bodies. See generally id. at chpt. XIV; Statute of the International Court of Justice, 59 Stat. 1055, T.S. No. 993 (1945), at chpts. II, IV. Each U.N. Member State "undertakes to comply" with any ICJ decision in a case to which it is a party. U.N. Charter, art. 94.

[27] Although TAC Article 17 describes the agreement's relationship with Article 33(1) of the U.N. Charter, it does not explain the agreement's relationship with the U.N. Charter as a whole. For discussion of the TAC's relationship with the U.N. Charter, see *infra* at "Right to Individual and Collective Self-Defense" and "Relationship Between the TAC and Other Agreements Concerning Human Rights, Trade, Terrorism, Transnational Crime, and Other Matters."

[28] See Vienna Convention, arts. 31-32 (describing general rules for treaty interpretation).

[29] See Declaration of Intention to Accede to the Treaty of Amity and Cooperation in Southeast Asia by Australia, signed July 28, 2005, available at http://www.aseansec.org/17624.htm; Declaration on Accession to the Treaty of Amity and Cooperation in Southeast Asia by Japan, signed December 25, 2003. As discussed *infra*, whether U.S. accession to the TAC requires further action by the legislative branch (e.g., the Senate providing advice and consent to accession) may be an issue for U.S. policymakers. In general, it appears that States with presidential or semi- presidential systems of government that have acceded to the TAC have not submitted the agreement to their legislative bodies for approval prior to accession.

[30] See Australian Department of Foreign Affairs and Trade, Report to the Joint Standing Committee on Treaties, "National Interest Analysis of Treaty of Amity and Cooperation with Southeast Asia" (2005) (hereinafter "Australian Analysis of TAC"), at 1, 3, available at http://www.aph.gov.au/House

[31] See, e.g., Instrument of Extension of the Treaty Of Amity And Cooperation In Southeast Asia to India, October 8, 2003, available at http://www.aseansec.org/15280.htm; Instrument of Extension of the Treaty of Amity and Cooperation in Southeast Asia to Republic of Korea, November 27, 2004, available at http://www.aseansec.org/ 16625.htm.

[32] See, e.g., Instrument of Accession to the Treaty Of Amity And Cooperation In Southeast Asia by the Russian Federation, November 29, 2004 , available at http://www.aseansec.org/16638.htm; Instrument of Accession to the Treaty Of Amity And Cooperation In Southeast Asia by the People's Republic of China, October 8, 2003, available at http://www.aseansec.org/15271.htm.

[33] Bliss, *supra* footnote 5, at 80-81.

[34] See generally id. See also Appendix A (side letter from United States to ASEAN Chairman); Appendix B (side letters between Australia and ASEAN).

[35] Vienna Convention, art. 31.

[36] Not every international agreement entered by the United States is intended to be legally binding. In some cases, the United States makes "political commitments" to foreign States. Although these commitments are non-legal, they may nonetheless carry significant moral and political weight. The Executive has long claimed the authority to enter such agreements on behalf of the United States without congressional authorization, asserting that the entering of political commitments by the Executive is not subject to the same constitutional constraints as the entering of legally binding international agreements. See generally Robert E. Dalton, Asst. Legal Adviser for Treaty Affairs, *International Documents of a Non-Legally Binding Character*, State Department, Memorandum, March 18, 1994, available at http://www.state (discussing U.S. and international practice with respect to non-legal, political agreements); Duncan B. Hollis and Joshua J. Newcomer, "'Political' Commitments and the Constitution," 49 VA. J. INT'L L. 507 (2009) (discussing U.S. political commitments made to foreign States and the constitutional implications of the practice). Obligations contained in political commitments may resemble those found in legally binding agreements. For example, the 1975 Helsinki Accords, a Cold War agreement signed by 35 nations, contains provisions concerning territorial integrity, peaceful settlement of disputes, implementation of confidence- building measures, scientific and economic cooperation, and cultural exchange that resemble provisions found in the TAC. However, whereas the obligations contained in the TAC are intended to be legally binding upon parties, those contained in the Helsinki Accords were intended to be political, rather than legal, commitments. See Dalton, *supra*, at 5 ("Clearly, the intent of the parties was that [the Helsinki Accords were] a politically binding not a legally binding document."); ASEAN Public Affairs Office, "ASEAN Knowledge Kit," March 2005, at 1, available at

http://www.aseansec.org/TAC-KnowledgeKit.pdf (describing the TAC as being "originally conceived as a legally- binding code of inter-State conduct among Southeast Asian countries").

[37] U.S. Const., art II, § 2; art. VI, § 2.

[38] There are three types of *prima facie* legal executive agreements: (1) congressional-executive agreements, in which Congress has previously or retroactively authorized an international agreement entered into by the Executive; (2) executive agreements made pursuant to an earlier treaty, in which the agreement is authorized by a ratified treaty; and (3) sole executive agreements, in which an agreement is made pursuant to the President's constitutional authority without further congressional authorization. The Executive's authority to promulgate the agreement is different in each case. For further discussion, see CRS Report RL32528, *International Law and Agreements: Their Effect Upon U.S. Law*, by Michael John Garcia, and CRS Report RL34362, *Congressional Oversight and Related Issues Concerning the Prospective Security Agreement Between the United States and Iraq*, by Michael John Garcia, R. Chuck Mason, and Jennifer K. Elsea.

[39] U.S. Const. art. II, § 1 ("The executive power shall be vested in a President of the United States of America ..."), § 2 ("The President shall be commander in chief of the Army and Navy of the United States ..."), § 3 ("he shall receive ambassadors and other public ministers ..."). Courts have recognized foreign affairs as an area of very strong executive authority. See United States v. Curtiss-Wright Export Corp., 299 U.S. 304 (1936); American Ins. Ass'n v. Garamendi, 539 U.S. 396, 413-417 (2003).

[40] The Foreign Assistance Act of 1961 (P.L. 87-195), as amended, and related legislation provide statutory authority for a broad range of executive agreements in matters including security and economic cooperation, and appear to serve as a legal authority supporting a substantial number of executive agreements entered into in recent decades. See Oona A. Hathaway, "Treaties' End: The Past, Present, And Future Of International Lawmaking In The United States," 117 YALE L.J. 1236at 1256 n. 40 (2008) (noting that a plurality of executive agreements entered since 1980 have involved matters addressed in the Foreign Assistance Act). Arguably, even if the TAC was interpreted to impose duties upon parties to positively engage in cooperative activities, such activities might already be authorized, in whole or in part, under existing U.S. law. The decision by the Executive to characterize accession to the TAC as being effectuated via sole executive agreement suggests an understanding that the TAC focuses upon external relations between TAC parties, and does not impose more concrete requirements upon States (e.g., legally obliging parties to provide financial or other types of assistance to ASEAN Members).

[41] See, e.g., Treaty of Friendship, Commerce and Navigation between the United States and the Republic of Korea, 8 U.S.T. 2217, November 7, 1957; Treaty of Friendship, Commerce and Navigation between the United States and Japan, 4 U.S.T. 2063, October 30, 1953; Treaty of Peace, Amity, Navigation, and Commerce between the United States and Colombia, 9 Stat. 881, June 10, 1848.

[42] See id. Other considerations may also affect the decision to enter an agreement of amity and cooperation as a treaty. For example, in 1976 the United States ratified a treaty with Spain entitled a "Treaty of Friendship and Cooperation," which included provisions concerning U.S. basing rights and the status of U.S. forces in Spain. The decision to enter the agreement as a treaty rather than as an executive agreement was primarily motivated by Senate concern over the Executive entering into a defense agreement with the unpopular Franco regime without first obtaining the advice and consent of the Senate. Following the end of the Franco regime and Spain's admittance into NATO, agreements between the United States and Spain covering the same issues as the 1976 treaty have been concluded as executive agreements. See *infra* at footnote 45.

[43] See, e.g., Agreement Relating to Scientific and Technological Cooperation between the United States and Turkey, TIAS 12185, June 14, 1994; General Agreement for Economic, Technical and Related Assistance between the United States and El Salvador, 13 U.S.T. 266, January 16, 1962; Agreement of Cooperation between the United States and Liberia, 10 U.S.T. 1598, July 8, 1959 (pledging cooperation in furthering economic development of Liberia and consulting on appropriate action in event that Liberia's security is threatened).

[44] Agreement on Friendship, Defense and Cooperation Between the United States and Spain, with Complementary Agreements, 34 U.S.T. 3885, entered into force May 14, 1983, at art. 1.

[45] In 1970, the Nixon Administration concluded an executive agreement with Spain which contained provisions concerning bilateral cooperation on economic, scientific, agricultural, and other matters, and also established more significant military ties between the countries than had previously existed. Agreement of Friendship and Cooperation between the United States America and Spain, 21 U.S.T. 1677, entered into force September 26, 1970. Several Senators voiced opposition to the agreement, arguing that it should have been presented to the Senate as a treaty. This opposition appeared to be primarily based on the military commitments made by the agreement, rather than with regard to the agreement's provisions concerning cooperation in non-military matters. The Senate thereafter passed S.Res. 469 (91st Cong.), which expressed the sense of the Senate that nothing in the executive agreement with Spain should be deemed a national commitment by the United States. In 1976, the Senate gave its advice and consent to ratification of the Treaty of Friendship and Cooperation with Spain, which included provisions relating to U.S. basing rights and the status of U.S. forces in Spain. Treaty of Friendship and Cooperation between the United States and Spain, with Supplementary Agreements, 27 U.S.T. 3005, entered into force September 21, 1976. Following the end of the Franco regime and Spain

becoming a member of NATO, the United States concluded the 1982 agreement, discussed *supra* at page 14. Despite taking a different form, the 1982 executive agreement is similar in scope to the 1976 treaty.

[46] 155 CONG. REC. S8029-S8030 (July 23, 2009) (hereinafter "Senators' Letter to State Department") (letter to Secretary of State Clinton from Senators Mitch McConnell, John Kerry, and Richard Lugar discussing State Department's congressional consultations regarding the TAC).

[47] This examination involved an electronic database search of several legal treatises and law review articles which discuss executive agreements, as well as a review of international agreements contained in the 2009 edition of *Treaties in Force*, a State Department publication compiling treaties and other agreements to which the United States is currently a party. It should be noted that this review did not consider any agreements that are classified or no longer in force.

[48] Budapest Treaty on the International Recognition of the Deposit of Microorganisms for the Purposes of Patent Procedure, with Regulations, 32 U.S.T. 1241, entered into force August 19, 1980. As the title of the agreement suggests, the Budapest Treaty is a technical agreement which established an international system for microorganism samples to be deposited for purposes of patent procedure. The Executive participated in the drafting of the Budapest Treaty with the intent to submit the final agreement to the Senate for its advice and consent to treaty ratification. However, following consultation with the Senate Committee on Foreign Relations, the agreement was instead handled as an executive agreement. See generally State Department, Digest of U.S. Practice in Int'l Law: 1977, at 788-789; Digest of U.S. Practice in Int'l Law: 1980, at 408-410. The State Department described the President's independent power with respect to the conduct of foreign relations and the statutory authority granted to the Patent and Trademark Office as providing the legal basis for the conclusion and implementation of the agreement. Digest of U.S. Practice in Int'l Law: 1980, *supra*, at 409. See also 35 U.S.C. § 6 (1980) (authorizing the Commissioner on Patents and Trademarks to "superintend and perform all duties required by law respecting the granting and issuing of patents and the registration of trademarks," and "to carry on programs and studies cooperatively with foreign patent offices and international intergovernmental organizations, or ... authorize such programs and studies to be carried out").

[49] Vienna Convention, art. 2.

[50] It should be noted, however, that the term "treaty" is not always interpreted under U.S. law to refer only to those agreements described in Article II, § 2 of the Constitution. See Weinberger v. Rossi, 456 U.S. 25 (1982) (interpreting statute barring discrimination except where permitted by "treaty" to refer to both treaties and executive agreements); B. Altman & Co. v. United States, 224 U.S. 583 (1912) (construing the term "treaty," as used in statute conferring appellate jurisdiction, to also refer to executive agreements).

[51] Senators' Letter to State Department, *supra* footnote 46 .

[52] Id. at S8030.

[53] Id.

[54] *See* Medellin v. Texas, 128 S.Ct. 1346, 1356 (2008) ("This Court has long recognized the distinction between treaties that automatically have effect as domestic law, and those that—while they constitute international law commitments— do not by themselves function as binding federal law.").

[55] TAC, arts. 4, 8.

[56] 22 C.F.R. § 18 1.2(a) (State Department regulation listing specificity of agreement as part of criteria used to determine whether it is legally binding); Congressional Research Service, *Treaties And Other International Agreements: The Role Of The United States Senate, A Study Prepared For The Senate Comm. On Foreign Relations* (Comm. Print 2001), at 52.

[57] Indeed, at least two TAC parties, Australia and France (as a member of the European Union), have imposed sanctions upon Burma, another TAC party and an ASEAN member, even though such measures could arguably be interpreted as being contrary to the principles of the TAC. See Australian Dept. of Foreign Affairs and Trade, *supra* footnote 24; European Union Council Regulation (EC) No.194/2008 (2008), available at http://eur-lex.europa.eu/ LexUriServ/LexUriServ.do?uri=OJ:L:2008:066:0001 :0087:EN:PDF. See also Senators' Letter to State Department, *supra* footnote 51, at S8030 ("the agreement is largely limited to general pledges of diplomatic cooperation and would not appear to obligate the United States to take (or refrain from taking) any specific action [with limited exception").

[58] Restatement, *supra* footnote 23, at § 321; Vienna Convention, arts. 26, 31.

[59] See generally Carin Kahgan, "*Jus Cogens* and The Inherent Right to Self-Defense," 3 ILSA J. Int'l & Comp. L. 767 (1997) (discussing State and scholarly views).

[60] Vienna Convention, art. 53.

[61] See, e.g., North Atlantic Treaty, 63 Stat. 2241, entered into force August 24, 1949; Security Treaty Between Australia, New Zealand and the United States of America, 3 U.S.T. 3420, entered into force April 29, 1952; Mutual Defense Treaty Between the United States of America and the Republic of the Philippines, 3 U.S.T. 3947, entered into force August 27, 1952; Mutual Defense Treaty Between the United States of America and the Republic of Korea, 5 U.S.T. 2368, entered into force November 17, 1954; Treaty of Mutual Cooperation and Security Between the United States of America and Japan, 11 U.S.T. 1632, entered into force June 23, 1960.

[62] See Appendix B (concerning mutual understandings reached between Australia and ASEAN members). In side letters exchanged prior to South Korea's accession to the TAC, the U.S.-South Korean military alliance was specifically listed as an example of an agreement that would not be affected by the TAC. A letter from the ASEAN Chairman that memorialized this understanding is on file with the authors of this chapter.

[63] Appendix A.

[64] See Bliss, *supra* footnote 5, at 77 (comparing principle of non-interference contained in TAC with the principles espoused in the U.N. Declaration on Principles of International Law Concerning Friendly Relations and Cooperation among States in accordance with the Charter of the United Nations).

[65] See id. (noting that Burma, a party to the TAC, had responded to criticism of its human rights record by the U.N. Human Rights Commission by claiming such criticism was a "blatant attempt to interfere" in Burma's domestic affairs).

[66] See Australian Analysis of TAC, *supra* footnote 30, at 6 (claiming that "the longstanding practice of States, including States Parties to the [TAC]... makes clear that nothing in the Treaty is to be interpreted as preventing a State Party from engaging on or commenting upon issues of international interest arising within another State Party to the Treaty").

[67] See *supra* footnote 57.

[68] See, e.g., Convention on the Prevention and Punishment of Genocide, entered into force January 12, 1951, 78 U.N.T.S. 277; Convention against Illicit Traffic in Narcotic Drugs and Psychotropic Substances, entered into force November 11, 1990, Treaty Doc. 101-4, 1988 U.S.T. LEXIS 194; Convention For Suppression of Financing Terrorism, entered into force April 10, 2002, Treaty Doc. 106-49, 2000 U.S.T. LEXIS 131; Agreement Establishing the World Trade Organization, entered into force January 1, 1995.

[69] U.N. Charter, art. 41.

[70] See International Convention on Civil and Political Rights, entered into force March 23, 1976, 999 U.N.T.S. 171, at Preamble (recognizing "the obligation of States under the Charter of the United Nations to promote universal respect for, and observance of, human rights and freedoms"); Australian Analysis of TAC, *supra* footnote 70, at 6-7 (suggesting that U.N. Charter Article 1 gives Member States that right and obligation to engage on and comment upon issues of international importance, including human rights concerns, occurring in the territory of another State).

[71] Appendix A ("the United States' accession to the Treaty does not affect the United States' rights and obligations under other bilateral or multilateral agreements ...").

[72] ASEAN Public Affairs Office, *supra* footnote 36, at 1. See also Bliss, *supra* footnote 5, at 79.

[73] See TAC, art. 6 ("The High Contracting Parties shall collaborate for the acceleration of the economic growth in the region in order to strengthen the foundation for a prosperous and peaceful community of nations in Southeast Asia.").

[74] See id., art. 10 ("Each High Contracting Party shall not in any manner or form participate in any activity which shall constitute a threat to the political and economic stability, sovereignty, or territorial integrity of another High Contracting Party.").

[75] See Appendix A ("The Treaty will not apply to, or affect, U.S. relationships with states outside of Southeast Asia."); Appendix B ("Further, the Treaty will not apply to, nor affect, Australia's relationships with states outside South-East Asia.").

[76] See Bliss, *supra* footnote 5, at 78-79.

In: Association of Southeast Asian Nations (ASEAN)... ISBN: 978-1-62417-982-2
Editor: Margaret E. Stamlin

Chapter 3

ASEAN: REGIONAL TRENDS IN ECONOMIC INTEGRATION, EXPORT COMPETITIVENESS, AND INBOUND INVESTMENT FOR SELECTED INDUSTRIES

United States International Trade Commission

ABSTRACT

This chapter describes trends in regional integration, export competitiveness, and inbound investment for six industries within the Association of Southeast Asian Nations (ASEAN): computer components, cotton woven apparel, hardwood plywood and flooring, healthcare services, motor vehicle parts, and palm oil. The six profiled industries are a subset of 12 priority sectors that ASEAN members identified in 2004 in order to promote regional integration. The members created a regional "Roadmap for Integration" (Roadmap) for each priority sector, and while these Roadmaps have promoted tariff reductions and streamlined certain administrative procedures, their success in promoting regional integration has been mixed. In general, economic factors and national government policies have had more influence than the Roadmaps over regional industrial structures. ASEAN members tend to view each other as competitors for inbound investment and jobs, and the members have no legally binding means to enforce compliance with the objectives of the Roadmaps.

The ease of importing and exporting varies widely among ASEAN members. Procedures for trading are relatively easy to complete in Singapore, Thailand, and Malaysia, but very difficult in Laos and Cambodia. The ASEAN Single Window (ASW) is ASEAN's most visible effort to facilitate trade among members. By enabling the rapid exchange of standardized data among members' customs agencies, it has the potential to bolster trade and support the emergence of intraregional supply chains. However, development of the ASW has proceeded slowly.

The quality of logistics services—such as customs brokerage, freight forwarding, and express delivery—also varies substantially from member to member. Logistics services are world-class in Singapore but poor in Laos, Cambodia, and Burma. In a number of ASEAN countries, restrictive regulations hamper the delivery of high-quality logistics services.

While large gaps persist among ASEAN members in e-commerce proficiency, the last decade has seen improvements throughout the region. Vietnam, in particular, has made significant progress in both e-commerce infrastructure development and legal reform. The ASEAN Secretariat has expressed a strong commitment to supporting wider adoption of e-commerce in the region and helping members to adopt new e-commerce laws, but has contributed less significantly to e-commerce infrastructure development.

ACRONYMS

3PL	Third-party logistics service provider
ACCSQ	ASEAN Coordinating Committee on Standards and Quality
ACDD	ASEAN Customs Declaration Document
ACE	ASEAN Competitiveness Enhancement
ACIA	ASEAN Comprehensive Investment Agreement
ADB	Asian Development Bank
AEC	ASEAN Economic Community
AEM	ASEAN Economic Ministers
AFAS	ASEAN Framework Agreement on Services
AFTA	ASEAN Free Trade Area
AFTEX	ASEAN Federation of Textile Industries
AHTN	ASEAN Harmonized Tariff Nomenclature
AIA	Framework Agreement on the ASEAN Investment Area
AICO	ASEAN Industrial Cooperation Scheme
APEC	Asia-Pacific Economic Cooperation
APWG	Automotive Product Working Group
ASEAN	Association of Southeast Asian Nations
ASW	ASEAN Single Window
ATC	Agreement on Textiles and Clothing
ATFWP	ASEAN Trade Facilitation Work Programme
ATIGA	ASEAN Trade in Goods Agreement
B2B	business to business
B2C	business to consumer
BMD	Bursa Malaysia Derivatives
C2C	consumer to consumer
CARB	California Air Resources Board
CBTA	Greater Mekong Subregion Cross-Border Transport Agreement
CEPT	Common Effective Preferential Tariff
CPC	Central Product Classification
DSM	Dispute Settlement Mechanism
EDI	electronic data interchange
ESCAP	Economic and Social Commission for Asia and the Pacific
EU	European Union
FAO	Food and Agriculture Organization of the United Nations
FDI	foreign direct investment
FLEGT	Forest Law Enforcement, Governance and Trade Action Plan

FSC	Forest Stewardship Council
FTA	free trade agreement
GATS	WTO General Agreement on Trade in Services
GDP	gross domestic product
GMS	Greater Mekong Subregion
HS	Harmonized System of Tariffs
HTS	U.S. Harmonized Tariff System
IAI	Initiative on ASEAN Integration
ICT	information and communication technology
IGA	investment guarantee agreement
IPR	intellectual property rights
ITTO	International Tropical Timber Organization
JCI	Joint Commission International
LEI	Lembaga Ekolabel Indonesia (Indonesian Ecolabeling Instititute)
LPI	World Bank Logistics Performance Index
MDF	medium density fiberboard
MFN	most-favored nation
MOU	memorandum of understanding
MRA	mutual recognition agreement
MTCC	Malaysia Timber Certification Council
NAFTA	North American Free Trade Agreement
NAP	National Automotive Policy
NGO	nongovernmental organization
NSW	National Single Window
NTM	nontariff measure
NTR	normal trade relations
OECD	Organisation for Economic Co-operation and Development
OEM	original equipment manufacturer
PEFC	Programme for Endorsement of Forest Certification
PIS	priority integration sector
QCD	quality, cost, and delivery
REIT	real estate investment trust
RSPO	Roundtable on Sustainable Palm Oil
SAFSA	Source ASEAN Full Service Alliance
SME	small and medium-sized enterprise
TEL	Temporary Exclusion List
UNECE	United Nations Economic Commission for Europe
USAID	U.S. Agency for International Development
VINATEX	Vietnam National Textile and Garment Group
VPA	voluntary partnership agreement
VVF	virtual vertical factories
WTO	World Trade Organization

EXECUTIVE SUMMARY

The Association of Southeast Asian Nations (ASEAN) was established as a regional organization in 1967 to accelerate economic growth, promote regional peace and stability, and enhance cooperation on economic, social, cultural, technical, and educational matters among Southeast Asian countries. The five founding countries— Indonesia, Malaysia, the Philippines, Singapore, and Thailand—were later joined by Brunei Darussalem (Brunei) in 1984, Vietnam (1995), Laos (1997), Burma (1997), and Cambodia (1999). Located in a region of the world with rapidly expanding economies, ASEAN has a combined population of 550 million people largely characterized by rising incomes. As such, the ASEAN nations are already an important market for U.S. companies, and they represent significant opportunities for expanded trade and investment. Trade between the United States and ASEAN countries has grown steadily in recent years, and in 2008 it totaled $177 billion.

This chapter has been prepared by the United States International Trade Commission (USITC, the Commission) at the request of the United States Trade Representative (USTR). The report provides an overview of regional trends in Southeast Asia in the areas of economic integration, export competitiveness, and inbound investment for 6 of 12 priority sectors identified in the ASEAN Economic Community Blueprint adopted in November 2007; the six sectors are agro-based products, automotives, electronics, healthcare, textiles and apparel, and wood-based products. The report also identifies and describes an industry within each of these priority sectors that has undergone either significant changes in regional economic integration with other ASEAN members, export competitiveness, or inbound investment in recent years; in some industries, there have been improvements in several of these areas. The six industries analyzed in greater detail are computer components, cotton woven apparel, hardwood plywood and flooring, healthcare services, motor vehicle parts, and palm oil. Information is provided about each industry's export competitiveness, trade flows, investment, and leading competitive factors, including trade facilitation, logistics services, and e-commerce. The major findings are summarized below.

Major Findings and Observations

ASEAN's manufacturing exports benefit from high productivity growth, proximity to large Asian markets, and free trade agreements, but challenges for future growth remain.

Low wages, high productivity growth in some countries, diverse production conditions, proximity to large Asian markets, and the region's liberalizing trade policy environment, including free trade agreements with a number of countries, have supported the growth of ASEAN's manufacturing exports. Challenges for ASEAN's export competitiveness remain in some countries, however, including a shortage of skilled labor and professionals; the lack of a developed system for setting product standards and conformity assessment procedures; unsophisticated consumer markets; and inadequate physical and institutional infrastructure, such as roads and other transport networks, communications, trade facilitation measures (e.g., customs procedures), and intellectual property rights (IPR) protection.

China is a major competitor to ASEAN for foreign investment and manufacturing jobs, but the new free trade agreement between China and ASEAN countries may create opportunities for integrating Asian production supply chains.

China is a major competitor of ASEAN countries in attracting foreign investment and in integrating regional production chains. However, ASEAN recently concluded an FTA with China. Because China has become an important hub in Asian supply chains, ASEAN countries hope the FTA will offer better opportunities to participate in these networks. ASEAN offers an alternative production location to China for multinational firms wanting to diversify their operations to reduce business and political risks.

A new ASEAN investment agreement was signed in February 2009, and is expected to offer significant improvements for investors.

The ASEAN Comprehensive Investment Agreement (ACIA) will enter into force once it is ratified by all ASEAN members, at which point it will replace the existing ASEAN Investment Area (AIA) and Investment Guarantee Agreement (IGA). Investment in most service industries is covered under the ASEAN Framework Agreement on Services (AFAS). The improvements over the AIA include new investor protections, such as the prohibition of performance requirements, more freedom for investors in appointing senior management, and a more comprehensive investor-state dispute settlement procedure; clear (though nonbinding) timelines for investment liberalization; and benefits extended to foreign-owned, ASEAN-based investors. Member countries' final offers under the agreement have not yet been published, so it is not yet possible to fully evaluate the ACIA's expected impact on intra-ASEAN investment.

The "ASEAN Single Window" could facilitate intra-ASEAN trade, but progress remains slow.

The ASEAN Single Window (ASW) is ASEAN's most visible effort to facilitate trade among members, by enabling the rapid exchange of standardized data among countries' customs agencies. Currently, the ease of importing and exporting varies widely among ASEAN members; procedural requirements are relatively easy to complete in Singapore, Thailand, Malaysia, but very difficult in Laos and Cambodia. Although the ASW has the potential to bolster trade and foster intra-regional supply chains, the development of the ASW has proceeded slowly since its creation in 2005. While there are some indications that work on the ASW is accelerating, faster progress and close consultation with ASEAN's business community on its design would ensure that the ASW contributes robustly to ASEAN economic integration.

Regulatory barriers inhibit improvements in logistics services.

ASEAN countries have committed to liberalize trade and investment in logistics services by 2013. However, while some members have made reforms, others (notably Indonesia) have introduced restrictive regulations that contradict their commitments and the quality of logistics services—such as customs brokerage, freight forwarding, and express delivery—

varies substantially from member to member. Logistics services are world- class in Singapore, but poor in Laos, Cambodia, and Burma.

Infrastructure and the legal framework for e-commerce among certain ASEAN members have improved.

Technical infrastructure (such as plentiful broadband connections), a sound legal framework for electronic transactions, and individuals' skills in using computers are prerequisites for the expansion of e-commerce. The ASEAN Secretariat is committed to supporting wider adoption of e-commerce in the region. It has helped members to adopt new e-commerce laws, but has contributed less significantly to infrastructure development. While large gaps persist among ASEAN members in technical infrastructure, legal frameworks, and computer skills, the last decade has seen improvements throughout the region. Vietnam, in particular, has made impressive progress in both infrastructure development and legal reform.

The ASEAN Roadmaps for Integration have achieved mixed success in promoting regional integration in all six industries.

While ASEAN Roadmaps for Integration (Roadmaps) have promoted tariff reductions and streamlined certain administrative procedures, they have achieved mixed success in promoting regional integration. In general, economic factors and national government policies have more influence over regional industrial structures. A number of different forces have contributed to greater regional integration, export competitiveness, and inbound investment in each of the six industries over the last five years, many of which came from outside the region. These forces include Japanese multinationals that invest and manufacture in more than one ASEAN country (motor vehicle parts), the removal of global quotas (cotton woven apparel), and global production sustainability standards (palm oil). ASEAN members often still view each other as competitors for inbound investment and jobs, and the six Roadmaps include no legally binding means to enforce member compliance with their objectives.

Computer components

While the ASEAN region is the world's second largest exporter of computer components, it faces a major challenge from China.

Production of most computer components is led by multinational firms, and ASEAN countries, especially Malaysia, Singapore, and Thailand, have been competitive investment destinations for over 30 years. In the past five years, Vietnam has also emerged as an important investment destination for the electronics sector, and its role in the computer industry in ASEAN is growing rapidly. The ASEAN region's total share of global computer components exports was 21 percent in 2008, the latest date for which data are available. But challenges remain from China. China has increased its share of global computer components production at the same time it has become the single largest assembler of finished computers. Many computer components firms are increasing production in China in order to be closer to their customer: the computer assembler.

Competition between ASEAN countries hampers efforts to create a more integrated regional computer components production base.

Using financial incentives and government policies, ASEAN countries that produce computer components compete to attract export-oriented investment and jobs from multinational firms and integrate their economies into the global computer supply chain. These incentives and policies are generally not coordinated between ASEAN members governments, hampering regional competitiveness.

Cotton woven apparel

Tariff reductions have helped facilitate a small but growing amount of integrated production of cotton woven apparel among ASEAN countries. However, trade programs such as free trade agreements have more heavily influenced regional integration.

While integrated production of cotton woven apparel among the ASEAN countries is still at a low level, it has been growing over the last few years. Intra-ASEAN trade in cotton woven fabrics, the primary input in cotton woven apparel, increased by 13 percent, from $168.4 million in 2004 to $190.8 million in 2008. To a certain degree, the reduction in duties on fabric inputs among member countries, as outlined in the ASEAN Roadmap for Integration of Textiles and Apparel, helped facilitate this integration. But in most cases, these duty reductions do not provide an incentive for regional integration when the final apparel is exported outside of the ASEAN region. Other factors have had a greater influence on regional integration in this industry. For example, the ASEAN-Japan FTA implemented in 2008 has been a driver for integrated ASEAN production because of the rules of origin in the agreement. Also stimulating integration are such factors as the increased competition among global suppliers created by the removal of quotas on textiles and apparel and the granting by the United States of normal trade relations (NTR) status to Vietnam.

Hardwood plywood and flooring

Access to legally sourced and sustainable wood supplies is an important competitive factor affecting integration, export competitiveness, and inbound investment in the ASEAN hardwood plywood and flooring industry.

To be competitive, ASEAN producers of hardwood plywood and flooring require access to wood raw materials that are becoming increasingly scarce, and they need to demonstrate to buyers in major markets that wood sources are legal and sustainable. In some fashion, each of the ASEAN countries regulates the volume of timber harvesting in natural forests, and some have imposed complete or partial bans on exports of wood raw materials. Amid mounting pressure to address illegal logging and deforestation, government (and other) buyers in consuming countries are increasingly requiring assurances that wood products are produced from legal and sustainable sources. ASEAN collaboration in responding to these concerns may be helping the industry to integrate in some respects, but ASEAN countries are also working independently in other bilateral and multilateral processes to protect their global competitiveness.

Despite tariff reductions, the region's competitive focus remains on exporting to major industrialized countries.

Intra-ASEAN trade of hardwood plywood and flooring is small compared with exports to non-ASEAN markets; tariff reductions coupled with other trade facilitation measures may encourage more intra-regional trade, but the region's hardwood plywood and flooring production is heavily weighted toward exports to Europe, Japan, and the United States. Over the past five years, China has become a formidable competitor to ASEAN in global markets, despite its own wood scarcity. Other significant challenges facing ASEAN hardwood plywood and flooring producers in major markets are product standards (such as formaldehyde emissions limits) and procurement policies requiring assurances of legal sourcing.

Healthcare services

Growth of private healthcare firms in the ASEAN market has generated increased trade and investment in healthcare services.

Malaysia, Singapore, and Thailand are the largest exporters of healthcare services, treating the largest number of foreign patients within their borders. Private healthcare firms from these countries have promoted their services in neighboring countries and expanded their presence by investing throughout the region to meet growing regional demand. Additionally, as these healthcare firms have grown, they have attracted increasing investment from investors outside of the ASEAN region.

In response to rising regional trade and investment in the healthcare services industry, member governments and the ASEAN Secretariat have launched efforts to support continued growth.

Governments of ASEAN member countries are encouraging investment in, and promoting exports of, healthcare services through programs that simplify procedures for the movement of foreign patients across borders and liberalize investment barriers. Similarly, the ASEAN Secretariat has facilitated regional healthcare investment through recent negotiations under the ASEAN Framework Agreement on Services, and initiated efforts to conclude mutual recognition agreements in certain medical professions.

Motor vehicle parts

ASEAN has been successful at meeting critical Roadmap targets and facilitating regional integration within the automotive sector. Despite these achievements, the regional automotive industry and market have yet to fully integrate.

ASEAN has eliminated duties or reduced nontariff measures for six members (Brunei, Indonesia, Malaysia, Philippines, Singapore, and Thailand) and implemented the ASEAN Industrial Cooperation Scheme (AICO); these initiatives have enhanced regional integration, export competitiveness, and inbound investment. Auto and parts manufacturers, in particular Japanese firms, have used the AICO system since the 1990s to develop an integrated production system in the region. ASEAN, however, has a history of national government intervention in support of local automotive industries through such measures as national car policies and local-content requirements that have protected local industries and encouraged the development of small-country markets. In addition, ASEAN member countries have made limited progress in harmonizing automotive safety and environmental standards and improving customs cooperation and coordination.

Further regional integration is key to improving the competitiveness of the regional motor vehicle parts industry by providing economies of scale.

ASEAN's large combined market provides it with the size necessary to manufacture components cost effectively. This is particularly important for the auto parts industry, which relies on production volume to lower costs and to supply auto assembly plants from a small number of focused production facilities.

Palm oil

The Roadmap for Integration of Agro-based Products has had far less impact on the structure of the ASEAN palm oil industry than have multinational corporations and international groups.

Two groups of influential actors have driven ASEAN regional actions in the palm oil industry over the last five years: multinational corporations, which produce and refine palm oil and process it into downstream goods, and international groups, such as the Roundtable on Sustainable Palm Oil (RSPO), through which the corporations seek to influence developments in the palm oil market. Investment and purchasing decisions by multinational palm oil producers and consumers are largely driven by the low price and abundant supply of palm oil (relative to other vegetable oils) and increasingly by sustainability concerns surrounding palm oil production.

ASEAN palm oil cultivation and initial processing can take place on a large scale in only two countries: Indonesia and Malaysia.

Because of climate requirements and the need for milling soon after harvest, palm oil can only be a large-scale, commercially viable agro-based industry in Indonesia and Malaysia, although small-scale palm oil cultivation exists in several other ASEAN countries. The prime growing conditions and developed processing infrastructure in Indonesia and Malaysia make them the most important palm oil suppliers to ASEAN and the world. The low cost of this vegetable oil, coupled with production that far exceeds ASEAN demand, results in most ASEAN palm oil production being exported to large markets such as China, the EU, and India.

1. INTRODUCTION

Purpose, Scope, and Organization of the Report

This chapter provides an overview of regional trends in economic integration, export competitiveness, and inbound investment for 6 of 12 priority sectors identified in the Association of Southeast Asian Nations (ASEAN) Economic Community Blueprint adopted in November 2007; the six sectors are agro-based products, automotives, electronics, healthcare, textiles and apparel, and wood-based products. The report also identifies and describes an ASEAN industry within each of these sectors that has undergone either significant changes in regional economic integration with other ASEAN members, export competitiveness, or inbound investment in recent years; in some industries, there have been improvements in several of these areas. The six industries are computer components, cotton woven apparel, hardwood plywood and flooring, healthcare services, motor vehicle parts, and palm oil. Information is provided about each industry's export competitiveness, trade flows, inbound investment, and leading competitive factors such as trade facilitation, logistics services, and e-commerce.

This chapter defines key terms used in the report and identifies information sources. Chapter 2 provides information on ASEAN and ASEAN's Economic Community Blueprint; trends in ASEAN economic integration, export competitiveness, and inbound investment; and regional improvements in trade facilitation, logistics services, and e- commerce. Chapters 3–8 describe these ASEAN regional trends more specifically for the six priority sectors. Chapters 3–8 also provide information about the six industries selected by the Commission, including (1) industry structure, size, and international competitiveness; (2) leading ASEAN exporting countries and major export markets; (3) leading ASEAN country recipients of inbound investment, the source countries of this investment, and ASEAN investment rules specific to the industry; (4) pairs or groups of ASEAN countries that have experienced significant economic integration in industry production and/or marketing; and (5) leading competitive factors affecting the industry's regional integration, export competitiveness, and inbound investment.

This chapter has been prepared by the United States International Trade Commission (USITC) at the request of the United States Trade Representative (USTR).[1] In its request letter, USTR noted that the 10 countries that are members of ASEAN have growing economies and a combined population of 550 million. As such, they are already an important market for U.S. companies, and they represent significant opportunities for expanded trade and investment. Trade between the United States and ASEAN countries has grown steadily in recent years, and in 2008 totaled $177 billion. ASEAN has a goal of achieving full regional economic integration by 2015. As a way to achieve this integration, ASEAN identified 12 priority sectors for accelerated economic integration. To assist U.S. trade policymakers in assessing progress toward this goal, USTR requested the USITC to prepare a report that would provide information about six of the priority sectors.

Definitions

Economic integration

Economic integration generally refers to a staged process through which a group of countries gradually coordinate or merge their economic policies over time. This coordination may be bilateral or carried out through a multilateral organization such as ASEAN. The purpose of economic integration is to lower trade barriers and other economic obstacles between countries, thereby expanding markets and trade, lowering prices, and improving the competitiveness of trade partners through lower costs and economies of scale. For some economic integration arrangements, the ultimate goal is a single market in which there is a free flow of goods, services, capital and labor, and harmonization of economic and monetary policies.[2] In other cases, member countries design the arrangement to be a free trade area, a customs union, or a common market, with no intentions to integrate further. The process of economic integration traditionally consists of four stages, each building on the other (table 1.1):

For the ASEAN Economic Community (AEC), the AEC Blueprint (Blueprint) defines the goal of economic integration as the free movement of goods, services, investment, and skilled labor, and freer flow of capital among member countries by 2015.[3] According to the Blueprint, AEC economic integration will create "(a) a single market and production base, (b) a highly competitive economic region, (c) a region of equitable economic development, and (d) a region fully integrated into the global economy."[4]

The net economic benefits of economic integration depend not only on trade created between member countries, but also on the economic costs associated with trade diverted away from lower-cost nonmember countries.[5] To the extent that economic integration arrangements are made between participating countries that are natural trading partners, it is considered probable that economic integration will benefit those countries.

Participating countries also gain additional long-term economic benefits through increased investment and harmonization of economic and monetary policies.

Studies of the economic integration process have generally focused on intraregional trade and investment flows, macroeconomic interdependence, and strong regional economic growth as indicators of increasingly integrated economies.[6] The USITC, in an earlier study of economic integration in East Asia, examined how economic policies and investment patterns in the region contributed to the process of merging the markets for goods, services, capital, and labor in East Asia.[7] The study highlighted the gradual convergence of economic policies, reductions in trade barriers, and improvements in infrastructure and intellectual property protection as factors facilitating economic integration among the regional economies.

The Asian Development Bank (ADB) has developed a number of specific quantitative indicators to gauge economic integration. These indicators include growth in regional gross domestic product (GDP) among member economies and changes in income gaps, intraregional trade and investment shares, production correlation (as a measure of macroeconomic linkages), correlation of equity prices, and interregional tourism flows.[8]

Table 1.1. Stages of Economic Integration

Arrangement	Economic integration actions
Free trade area or agreement (including preferential trade agreements)	Tariffs among member economies are reduced or eliminated, but the members maintain their own external tariffs against the rest of the world.
Customs union	A common external tariff is set among members, in which the same tariffs are applied to third-country nonmembers.
Common market	Labor, capital, persons, and services are allowed to move freely within and between member countries.
Economic and monetary union	Monetary and fiscal policies among members are harmonized, including the use of a common currency. Requires the existence of supra-national institutions to coordinate policies.

Source. United Nations University, "Different Forms of Integration."

This chapter calculates two measures that are commonly used to show trends in regional interdependence.[9] The intraregional trade share shows a region's trade with itself as a share of the region's total trade, which in this case is the share of ASEAN's total trade that is intra-ASEAN. The other measure is the intraregional trade intensity, which shows the intensity with which a region trades with itself compared with its trade with the world.[10] It is calculated by dividing the share of intraregional trade in its overall trade by the share of its trade in world trade. If the index measures 1, there is no regional bias in trade; if it is more than 1, the region's share of trade with itself is greater than would be expected given its importance in world trade. An increase in the index over time indicates that regional trade is increasing more rapidly than the region's share in global trade. Unlike the intraregional trade share indicator, trade intensity does not suffer from a size bias; that is, it does not rise simply because the region's relative weight in the world economy is increasing.

Export competitiveness

USTR requested that the USITC analyze the export competitiveness of the profiled industries in ASEAN member countries. In referring to competitiveness as a concept, theorists distinguish between "competitiveness" at the national level (competitive economies) and export competitiveness. The former is based on macroeconomic factors, such as access to healthcare, education, macroeconomic stability, innovation, technological readiness, market size, the efficiency of the goods market, and the sophistication of the business and financial markets, that raise long-term economic growth rates and standards of living within an economy.[11] On the other hand, export competitiveness focuses more narrowly on specific industries and the factors that affect competition for global market share and exports. For example, in past reports, the USITC has defined export competitiveness "as the ability of producers to sell goods in foreign markets at price, quality, and timeliness comparable to competing foreign producers."[12] Changes in export competitiveness indicate the relative positions of firms in international markets and their potential opportunities for growth and investment.

To the extent that economic integration results in trade creation and overall gains to participating economies, it contributes to the long-term competitiveness of those countries. It also contributes to export competitiveness to the extent that firms located within the member countries are able to take advantage of economies of scale, improved market efficiencies, the creation of larger markets, and other opportunities from economic integration to expand exports.

Export Competitiveness Factors

A number of common supply and demand factors affect the ability of firms to export and increase market share (table 1.2). In general, factors that improve firm productivity and lower costs (supply side factors) improve industry competitiveness. Additionally, increasing demand or growth in demand for enhanced quality or specific product characteristics can also benefit specific firms, including exporters. Macroeconomic factors, such as exchange rates, also influence export competitiveness. A change in the "real" exchange rate affects the cost competitiveness of all industries in a country, while exchange rate volatility could reduce export demand to the extent buyers are faced with uncertain prices. These factors and their relationship to ASEAN industry competitiveness in the six sectors are discussed more fully in Chapters 3–8.

Table 1.2. Export Competitiveness Factors

Demand and competitiveness:	**Demand-side factors:**
Growing demand in a group of countries engaged in an economic arrangement may benefit imports from outside the region. But it may also benefit specific exporters located inside the region, to the extent that those exporters have an advantage in supplying products with specific attributes that consumers value.	• Growing demand for final products that drives higher demand for intermediate products • Changes in consumer tastes and preferences • New standards or quality requirements • Consumer concerns, such as the • environment, fair trade, and labor protections • Marketing practices • Channels of distribution
Supply and competitiveness:	**Supply-side factors:**
Lower production costs and/or increased firm productivity improve the ability of exporters to supply specific products at the lowest cost.	• Production costs (such as purchased inputs, labor, and capital costs) • Labor productivity • Investment • Capacity utilization • Infrastructure • Proximity to the market • Product innovation and entrepreneurial • ability • Industry structure (such as concentration or vertical and horizontal integration)
Exchange rates:	
Exchange rate volatility and changes in the "real" exchange rate can affect export competitiveness.	• Exchange rate volatility increases risk from exporting and may deter producers from entering the export market. It may also induce buyers to switch sources of supply. • An increase in a country's real exchange rate, observed in changes in the relationship between domestic and foreign prices for the same goods, means that its exports may become more expensive than competing products in foreign markets.

Sources. WEF, *Global Competitiveness Report 2009-2010;* USITC, *Wood Flooring and Hardwood Plywood, 2008*, 6-2; other factors compiled by USITC staff.

Inbound Investment

According to the AEC Blueprint, "[s]ustained inflows of new investments and reinvestments will promote and ensure dynamic development of ASEAN economies."[13] In general, inbound investment is calculated by measuring direct investment (which indicates a lasting interest in a foreign business enterprise), as opposed to portfolio investment (box 1.1).

BOX 1.1. DEFINITIONS OF FOREIGN DIRECT INVESTMENT (FDI) AND PORTFOLIO INVESTMENT

The Organisation for Economic Co-operation and Development (OECD) has developed the globally recognized definition of FDI. This definition is used by both the OECD and the International Monetary Fund (IMF) as part of their guidance for national statistical offices in the preparation of FDI statistics.

> Direct investment refers to cross-border investment made by a resident in one economy (the *direct investor*) with the objective of establishing a lasting interest in an enterprise (the *direct investment enterprise*) that is resident in an economy other than that of the direct investor. The motivation of the direct investor is a strategic long-term relationship with the direct investment enterprise to ensure a significant degree of influence by the direct investor in the management of the direct investment enterprise. The lasting interest is evidenced when the direct investor owns at least 10% of the voting power of the direct investment enterprise. (OECD, 17)

Total direct investment in a given company consists of the sum of equity investment, reinvested earnings, and inter-company debt.

Investment in a foreign enterprise that falls below the threshold of 10 percent voting power is defined as portfolio investment. The underlying difference between direct and portfolio investment is the investor's motivation to significantly influence or control an enterprise. Portfolio investors "do not generally expect to influence the management of the enterprise" and are assumed to be focused on earnings resulting from the acquisition and sales of securities, rather than on long-term management control of the enterprise (OECD, 53).

Source: *OECD, Benchmark Definition of Foreign Direct Investment,* 2008, 17, 22–23, 48–49, 53.

Official government FDI statistics are commonly reported as FDI position (cumulative investment over time) or FDI flows (new investment recorded during a specific time period, generally annually or quarterly). Many ASEAN governments report FDI position and flows by country and by industry, but none, unfortunately, report FDI statistics at the level of the specific industries needed for the industry profiles discussed in chapters 3–8. Therefore, FDI data throughout this chapter are reported at a higher level of aggregation. Direct investment can take place through either "greenfield" FDI projects (new investment, such as construction of a new factory) or through M&A (merger with, or acquisition of, an existing company). Official data include both types of investment.

In an effort to estimate FDI in certain specific industries within ASEAN, the USITC relies on databases that show particular FDI transactions, either greenfield or M&A

investment projects.[14] These databases compile available information on particular investment transactions. The information can provide insight into the companies and countries involved in FDI in particular industries, but it cannot substitute for official FDI data, since the transaction values are not compiled in the same way as official FDI statistics.[15] In particular, FDI position and flows data include three components: equity transactions, reinvested earnings, and inter-company debt. In general, the available commercial databases only reflect the equity transactions component of the total.[16]

Trade Facilitation, Logistics Services, and E-Commerce

Trade facilitation, logistics services, and e-commerce are specific work plan elements in the ASEAN economic integration agenda. Improvements in these activities contribute to increased export competitiveness by raising firm productivity and by reducing the cost and time to export, as well as communication and information costs (table 1.3). Enhancements in these areas involve legal and regulatory reforms; customs modernization; improvements to roads, ports, and other infrastructure; and increased access to the Internet. These activities are discussed in more detail in chapter 2.

INFORMATION SOURCES

Information on ASEAN for this chapter was obtained from a number of sources, including written submissions received in response to the Commission's *Federal Register* notice announcing the investigation;[17] testimony at the public hearing held by the Commission in connection with this investigation on February 3, 2010;[18] and interviews during factfinding travel to certain ASEAN member countries, including Indonesia, Malaysia, Singapore, Thailand, and Vietnam.

Table 1.3. Trade Facilitation, Logistics Services, E-Commerce, and Competitiveness

Factor	Effect on export competitiveness
Trade facilitation: "The simplification, standardization, and harmonization of procedures and associated information flows required to move goods from seller to buyer and to make payments."[a]	Enhances productivity and efficiency, lowering trade costs. Reduces risks associated with undue delay and loss of goods during shipment.
Logistics services: "A range of related activities intended to ensure the efficient movement of production inputs and finished products."[b]	More efficient transportation and freight services reduce the cost of getting goods to markets.
E-commerce: "The buying, selling, or exchanging of goods, services, and information through electronic networks."[c]	High-speed connectivity and the ability to complete transactions electronically facilitate communication and reduce information costs.

Sources:

a United Nations Centre for Trade Facilitation and Electronic Business, cited in Asian Development Bank (ADB), *Designing and Implementing Trade Facilitation,* 2009, 3.

b USITC, *Logistic Services*, 2005, vii.

c Acoca, *Empowering E-Consumers: Strengthening Consumer Protection in the Internet Economy,* 2009, 6.

Key sources of published information on ASEAN were international organizations, including the OECD, International Monetary Fund (IMF), and the World Trade Organization (WTO); numerous scholarly and peer-reviewed academic publications; business and industry publications; government print documents and information obtained from ASEAN and member government Web sites; and reports by U.S. government agencies. These sources are cited in bibliographies at the end of each chapter.

2. ASEAN AND SELECTED REGIONAL TRENDS AND IMPROVEMENTS

Introduction to ASEAN

ASEAN was established in 1967 to accelerate economic growth, promote regional peace and stability, and enhance cooperation on economic, social, cultural, technical, and educational matters.[19] The five founding countries—Indonesia, Malaysia, the Philippines, Singapore, and Thailand—were later joined by Brunei Darussalam (Brunei) in 1984, Vietnam (1995), Burma (1997), Laos (1997), and Cambodia (1999).[20]

Since its founding, ASEAN's progress on economic integration has been affected by various factors. As a largely voluntary, consensus-based institution, [21] with an economically and politically diverse membership, ASEAN has generally followed a slow, step-by-step approach to building regional cooperation and has progressively entered into more legally binding and institutionalized agreements.[22] However, certain external events have stimulated faster and deeper progress on integration among member countries: a growing international trend toward regionalism and free trade agreements (FTAs), especially those involving ASEAN's important trading partners; the Asian financial crisis of 1997; the rise of emerging economies that compete with ASEAN countries, especially China and India; and the 2008–09 global economic slowdown.[23]

Although political security was ASEAN's initial focus, economic cooperation grew in the 1970s with agreements on joint industrial projects and preferential trading arrangements (box 2.1).[24] The first substantial step toward integrating the ASEAN market came in 1992 when the ASEAN-6 agreed to establish the ASEAN Free Trade Area (AFTA).[25] The AFTA provided for the reduction or elimination of tariffs under a Common Effective Preferential Tariff scheme (CEPT) and the removal of quantitative restrictions and other nontariff measures (NTMs). It also addressed other cross-border measures, such as trade facilitation and standards harmonization.[26] ASEAN leaders signed agreements to liberalize services trade in 1995 (ASEAN Framework Agreement on Services, or AFAS) and investment flows in 1998 (Framework Agreement on the ASEAN Investment Area, or AIA).

In 2003, ASEAN officials agreed to establish an ASEAN Community by 2020, a goal that was subsequently accelerated to 2015. The ASEAN Community consists of three pillars: the ASEAN Political-Security Community, ASEAN Economic Community, and ASEAN Socio-Cultural Community. To guide the creation of the ASEAN Community, in 2009 ASEAN leaders published the *Roadmap for an ASEAN Community*, which included blueprints for achieving each of the pillar communities as well as an updated work plan for the Initiative for ASEAN Integration (IAI), a program that aims to narrow the development gap between members.

Box 2.1. Timeline of Important Milestones in ASEAN Economic Integration

Date	Milestone	Description
1967	Bangkok Declaration	ASEAN founded by Indonesia, Malaysia, Philippines, Singapore, and Thailand.
1977	Agreement on ASEAN preferential trading arrangements	One of the earliest ASEAN agreements to carry some legal obligation. Members agreed to apply preferential tariff rates based on a margin of preference over MFN rates on basic commodities, products of ASEAN industrial projects, and others of interest.
1984	Brunei joins ASEAN	
1987	Enhanced preferential trading arrangements	Improved the preferential trading arrangements by, e.g., reducing exclusion lists, further reducing tariffs, and relaxing ASEAN content requirements in the rules of origin.
1987	Investment Guarantee Agreement (IGA)	The IGA provides investment protections for FDI between ASEAN member countries, including compensation in case of expropriation; guarantees of an investor's right to repatriate earnings, subject to local laws; and provision for arbitration between parties in cases of disputes.
Jan. 1, 1993	AFTA implemented	Members agreed to establish the ASEAN Free Trade Area (AFTA) and a Common Effective Preferential Tariff (CEPT) scheme, where 99 percent of product categories will have intra-ASEAN tariff rates reduced to 0–5 percent.
1995	Vietnam joins ASEAN	
1995	ASEAN Framework Agreement on Services (AFAS)	Based closely on the General Agreement on Trade in Services (GATS). Aims to eliminate restrictions on trade in services, enhance intra-ASEAN services cooperation, and liberalize services trade based on the GATS-plus principle. Mandates successive negotiations to progressively liberalize services trade.
1996	ASEAN Industrial Cooperation Scheme (AICO)	Replaced earlier ASEAN industry project cooperation programs. Promotes joint manufacturing industrial activities between ASEAN-based companies. AICO products enjoy preferential tariff rates of 0–5 percent.
1997	Burma and Laos join ASEAN	
1997	ASEAN Vision 2020	Laid out a vision of ASEAN in 2020, including closer economic integration and a commitment to create "a stable, prosperous and highly competitive ASEAN Economic Region in which there is a free flow of goods, services and investments, a freer flow of capital, equitable economic development and reduced poverty and socio-economic disparities."

(Continued)

Date	Milestone	Description
1998	Framework Agreement on the ASEAN Investment Area (AIA)	Aims to ensure a free flow of investment (in manufacturing, fisheries, forestry, mining, agriculture, and services) by 2020. Reservations made by members are scheduled to be eliminated in 2015 for ASEAN investors and 2020 for non-ASEAN invest-tors. The ASEAN-6 countries (original ASEAN members and Brunei) agreed to accelerate this process by eliminating reservations in manufacturing for ASEAN investors by 2003 and for all investors by 2010.
1998	Hanoi Plan of Action	First of a series of action plans to help implement the ASEAN Vision 2020. It lays out steps to promote economic integration over the period 1999–2004.
1999	Cambodia joins ASEAN	
2000	Initiative on ASEAN Integration (IAI)	Goal is to address the development gap between member states through soft infrastructure projects (such as training, technical studies, and capacity building) and physical transport and communication infrastructure projects, and to mobilize funding from international financial institutions and developed countries for support. About 258 projects have been completed to date.
October 2003	Declaration of ASEAN Concord II: ASEAN Community by 2020 (9th ASEAN Summit or Bali Concord II)	Agreed to establish an ASEAN Community by 2020 that consists of three pillars or communities based on political and security cooperation, economic cooperation, and socio-cultural cooperation. The ASEAN Economic Community (AEC) is the end goal of the economic integration process described in the ASEAN Vision. Eleven priority sectors are identified for accelerated integration.
November 2004	Vientiane Action Program	Successor to the Hanoi Plan of Action to help realize the ASEAN Vision and the ASEAN Community. It covers the period 2004- 2010.
November 2004	ASEAN Framework Agreement for the Integration of Priority Sectors	Includes roadmaps for each priority sector that identify measures to be implemented and timelines for their implementation.
January 2007	ASEAN Community by 2015	Leaders at the 12th ASEAN Summit agreed to accelerate the establishment of an ASEAN Community; the target date is now 2015.
November 2007	ASEAN Economic Community Blueprint	Leaders at the 13th ASEAN Summit adopted the ASEAN Economic Blueprint, which provides the framework for achieving the AEC by 2015.

Date	Milestone	Description
December 15, 2008	ASEAN Charter implemented	Establishes the legal and institutional framework for ASEAN.
February 26, 2009	ASEAN Comprehensive Investment Agreement (ACIA) signed	ACIA adds to investor protections under the AIA in several ways: includes comprehensive investment liberalization and protection provisions, including prohibition of performance requirements; includes an investor-state dispute settlement process; and extends benefits to foreign-owned, ASEAN-based investors. The ACIA framework is a "negative list" framework; each member state also compiles a list of reservations, or exclusions to the agreement.
March 2009	Roadmap for the ASEAN Community, 2009–2015	Consists of the Economic Community Blueprint (approved in 2007), the Political-Security Community Blueprint, the SocioCultural Community Blueprint, and the Second IAl Work Plan. Replaces the Vientiane Action Program.
May 17, 2010	ASEAN Trade in Goods Agreement (ATIGA) enters into force	ATIGA builds on existing initiatives related to trade in goods (e.g., CEPT-AFTA, nontariff measures, customs, ASEAN single window, mutual recognition agreements, e-ASEAN, integration of priority sectors, etc.). Goal is to achieve the free flow of goods to establish a single market and production base, making it possible to realize the AEC by 2015.

Source. Compiled by USITC from ASEAN documents, http://www.aseansec.org.

According to the 2003 declaration on the ASEAN Community, the ASEAN Economic Community (AEC) will be the "realization of the end-goal of economic integration ... to create a stable, prosperous and highly competitive ASEAN economic region in which there is a free flow of goods, services, investment and a freer flow of capital, equitable economic development and reduced poverty and socio-economic disparities."[27] The AFTA, as well as the agreements on services trade (AFAS) and investment (AIA), form the basis of the AEC.[28] Although the AEC is similar to the European Union (EU) in some respects, it is a "hybrid" FTA-plus arrangement; it includes some elements of a common market, but not a common external tariff.[29]

The AEC Blueprint, which was adopted by ASEAN leaders in November 2007, lists the steps to achieving the AEC by 2015 and a timeline for their implementation. The Blueprint categorizes the AEC goals into four areas (box 2.2). To help form a single market and production base, the Blueprint includes a commitment to more rapidly integrate 12 sectors: agro-based products, air travel, automotives, e-ASEAN, electronics, fisheries, healthcare, logistics, rubber-based products, textiles and apparel, tourism, and wood-based products. These so-called priority sectors (with the exception of logistics, which was added in 2006) were originally identified for accelerated integration in 2003. They were chosen "on the basis of comparative advantage in natural resource endowments, labor skills and cost competitiveness, and value-added contribution to ASEAN's economy," and together accounted for more than 50 percent of intra-ASEAN trade.[30] In 2004, Roadmaps for each priority sector, outlining the path to full integration, were adopted, and country coordinators were selected to oversee progress.[31] Today the priority sectors are considered the "catalysts" to realizing the goals of the AEC.[32] This study examines industries within 6 of the 12 priority sectors, as discussed below.

BOX 2.2. AEC BLUEPRINT: GOALS AND MEASURES

Goal	Measures/actions
A single market and production base	Measures to ensure the free flow of goods, services, investment, and skilled labor, as well as the freer flow of capital, within ASEAN. Measures on the free flow of goods cover tariffs, NTMs, rules of origin, trade facilitation, customs integration, and standards and technical regulations.
A highly competitive economic region	Actions on competition policy, consumer protection, intellec-tual property rights, taxation, e-commerce, and infrastructure development.
A region of equitable economic development	Actions to develop small and medium-sized enterprises and to implement and enhance the IAI.
A region fully integrated into the global economy	Actions to develop a coherent approach among ASEAN members on external economic relations and to enhance participation in global supply networks.

Source. ASEAN, "ASEAN Economic Community Blueprint," November 20, 2007.

The ASEAN countries differ from each other in a number of important ways, and member states recognize that this poses a challenge to implementation of the blueprint. For example, member countries vary by size, level of development, political system, investment environment, and economic structure. In particular, the last four countries to join ASEAN—Burma, Cambodia, Laos, and Vietnam (often referred to as the ASEAN4)—are at a lower stage of economic development than the other six member countries (table 2.1). Consequently, an integral part of the AEC Blueprint is a commitment to narrow the development gap among the ASEAN member states, primarily through the IAI work program, in order to facilitate regional integration.[33]

Table 2.1. ASEAN member states, selected indicators, 2008

	Population (million)	GDP per capita, PPP	Share of total ASEAN trade (%)	Intra-ASEAN trade (% global) 2004	2008
Brunei	0.4	(a)	0.7	23.0	29.8
Burma	49.6	435	0.6	49.7	53.6
Cambodia	14.6	1,921	0.5	16.7	21.8
Indonesia	237.5	3,824	15.6	20.9	25.6
Laos	6.2	(a)	0.2	74.4	84.4
Malaysia	27.7	13,852	19.8	20.7	25.1
Philippines	92.7	3,457	6.2	18.1	20.3
Singapore	4.8	40,319	27.6	29.5	36.3
Thailand	66.3	8,235	20.6	19.2	19.7
Vietnam	86.1	2,793	8.3	24.1	20.9

Source. Country profiles (1st two indicators), app. E; ASEAN Secretariat.
[a] Not available.

The AEC Blueprint also includes a commitment to global integration, including a review of its FTA commitments vis-à-vis ASEAN's internal integration commitments. Because nearly three-quarters of ASEAN's trade is with external partners (figure 2.1), ASEAN has taken a two-pronged strategy to boost its economic performance, namely to foster both internal integration and global relationships, and to ensure their complementary relationship.[34] To improve global market opportunities and attract investment, ASEAN has implemented FTAs with Australia and New Zealand (in a single FTA), China, India, Japan, and Korea (box 2.3).[35] Individual ASEAN countries have also implemented FTAs with various partners, resulting in an overlapping web of FTAs.[36]

According to the AEC Blueprint, the ASEAN Economic Ministers (AEM) are in charge of economic integration and accountable for the overall implementation of the AEC Blueprint. Relevant ASEAN sectoral bodies are responsible for coordinating, implementing, and monitoring commitments under their respective purview. The ASEAN Secretariat (Secretariat) reviews and monitors compliance of implementation.[37] The Secretariat was tasked with developing the AEC Scorecard to monitor progress. According to the first AEC Scorecard, which was published in 2010 and covers the first two-year period (2008–09), 73.6 percent of AEC targets were achieved, including 82 percent of the measures and activities included under the goal of achieving a single market and production base; 50 percent of measures under that of creating a competitive economic region; and 100 percent of measures under the goals of attaining equitable economic development and integration with the global economy. Actions that still need to be addressed include ratification of various agreements, including the ASEAN Comprehensive Investment Agreement (ACIA) and several agreements covering goods in transit and air services; implementation of National Single Windows (see discussion below); and determination of final tariff reduction rates on highly sensitive products, such as rice and sugar.[38]

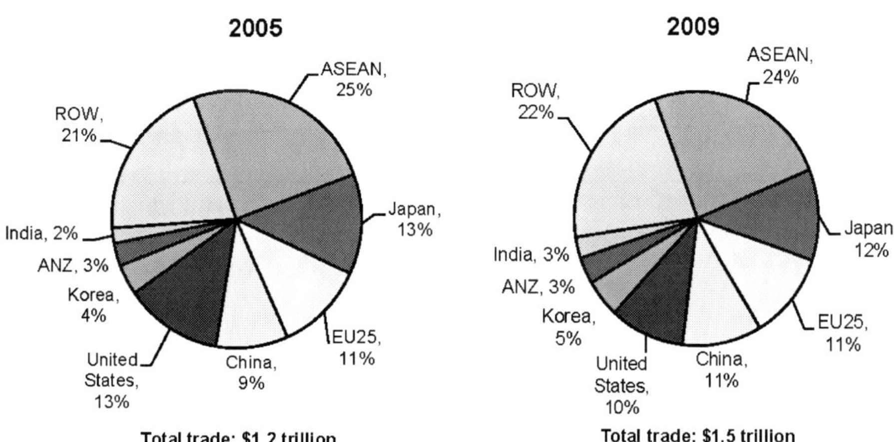

Source: ASEAN Secretariat Yearbook.

Note. ROW = rest of the world; ANZ = Australia and New Zealand; EU25 = Austria, Belgium, Cyprus, Czech Republic, Denmark, Estonia, Finland, France, Germany, Greece, Hungary, Ireland, Italy, Latvia, Lithuania, Luxembourg, Malta, the Netherlands, Poland, Portugal, Slovak Republic, Slovenia, Spain, Sweden, and the United Kingdom. Figures may not total to 100 percent due to rounding.

Figure 2.1. ASEAN trade, by partner, 2005 and 2009.

Following a meeting of the AEM in February 2010, the Malaysian Trade Minister, who hosted the meeting, reported that there were gaps in the implementation of some commitments, particularly in the areas of eliminating nontariff barriers, streamlining customs clearance, harmonizing standards, and removing duplication of testing and certification procedures. The Minister noted that among the priority sectors, "satisfactory progress has been only achieved in the automotive, textiles, air travel and tourism sectors."[39]

Box 2.3. ASEAN-Wide Free Trade Agreements, 2010

FTA partner	Description	Status
Signed FTAs		
Australia and New Zealand	Trade in goods and services; investment; economic cooperation; and legal and institutional issues	Framework Agreement signed in November 2004. FTA with Australia and New Zealand was a single undertaking (covering goods, services, and investment): signed on February 27, 2009; implementation on January 1, 2010.
China	Trade in goods and services; investment; trade facilitation measures; and economic cooperation in areas of interest	Early Harvest Programme (EHP) entered into effect on January 1, 2004 for ASEAN-6 and China. Duties for EHP products were fully eliminated as of 2006. Trade in Goods Agreement signed on November 29, 2004; entered into effect on July 1, 2005 to reduce tariffs in 2005, 2007, 2009, and January 1, 2010. Trade in Services Agreement and commitments under the First Package signed on January 14, 2007; entered into effect on July 1, 2007. Negotiations are ongoing f r the Second Package. Investment Agreement signed on August 15, 2009.
India	Trade in goods and services; investment; and economic cooperation	Framework Agreement signed on October 8, 2003. Trade in Goods Agreement signed in August 2009, effective January 1, 2010. Negotiations on trade in services and investment ongoing (target date of August 2010 for conclusion).
Japan	Trade in goods and services; investment; rules of origin; sanitary and phytosanitary (SPS) measures; technical barriers to trade; dispute settlement mechanism; and economic cooperation	Framework Agreements signed on October 8, 2003. Trade in Goods Agreement signed on April 14, 2008; implemented on December 1, 2008. Negotiations on trade in services are yet to commence. Negotiations on investment are yet to commence.

FTA partner	Description	Status
Korea	Trade in goods and services; investment; and economic cooperation in areas of interest	Trade in Goods Agreement signed on August 26, 2006; entered into effect on June 1, 2007. Agreement on Trade in Services signed on November 21, 2007; effective May 1, 2009. Investment Agreement signed on June 2, 2009.
FTAs under negotiation		
European Union	Trade in goods and services; government procurement; competition policy; intellectual property rights; industrial and commercial property rights; sustainable development including labor and environmental issues; cooperation on trade related issues; and dispute settlement mechanism	Negotiations launched on May 4, 2007, on a region-to-region basis, but stalled in early 2009. In December 2009, EU agreed to pursue negotiations towards FTAs with interested individual ASEAN countries, which could become building blocks for an ASEAN-wide regional agreement. Negotiations with Singapore began in March 2010. Vietnam has shown interest in a bilateral FTA.

Source. ADB, Free Trade Agreement Database for Asia (accessed February 2, 2010); WTO, Regional Trade Agreements portal (accessed February 2, 2010); European Commission, "Overview of FTA and other Trade Negotiations," April 21, 2010.

Note. On the implementation date indicated, the ASEAN FTA enters into force only with respect to those ASEAN countries that have completed their respective legal procedures. The FTA will become effective with respect to the other ASEAN countries once they notify the completion of their respective legal process.

In December 2008, the ASEAN Charter entered into effect, which provides the legal and institutional framework for ASEAN to be a more rules-based organization. According to ASEAN officials, the Charter strengthens ASEAN institutions and supports ASEAN's community-building and integration efforts by enhancing the formal nature of ASEAN integration.[40] To ensure compliance with agreed rules and obligations, the Charter provides that "disputes that concern the interpretation or application of ASEAN economic agreements shall be settled in accordance with the ASEAN Protocol on Enhanced Dispute Settlement Mechanism,"[41] which was adopted in 2004. The enhanced dispute settlement process includes three stages: advisory, consultative, and adjudicatory. As of February 2009, no disputes had been raised for adjudication.[42]

ASEAN Trends in Economic Integration, Export Competitiveness, and Inbound Investment

Strengthening economic integration, improving export competitiveness, and attracting foreign direct investment are all goals of the formation of a single market and production base, as envisaged by the AEC. Broad ASEAN trends in these three areas over the past five years are presented below.

Economic Integration

Between 2004 and 2008, intra-ASEAN trade increased slightly. Although intraregional trade is important in each of the six priority sectors covered in this chapter, none of the sectors exhibited a clear trend towards increased integration during this period.

ASEAN agreements affecting integration

Efforts to form a single market and production base have met with the most success in the area of tariff liberalization under AFTA.[43] AFTA is considered a "deep" FTA relative to others among developing countries because of its comprehensive coverage, ambitious liberalization to zero or near-zero rates, and timely implementation.[44] Intraregional tariff reductions and eliminations to 0–5 percent rates under the CEPT scheme of AFTA are supposed to be completed in 2010 for the ASEAN-6 and in 2015 for the ASEAN-4, and by 2018 for the ASEAN-4 on some sensitive products (unprocessed agricultural products). On January 1, 2010, the ASEAN-6 reduced intraregional duties on an additional 7,881 tariff lines to zero, bringing the total number of intraregional duty-free lines to over 54,000, or 99.11 percent of all lines.[45] About 70 percent of ASEAN-4 intraregional tariff lines are now free of duty.[46]

ASEAN rules governing the free flow of goods are largely contained in the ASEAN Trade in Goods Agreement (ATIGA), which consolidates and builds on existing initiatives related to trade in goods to provide a comprehensive legal framework (box 2.4). ATIGA entered into force on May 17, 2010, and by strengthening the rules governing the liberalization and facilitation of goods trade, it provides an important foundation for more rapidly achieving the AEC.

Services trade liberalization among ASEAN countries has been conducted through successive rounds of negotiations aimed at progressively higher-level commitments, as mandated under the 1995 AFAS.[47] Services liberalization is undertaken according to the "ASEAN minus X" principle. Under this approach, two or more ASEAN countries that are ready may proceed to liberalize a services sector, while others can join at a later time.[48] To date, seven packages (or schedules) of commitments have been concluded, covering trade in a wide variety of services, including business and professional services (e.g., accounting, auditing, architecture, and engineering), construction, distribution, education, environment, healthcare, maritime transport, telecommunications, and tourism services. Negotiations to liberalize financial services and air transport services are conducted separately and have resulted in separate packages of commitments for each.[49]

To facilitate the flow of foreign professionals in ASEAN, member countries have signed seven MRAs on architectural, accountancy, engineering, dental practitioner, medical practitioner, nursing, and surveying qualifications.[50]

Box 2.4. What is ATIGA?

ATIGA builds on the commitments under the following ASEAN economic initiatives: the Agreement on ASEAN Preferential Trading Arrangements (1977), the Agreement on the Common Effective Preferential Tariff Scheme for the ASEAN Free Trade Area (1992), the ASEAN Agreement on Customs (1987), the ASEAN Framework Agreement on Mutual Recognition Arrangements (1998), the e-ASEAN Framework Agreement (2000), the Protocol Governing the Implementation of the ASEAN Harmonized Tariff Nomenclature (2003), the ASEAN Framework Agreement for the Integration of Priority Sectors (2004), and the Agreement to Establish and Implement the ASEAN Single Window (2005). Member states are to agree on the list of agreements to be superseded by ATIGA within six months from the date of ATIGA's entry into force (by November 17, 2010).

Among the provisions of ATIGA are those covering tariff liberalization schedules and commitments, rules of origin (which require regional value content of 40 percent or a change in tariff heading),[a] NTMs, trade facilitation, customs procedures, standards and conformance, sanitary and phytosanitary measures, and trade remedy measures (as provided for in the WTO). To improve transparency, ATIGA requires member state notification of actions and measures they intend to take with respect to tariffs, NTMs, standards, SPS, and other measures;[b] requires publication of trade laws, regulations, and judicial decisions in accordance with article X of GATT 1994; and establishes a trade repository, which will publish on the Internet all member state trade and customs laws, measures, and procedures, including tariff schedules. The trade repository will include a database of NTMs, with a view to identifying NTMs for elimination.[c] Relevant ASEAN bodies are required to review any NTM notified or reported by any other member state or by the private sector with a view to determining whether the measure constitutes a nontariff barrier. ATIGA specifies deadlines for the elimination of identified nontariff barriers: by 2010 for the ASEAN-5, 2012 for the Philippines, and 2015 (with flexibilities to 2018) for the ASEAN-4. ATIGA also seeks to prevent member states from introducing new NTMs that are inconsistent with WTO obligations.[d]

ATIGA provides guidance on the development and application of standards, technical regulations, and conformity assessment procedures, which are to be transparent and consistent with international standards and practice. It also directs member states to develop and implement sectoral Mutual Recognition Arrangements (MRAs) and harmonized regulatory regimes. ATIGA includes an MRA on conformity assessment with respect to electrical and electronic equipment, and two agreements on harmonized regulatory regimes (for electrical equipment and cosmetics).[e]

[a] ATIGA, 22.

[b] See ATIGA, Annex 1.

[c] According to the Secretariat, there is an outdated list of NTMs, which currently contains 7,378 NTMs, of which 43.9 percent are non-automatic licenses. Other categories are automatic licenses, quotas, prohibitions and quantitative restrictions, and tariff-rate quotas. Government official, interview by USITC staff, Singapore, March 1, 2010. The database can be found at http://www.aseansec.org/16355.htm.

[d] ATIGA, articles 40–42.

[e] ASEAN economic ministers signed an MRA for pharmaceutical products in 2009.

Intra-ASEAN trade trends

Two measures are typically cited to show trends in regional trade integration—the intraregional trade share and the intraregional trade intensity.[51] The intraregional trade share for ASEAN shows the share of total ASEAN trade that is accounted for by intraASEAN trade.[52] As shown in figure 2.1, intra-ASEAN trade decreased slightly as a share of total ASEAN trade between 2005 and 2009, declining from 25 to 24 percent. It remains smaller than that of other regional groups, such as the EU- 15, the members of the North American Free Trade Agreement (NAFTA), and broader East Asia groupings that include China. The trend and size of ASEAN's intraregional trade share reflects its growing links with China and its continued reliance on trade with outside partners.[53] However, an examination of longer-term trends shows that the share of intra-ASEAN trade has increased substantially from the 19 percent share it held when AFTA entered into effect in 1993. During the first decade of the 2000s, it gradually increased from 22 percent in 2001–02, to 24–25 percent in 2003–07 (figure 2.2). In 2008, intra-ASEAN trade reached its highest share—nearly 27 percent—but it dropped back in 2009, at the same time that ASEAN's shares with each of its other top partners (Japan, EU-25, China, and the United States) also declined. The global economic downturn reduced demand both absolutely and relatively from ASEAN's major partners and hurt global production networks as well, which likely disproportionately affected intraregional trade.[54] In four of the past five years, the share of intra-ASEAN exports in total ASEAN exports was larger than the share of intra-ASEAN imports in total ASEAN imports, but the difference was small (table 2.2).[55]

The share of intra-ASEAN trade in total ASEAN trade varies by member state, but generally measures between 20 and 30 percent (table 2.1). Singapore's 36 percent share reflects its importance as a port and transshipment point. Two member states have intraregional trade shares far above the average—Burma and Laos—because of the small size of their economies and their small overall level of trade (each accounting for less than 1 percent of ASEAN trade).

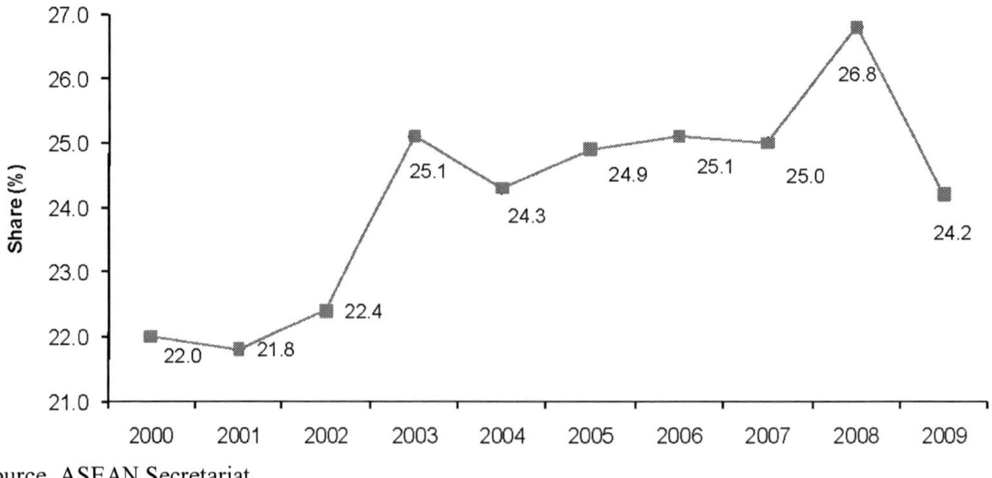

Source. ASEAN Secretariat.

Figure 2.2. Intra-ASEAN trade as a share of total ASEAN trade, 2000–09.

Table 2.2. Intra-ASEAN Imports and Exports, Share of Total, by Priority Sector, 2004–08 (Percent)

	2004	2005	2006	2007	2008
Intra-ASEAN imports					
Agro-based products	39.0	28.2	31.7	29.3	31.9
Automotives	17.3	19.6	21.3	23.4	23.6
Electronics	27.9	27.9	26.4	24.3	24.1
Healthcare	15.3	14.0	15.6	16.8	17.6
Textiles and apparel	16.0	16.5	16.7	15.5	14.7
Wood-based products	61.2	60.5	55.0	50.2	50.8
Total, all products	23.8	24.5	25.0	24.6	25.9
Intra-ASEAN exports					
Agro-based products	12.6	11.4	11.9	9.9	9.7
Automotives	36.3	35.7	30.9	29.2	31.4
Electronics	25.0	24.7	24.0	23.3	22.6
Healthcare	29.8	23.5	21.6	20.1	24.6
Textiles and apparel	10.8	12.8	13.0	11.9	12.4
Wood-based products	8.2	8.1	8.1	8.2	8.2
Total, all products	24.8	24.0	25.2	25.3	27.6

Source. Specific products: WITS, Integrated Warehouse (accessed March 29, 2010). All products: ASEAN Secretariat.

Note. Brunei, Indonesia, Laos, and Burma do not report trade data for priority sectors, and are not included in product- specific shares. These countries are included in the "all products" shares.

The intraregional trade intensity—the intensity with which a region trades with itself, compared with its trade with the world—is shown in table 2.3.[56] The intra-ASEAN trade intensity, ranging between 4.1 and 4.6 percent during 2004–08, indicates that there is a regional bias in ASEAN trade (i.e., ASEAN's share of trade with itself is greater than would be expected given ASEAN's importance in world trade).[57] The ASEAN score on this index is also significantly higher than for other regional groups, including members of NAFTA, the EU- 15, and broader Asian groupings, but this could result from the greater weight of these other groups in world trade.[58] The slight increase in ASEAN's intraregional trade intensity over the past five years shows that intra-ASEAN trade is increasing relative to the world's share of trade with ASEAN, but the change is small.

Table 2.3. Intraregional Trade Intensity Index, by Priority Sector, 2004–08 (Percent)

	2004	2005	2006	2007	2008
Agro-based products	2.1	1.7	1.7	1.5	1.6
Automotives	13.1	12.7	12.8	11.7	10.7
Electronics	2.3	2.3	2.3	2.2	2.3
Healthcare	13.8	9.9	8.9	8.6	10.2
Textiles and apparel	2.4	2.7	2.6	2.3	2.3
Wood-based products	2.6	2.5	2.4	2.4	2.4
Total, all products	4.2	4.1	4.3	4.3	4.6

Source. Specific products: WITS, Integrated Warehouse (accessed March 29, 2010). All products: ASEAN Secretariat, IMF Direction of Trade Statistics

Intra-ASEAN trade trends in six priority sectors

Table 2.4 shows both intra- and extra-ASEAN trade in the six priority sectors covered in this chapter. The electronics sector is by far the largest in terms of trade. Table 2.2 shows the shares of intra-ASEAN imports and exports in each sector. Unlike intra-ASEAN trade in all products, intra-ASEAN trade in most of the six sectors is asymmetrical. In both the agro-based products and wood-based products sectors, the vast majority of the intraASEAN trade share is accounted for by intra-ASEAN imports, indicating a strong reliance on outside markets for their exports. In the automotives sector, the majority of the intra-ASEAN trade share is accounted for by intra-ASEAN exports, reflecting the continued importance of ASEAN as a market for automobiles and parts. Notably, in the electronics sector, both the intra-ASEAN import and export shares steadily declined over the 2004–08 period, reflecting China's growing role in electronics production and ASEAN's increased participation in global electronics production chains.

The intra-ASEAN trade intensity shows a regional bias in trade in each of the six priority sectors, but the index varies widely among the sectors. The index is largest (although on a declining trend) in the automotives sector, followed by the healthcare sector, indicating these are the most regionally integrated sectors. In the remaining four sectors, the index is smaller than the index for all products. The index is very similar and fairly stable during 2004–08 in the electronics, textiles and apparel, and wood-based products sectors, indicating a similar level of regional integration. None of the sectors shows a clear trend towards increased integration over the past five years.

Export Competitiveness

ASEAN's export competitiveness, as measured by its share of global export markets, has remained fairly stable over the past five years. During the same period, world market shares in each of the six priority sectors have generally increased. The region's outward orientation in trade and investment has been particularly important in driving ASEAN's international trade, growth, and development.[59] Although several factors continue to limit the growth of ASEAN's exports to world markets, the AEC was designed to address a number of them.

ASEAN's openness to international trade and investment has contributed to "vibrant export sectors" and the development of regional manufacturing production networks.[60] Because ASEAN markets are relatively small, a major goal of the creation of the AEC is to reduce transaction costs and attract investment so as to better exploit opportunities to participate in global supply networks and serve as a regional production center.[61]

According to the Asian Development Bank (ADB), ASEAN's participation in Asian production networks has been growing, particularly in the automotive and electronics sectors. The share of parts and components in ASEAN's manufacturing trade increased from 35 percent in 1996 to 43 percent in 2006.[62] The ADB attributes this increase to several factors:

> As a share of GDP, parts and components trade is among the highest in the world in the ASEAN (especially in Malaysia, the Philippines, Singapore, and Thailand), perhaps because the relatively small size of their economies makes specializing in small niches of comparative advantage particularly important. Broadly speaking, the success of these economies is based on policies that welcome foreign companies, encourage technological upgrading, and build strong connections with world markets, as well as on their proximity to Asian neighbors following similar strategies.[63]

Table 2.4. Total Imports, Exports, and Intra-ASEAN Exports by Priority Sectors, 2004–08

	2004	2005	2006	2007	2008
			thousand $		
Intra-ASEAN exports					
Agro-based products	1,360,040	933,008	1,144,436	1,687,945	2,788,211
Automotives	3,283,824	4,191,755	4,690,222	6,416,477	7,987,317
Electronics	47,876,167	52,268,178	56,110,336	57,982,542	56,928,581
Healthcare	869,375	987,092	1,235,436	1,685,250	2,000,350
Textiles and apparel	1,932,309	1,912,064	2,122,545	2,420,349	2,431,291
Wood-based products	1,143,053	1,230,734	1,237,746	1,394,430	1,424,370
ASEAN exports to rest of world					
Agro-based products	12,103,978	12,486,983	14,057,296	19,313,273	28,435,800
Automotives	11,893,715	15,714,044	18,579,336	25,174,340	27,124,044
Electronics	215,482,996	239,608,257	265,615,439	266,126,413	267,278,788
Healthcare	3,506,738	5,692,372	8,071,235	10,049,727	10,167,219
Textiles and apparel	28,252,081	29,279,163	33,140,705	36,301,953	38,527,881
Wood-based products	14,874,195	16,074,553	17,000,980	18,327,184	17,661,561
ASEAN imports from rest of world					
Agro-based products	2,858,852	3,280,498	3,798,554	5,090,003	6,695,704
Automotives	20,993,758	22,482,370	21,549,767	26,486,024	33,332,207
Electronics	112,712,493	115,987,928	135,143,296	140,732,507	146,680,531
Healthcare	5,299,171	6,456,913	7,290,542	8,359,650	9,339,715
Textiles and apparel	13,597,104	14,690,394	16,692,532	21,385,662	22,541,775
Wood-based products	865,724	989,336	1,255,923	1,615,914	1,543,718

Source. WITS, Integrated Data Warehouse (accessed March 29, 2010).

Note. Data are based on partner country trade data, rather than reporter country trade data, because Brunei, Burma, Cambodia, Laos, Phillipines, and Vietnam do not report trade data in all years. Intra-ASEAN exports between these members may be understated.

Low wage rates, high productivity growth rates in some countries,[64] diverse production conditions, proximity to large Asian markets, and the region's trade policy environment, including free trade agreements with a number of countries (box 2.2 and app. E), have supported the growth of ASEAN's manufacturing exports, particularly through global production chains.[65] However, challenges for ASEAN's export competitiveness remain, particularly in some of the member states. These impediments include a shortage of skilled labor and professionals; the lack of a developed system for setting product standards and conformity assessment procedures; unsophisticated consumer markets; and inadequate physical and institutional infrastructure, such as roads and other transport networks, communications, trade facilitation measures (e.g., customs procedures), and intellectual property rights (IPR) protection.[66] USITC hearing witnesses note that many of the member countries continue to implement domestic polices that restrict trade and provide an unfavorable climate for foreign investment.[67]

China is often viewed as a major competitor of ASEAN countries in attracting foreign investment and in integrating regional production chains.[68] However, ASEAN recently concluded an FTA with China. According to a written submission, because China has become an important hub in Asian supply chains, ASEAN countries hope the FTA will offer better opportunities to participate in these networks.[69] In addition, ASEAN offers an alternative production location to China for firms wanting to diversify their operations to reduce business and political risks.[70]

Table 2.5 shows the trend in ASEAN's share of the world market over the past five years. During this period, ASEAN's share of world exports was fairly stable, but declined in 2008, probably because of the economic downturn that reduced demand from its top markets. In the six priority sectors, world export shares have generally risen over the 2004–08 period. For more information on export competitiveness in these sectors, see the discussion of each priority sector in chapters 3–8.

Longer-term analysis shows that ASEAN's world export share grew steadily until the 1997 Asian financial crisis, and subsequently has remained fairly stable during a period when China emerged as a major exporter.[71] Within ASEAN, the ASEAN-4 (Burma, Cambodia, Laos, and Vietnam) have shown the highest rates of export growth. Although some studies have shown that the economic rise of China and India has not reduced ASEAN's exports or growth rates,[72] ASEAN exporters continue to face stiff competition from these countries.[73]

The AEC addresses many of the competitive challenges facing the region. Many ASEAN countries still specialize in only a limited number of sectors, and exports are usually destined for a few partners.[74] Completing the AEC is expected to create opportunities for greater industrial efficiency and cost competitiveness in a much broader range of products and services, which would benefit ASEAN on the global market.[75]

Inbound Investment

Once it enters into force, the new ASEAN Comprehensive Investment Agreement (ACIA) will replace the two existing ASEAN investment agreements. The ACIA offers strengthened protections for foreign investors and is likely to contribute to increased direct investment between ASEAN members.

Investment agreements

The ASEAN Investment Area (AIA) and Investment Guarantee Agreement (IGA)

Table 2.5. ASEAN World Market Shares, by Priority Sector, 2004–08

	2004	2005	2006	2007	2008
	Share of world merchandise exports (%)				
Agro-based products	8.8	7.9	8.5	10.1	10.4
Automotives	1.5	1.8	2.0	2.3	2.6
Electronics	12.3	12.3	12.2	12.7	11.8
Healthcare	1.5	1.9	2.4	2.5	2.1
Textiles and apparel	4.4	3.9	3.9	4.3	4.5
Wood-based products	6.7	6.7	7.1	7.4	7.7
Total, all products	6.2	6.6	6.3	6.2	5.5

Source: Specific products: WITS, Integrated Warehouse (accessed March 29, 2010). All products: ASEAN Secretariat; IMF, *Direction of Trade Statistics*

As part of its Blueprint, the ASEAN Economic Community calls for "a free and open investment regime" that is designed to serve as a crucial step in attracting FDI to the ASEAN economies.[76] As of April 2010, the AIA agreement is the vehicle for this integration of ASEAN member investment regimes. The AIA was signed in October 1998 and is still in force, pending the ratification of the new ACIA by all members (discussed in further detail below).

The AIA was the first agreement to promote ASEAN as a single investment area, increase regional cooperation on investment issues, and provide guarantees of national treatment and transparency in investment regulations to investors. All industries in the areas of manufacturing, agriculture, fishery, forestry, and mining and quarrying, including services incidental to those sectors, are covered, except as outlined in each country's Temporary Exclusion and Sensitive Lists. A General Exception List also excludes industries that remain closed to investment for reasons of national security, public morals, public health, or environmental protection.[77] Investment in services is covered through the separate services negotiations conducted under the Coordinating Committee on Services, discussed below.[78]

The AIA operates in tandem with the IGA, signed in 1987, which provides investment protections. The IGA applies only to FDI between ASEAN member countries. The agreement offers most-favored-nation (MFN) treatment to investors, but not national treatment. [79] It provides for compensation in case of expropriation; guarantees an investor's right to repatriate earnings, subject to local laws; and provides for arbitration in cases of dispute, but does not provide for investor-state arbitration. The latter is guaranteed to investors under the recently signed ACIA agreement.[80]

The ASEAN Comprehensive Investment Agreement (ACIA)

Signed in February 2009, the ACIA provides enhanced investor protections as compared with the AIA and the IGA. The agreement is organized around four pillars— liberalization, protection, facilitation, and promotion of investment—and follows international standards for investment protection agreements, including the U.S. Model Bilateral Investment Treaty and the OECD Guidelines for Multinational Enterprises.[81] The ACIA applies to the same list of industries as the AIA, basically covering the entire economy except for most services, which are covered under the AFAS, discussed below.

The ASEAN Secretariat has identified a number of ACIA provisions that mark significant improvements in investor protections, compared with the AIA.

- Comprehensive investment liberalization and protection provisions:
  - Performance requirements are prohibited (Article 7).[82]
  - Member countries may not require investors to appoint senior management from the host country, although they may require that a majority of the Board of Directors be nationals of the host country (Article 8).[83]
  - Investor-state dispute settlement procedures are more comprehensive (Articles 28–41).
- Clear timelines for investment liberalization, although these are not binding.[84]
- Benefits extended to foreign owned (non ASEAN) investors, for investments that are based in an ASEAN member country, from entry into force of the agreement.

- Preservation of the special and differential treatment granted under the AIA for the newer members of ASEAN, allowing those countries to delay commitments under the ACIA in deference to their state of development (Article 23).
- A more liberal and transparent investment environment.[85]

The provisions of the ACIA agreement apply to all covered investments through a negative list framework,[86] but each member state also compiles lists of reservations, or exclusions to the agreement. Such reservation lists typically exempt certain industries considered strategic or in need of protection from coverage by the overall agreement. Because the ACIA has not yet entered into force, the reservation lists that will be attached to the agreement have not been made public, so it is difficult to assess the final impact of the ACIA.[87] Like the AIA, the ACIA includes a list of general and security exceptions, whereby measures that are needed to protect public morals, public order, the environment, protection of national treasures, and national security can be excluded.[88]

According to a representative of the ASEAN Secretariat, the reservation lists are likely to reflect current law in most ASEAN member countries, and to retain most of the reservations that existed under the AIA.[89] If that is the case, the immediate benefits of the ACIA to foreign investors will consist of the expanded protections offered to covered investments, such as the investor-state dispute settlement process, but will not expand the scope of those covered investments.

Investment-related Provisions of the ASEAN Framework Agreement on Services (AFAS)

For services industries, other than services incidental to the goods sectors covered by the ACIA and its predecessor agreements, liberalization of FDI regulations is also covered under the AFAS and is also addressed by successive rounds of negotiations primarily under the aegis of the Coordinating Committee on Services. The AFAS recognizes the AEC Blueprint's goal of facilitating the free flow of services within the region by 2015.[90] Included in the overall liberalization of services trade are several provisions that apply particularly to investment. The AEC Blueprint sets goals under which ASEAN members will change their regulations to permit foreign investors to hold a minimum equity ownership in investments in certain industries (table 2.6), as well as a broad goal to progressively remove, by 2015, other provisions that limit market access to commercial establishment in an ASEAN market.[91] In addition, the investment protections extended to covered investments under the ACIA are also extended to investment in the service sectors covered under the AFAS.[92] Because the AFAS operates through a positive list framework, only services covered under each country's Schedule of Commitments receive these protections.

Expected effects of ACIA on intra-ASEAN FDI[93]

Because the ACIA has not yet entered into force, it is difficult to predict the effects of the new agreement on intra-ASEAN flows of FDI in coming years. However, data for 2008 show that intra-ASEAN FDI flows recorded a substantial increase of 13.4 percent, even in the face of overall declines in FDI flows from developed countries into ASEAN. As a result, the share of intra-ASEAN FDI flows increased to 18.2 percent of the total in 2008, up from 13.5 percent in 2007, and close to the 21.9 percent share of the European Union in overall FDI flows to ASEAN (table 2.7).

Table 2.7. FDI Inflows to ASEAN, 2004–08

	FDI Inflows (million $)					Share of total FDI Inflows (percent)				
Country	**2004**	**2005**	**2006**	**2007**	**2008**	**2004**	**2005**	**2006**	**2007**	**2008**
European Union (EU)-25	8,630	8,187	10,672	18,481	12,942	25	20	19	26	22
ASEAN	2,804	6,242	7,596	9,462	10,727	8	12	14	14	18
Japan	5,732	7,235	10,230	8,382	7,157	16	21	19	12	12
United States	5,232	3,865	3,419	6,346	3,013	15	7	6	9	5
All other	12,719	15,539	23,063	27,269	25,169	36	40	42	39	43
Total FDI inflows to ASEAN	35,117	41,068	54,980	69,939	59,008	100	100	100	100	100

Sources: ASEAN Foreign Direct Investment Statistics Database, http://www.asean.org/18144.htm (accessed March 25, 2010); ASEAN Investment Report 2007, http://www.aseansec.org/21406.pdf (accessed April 22, 2010).

Note: For 2004 and 2005, EU data reflect the United Kingdom, the Netherlands, and Germany only, which are believed to account for the great majority of total EU FDI inflows into ASEAN. Data for EU-25 are not available prior to 2006, and separate data for the United Kingdom, the Netherlands, and Germany are not available for 2004 and 2005. Investment in Burma by U.S. persons is a violation of U.S. law. See U.S. Department of the Treasury, "What You Need to Know About U.S. Sanctions Against Burma (Myanmar)," December 5, 2008, http://www.treas.gov/offices/ enforcement. (accessed July 22, 2010); Congressional Research Service, "Burma: Economic Sanctions," August 3, 2009, http://www.fas.org/sgp/crs/row/RS22737. pdf (accessed July 22, 2010).

The ASEAN Secretariat attributes the rise of intra-ASEAN FDI flows in part to the rising confidence of ASEAN investors based on shared cultural and geographic similarities, and expects continued growth in intra-ASEAN FDI flows as integration efforts continue. The increased share of intra-ASEAN FDI may also reflect the drop in FDI flows from most developed countries, particularly the EU and the United States, due to the 2008–09 global financial crisis.[94]

Indonesia, Thailand, Vietnam, and Malaysia received the largest shares of intra-ASEAN FDI flows in 2008. The largest investors were Singapore and Malaysia, with Singapore accounting for 57.2 percent of total intra-ASEAN FDI flows that year.[95]

Regional Improvements in Trade Facilitation, Logistics Services, and E-Commerce

USTR specifically requested analysis of the impact of trade facilitation, logistics services, and e-commerce on regional integration, export competitiveness, and inbound investment for the six industries analyzed in this chapter. The following discussion describes selected developments in these three areas that have affected multiple industries and ASEAN member countries.

Trade Facilitation

Trade facilitation is "the simplification, standardization, and harmonization of procedures and associated information flows required to move goods from seller to buyer and to make payments."[96] This chapter distinguishes improvements in trade facilitation from logistics services by treating the former as reforms to policies and procedures for importing and exporting, and the latter as discrete services that importers and exporters pay for in order to move goods from sellers to buyers.

The time and cost required to import and export vary widely among ASEAN members. The World Bank's Ease of Trading Across Borders index for 2009 ranked Singapore first and Laos 168th among 183 countries, with all other ASEAN members falling between them.[97] While importing and exporting in several of the ASEAN countries—led by Singapore and Thailand—is generally cheaper, faster, and less complicated than the average for countries in East Asia and the Pacific, only Singapore consistently outperforms the wealthy nations in the OECD (table 2.8).[98]

ASEAN members addressed trade facilitation to a limited extent in the 1990s,[99] but increased their focus on it in the next decade[100] when they recognized that they would need to reduce NTMs in order to enjoy the maximum benefits of tariff liberalization.[101] The ASEAN Trade Facilitation Work Programme (ATFWP), adopted in August 2008,[102] calls for an ambitious set of trade facilitation measures in areas including customs modernization, simplified rules of origin, and harmonization of product standards and technical regulations.[103] It also includes ASEAN's most visible effort to facilitate trade: establishment of the ASEAN Single Window (ASW).[104]

Table 2.8. Ease of Trading Across Borders, 2009

Country or Region[a]	**Global Rank (of 183)**	**Documents to export (number)**	**Time to export (days)**	**Cost to export (US$ per container)**	**Documents to import (number)**	**Time to import (days)**	**Cost to import (US$ per container)**
Singapore	1	4	5	456	4	3	439
Thailand	12	4	14	625	3	13	795
Malaysia	35	7	18	450	7	14	450
Indonesia	45	5	21	704	6	27	660
Brunei	48	6	28	630	6	19	708
Philippines	68	8	16	816	8	16	819
Vietnam	74	6	22	756	8	21	940
Cambodia	127	11	22	732	11	30	872
Laos	168	9	50	1,860	10	50	2,040
United States	18	4	6	1,050	5	5	1,315
East Asia and Pacific (average)		7	23	909	7	24	953
OECD (average)		4	11	1,090	5	11	1,146

[a] No data available for Burma.

ASEAN Single Window

ASEAN members' senior economic officials have deemed the ASW "the single most important initiative of customs that will ensure expeditious clearance of goods and reduce the cost of doing business in ASEAN."[105] ASEAN members signed the Agreement to Establish and Implement the ASEAN Single Window (ASW Agreement) in December 2005. The agreement defines the ASW as "the environment where National Single Windows (NSWs) of Member Countries operate and integrate," and says that National Single Windows enable:

a. a single submission of data and information;
b. a single and synchronous processing of data and information; and
c. a single decision-making point for customs release and clearance.[106]

Members agreed that the ASEAN-6 (Brunei, Indonesia, Malaysia, the Philippines, Thailand, and Singapore) would activate their NSWs by 2008, while the ASEAN-4 (Cambodia, Laos, Burma, and Vietnam) would do so no later than 2012. The effective deadline for completion of the ASW is 2015.[107] As of March 2010, Singapore was the only member with a fully operational NSW. Brunei, Indonesia, Malaysia, the Philippines, and Thailand had partially completed their NSWs, while Vietnam had completed a National Single Window Master Plan and was piloting an "E-Customs" system intended to be a core element of its NSW. Cambodia, Laos, and Burma have made much less progress.[108]

The ASW will enable ASEAN member countries' customs agencies to exchange information required for the ASEAN Customs Declaration Document (ACDD), certificates of origin for preferential trade agreements, and other key documents for trading and transporting goods. The ASW will use a "federated approach"[109] that permits the exchange of information among ASEAN members' customs agencies without storing the information on a central server. Design of an ASW prototype (the ASW Pilot Project) is scheduled for completion in fall 2010.[110]

Private sector representatives consulted for this study were enthusiastic about the ASW's potential to produce significant benefits for companies and member economies.[111] For example, the director of a multinational manufacturer in Vietnam suggested that the ASW, as part of a broader regional program to eliminate barriers to cross-border trade, could enable his firm to concentrate production at the most efficient locations and then ship the products throughout the region.[112]

However, firms had a number of concerns. First, many felt that work on the ASW and NSWs had proceeded too slowly. Observers attributed delays to disagreements among ASEAN member countries over the ASW's architecture rather than technical challenges. Sources suggested that one country in the region feared that the ASW could compromise the security of traders' information and could undermine the competitive advantages that the country enjoys due to the strength of its own NSW.[113] Secondly, private sector representatives were uncertain whether the ASW would include features that would make it maximally useful, such as the ability to file an import declaration electronically from a point outside the country of importation[114] or the opportunity to register with an "integrated trader database" that would make traders eligible for expedited clearance throughout the region.[115] Finally, while firms appreciated the opportunities that they had had for dialogue with ASEAN member countries and the ASEAN Secretariat, they wanted the discussions to occur more regularly, perhaps through a dedicated body through which the private sector could provide suggestions for the ASW. They also hoped for clearer indications that the dialogue was leading to tangible progress.[116]

There are signs that these concerns are being addressed. The launch of the aforementioned ASW Pilot Project and completion of a recently expanded "ASEAN Data Model" specifying standard data elements for key trade documents[117] suggest that work on the ASW is accelerating. Members are also intensifying their efforts to develop a legal framework for the ASW. [118] For example, in August 2009, members adopted a memorandum of understanding on the legal terms for the exchange of data through the ASW Pilot Project, providing a foundation upon which to base the ASW's comprehensive legal framework.[119] In regard to private sector engagement, ASEAN Secretariat staff are increasing their efforts to exchange ideas with the private sector, such as through a roundtable discussion with the US-ASEAN Business Council in Singapore in March 20 10.[120] Continued public-private engagement and consistent demonstration of progress will be vital to sustaining the business community's enthusiasm for the ASW.[121]

Trade Facilitation Activities Sponsored by Other Entities

Several ASEAN member countries participate in trade facilitation programs that are not coordinated by ASEAN. For example, Burma, Thailand, Cambodia, and Vietnam (along with China's Yunnan province) participate in the Greater Mekong Subregion (GMS) program, which is sponsored principally by the Asian Development Bank. Its agenda includes "hard" infrastructure improvements (e.g., road and rail infrastructure), as well as facilitation of the cross-border transport of goods and people. The GMS Cross-Border Transport Agreement (CBTA) aims to permit trucks to make a single stop to complete all border procedures (as opposed to stopping at multiple agencies' posts on each side of the border). As of 2010, implementation of the CBTA's single-stop procedures had not progressed beyond pilot applications at select locations (e.g., the crossing between Lao Bao, Vietnam, and Dansavanh, Laos). Observers noted the challenges of harmonizing countries' regulations and reaching agreement for procedures at the single-stop sites.[122]

The ASEAN Secretariat provides little direct assistance for modernization of member countries' customs agencies because of the Secretariat's limited financial resources (as noted earlier, its total budget for 2010 is $14.5 million).[123] Multilateral and bilateral donors sponsor most such assistance. Examples include the World Bank's Laos Customs and Trade Facilitation Project [124] and Vietnam Customs Modernization Project[125] and the U.S. Agency for International Development's Local Implementation of National Competitiveness for Economic Growth (LINC-EG) Program for the Philippines.[126]

Logistics Services

Logistics services are "a range of related activities that ensure the efficient movement of production inputs and finished products." [127] Companies that specialize in logistics, known as third-party logistics service providers (3PLs), offer services such as supply- chain consulting, transportation management (e.g., warehousing, cargo handling, and customs brokerage), freight transport, and express delivery.[128] Large 3PLs operating in ASEAN include firms from outside the region (e.g., DB Schenker, DHL, Kuehne + Nagel, and Nippon Express), as well as firms based within it (e.g., Singapore's APL Logistics and Toll Global Logistics).[129]

In August 2007, ASEAN members named logistics services (like the six sectors discussed in detail in this study) a Priority Integration Sector. The Roadmap for the Integration of Logistics Services commits members to "substantial liberalization" of logistics services by 2013.[130] The Roadmap names specific logistics services to be liberalized, including customs brokerage, storage and warehousing, packaging, freight forwarding, cargo handling, international freight transport, and express delivery. It also lists numerous measures in the domains of "trade and customs facilitation" and "logistics facilitation," including the completion of new ASEAN agreements (e.g., the ASEAN Framework Agreement on Facilitation of Inter-State Transport) and implementation of existing ones (e.g., the ASEAN Framework Agreement on Facilitation of Goods in Transit).

The quality of logistics services varies substantially among the ASEAN countries. The World Bank's Logistics Performance Index (LPI) for 2010 ranks Singapore second and Burma 133rd among 155 countries (table 2.9). Recent research points to a strong inverse relationship between the restrictiveness of the logistics policies of ASEAN countries and perceptions of logistics services quality, as reflected in the LPI.[131] This finding suggests that implementation of the Roadmap could lead to improvements in the perceived quality of logistics services in member countries.

Trade facilitation enables 3PLs to deliver their services more cheaply and quickly—and equally importantly, with more consistent costs and delivery times. For this reason, trade facilitation is a prerequisite for high-quality logistics services. The wide variation in ASEAN members' LPI scores for efficiency of customs clearance is notable because logistics services providers consistently rank customs procedures and inspections as the most significant impediments to the delivery of logistics services.[132]

Table 2.9. ASEAN in the World Bank's 2010 Logistics Performance Index (LPI)

Country or Region[a]	Global Rank (of 155)	LPI aggregate[b]	Cust-oms	Infrastructure	International shipments	Logistics competence	Tracking & tracing	Time-liness
Singapore	2	4.09	4.02	4.22	3.86	4.12	4.15	4.23
Malaysia	29	3.44	3.11	3.50	3.50	3.34	3.32	3.86
Thailand	35	3.29	3.02	3.16	3.27	3.16	3.41	3.73
Philippines	44	3.14	2.67	2.57	3.40	2.95	3.29	3.83
Vietnam	53	2.96	2.68	2.56	3.04	2.89	3.10	3.44
Indonesia	75	2.76	2.43	2.54	2.82	2.47	2.77	3.46
Laos	118	2.46	2.17	1.95	2.70	2.14	2.45	3.23
Cambodia	129	2.37	2.28	2.12	2.19	2.29	2.50	2.84
Burma	133	2.33	1.94	1.92	2.37	2.01	2.36	3.29
East Asia & Pacific (average)		2.73	2.41	2.46	2.79	2.58	2.74	3.33

Source. World Bank, Logistics Performance Index database, accessed April 20, 2010.

[a] No data available for Brunei.

[b] For the LPI aggregate and each component, the range of possible scores is 1 (worst) to 5 (best).

Restrictions on foreign ownership also affect logistics services providers' ability to offer high-quality services. For example, if regulations require a foreign company to operate as a minority partner in a joint venture, the foreign company may have less power to ensure that its standards of service quality are maintained.[133] While the ASEAN Economic Blueprint called upon all ASEAN members to allow 51 percent foreign ownership in logistics services by 2010 and 70 percent by 2013, several countries have not complied with this commitment (see below). The scope for enforcement of the commitments is unclear. The AEC Blueprint "recommends" that members use the ASEAN Enhanced Dispute Settlement Mechanism (DSM) "in enhancing the implementation arrangement ... identified in the Blueprint" but does not specifically state that liberalization commitments are subject to dispute resolution. Furthermore, members' willingness to use the Enhanced DSM is questionable. Members have never used the Enhanced DSM since its creation in 2004, nor did they use the original ASEAN DSM created in 1996.[134]

Industry representatives were particularly concerned about foreign ownership restrictions in Thailand, the Philippines, and Indonesia. Thailand prohibits majority foreign ownership of "domestic transportation" activities, which includes movements of goods within Thailand that may be part of an international shipment.[135] Service providers expressed confusion about the rules in the Philippines. While foreign participation in logistics services is capped at 40 percent, the service providers reported that the Philippine Department of Justice had ruled the limit nonapplicable to air freight forwarders. However, when one express delivery firm applied for a license, it was told that the application would have to be reviewed in court, where delays of several years are common.[136]

Indonesia's 2009 Postal Law defines all logistics services as "postal services" and reserves these activities for Indonesians.[137] The law allows foreign firms to offer services in collaboration with domestic partners in "province[s] or capital[s] with an international airport and/or harbour only."[138] The law does not preserve the distinctions made in the ASEAN logistics sector's Roadmap between postal services and other logistics services and ignores the liberalization commitments in the AEC Blueprint.[139]

Malaysia has long required that a certain percentage of companies' ownership shares be reserved for *bumiputera* (indigenous Malays) in various industries. In April 2009, the government removed those requirements for 27 service industries, including several logistics services (e.g., road freight transport and maritime agency services).[140] However, 3PL industry representatives said that the requirement for majority *bumiputera* ownership of customs brokerages, which was not eliminated, prevents some firms from offering their standard services.[141]

Vietnam and Cambodia liberalized many logistics services as part of their World Trade Organization (WTO) accession commitments. In Vietnam, foreign ownership for express delivery firms is presently capped at 51 percent; the cap is to be lifted entirely in 2012. Road freight transport opened to majority (51 percent) foreign ownership in 2010, although Vietnam reserved the right to screen investments using "economic needs tests."[142] Cambodia's WTO accession package included commitments not to restrict foreign ownership for road freight transport and courier services (generally understood to include express delivery).[143]

The Commission did not find evidence of notable foreign ownership restrictions on logistics services in Brunei or Singapore. The latter is recognized as one of the most open markets for logistics services in the East Asia and Pacific region.[144] Limited information was available for restrictions in Laos and Burma. One industry source noted that full foreign ownership of express delivery service firms in Burma is permitted, but described licensing procedures as "restrictive."[145]

E-Commerce

E-commerce is "the buying, selling, or exchanging of goods, services, and information through electronic networks."[146] It occurs through three channels: business-to-business (B2B), business-to-consumer (B2C), and consumer-to-consumer (C2C). Use of e- commerce varies by country and industry. Its prevalence depends on a combination of factors, including the quality and extent of infrastructure; the existence of a sound legal framework; and individuals' access to information and communication technologies (ICTs), facility with the use of ICTs, and willingness to make electronic transactions.

Broadband service is a critical element of the infrastructure required for e-commerce.[147] People who subscribe to broadband are more likely to shop online;[148] the high-speed Internet connections enabled by broadband service permit shoppers to complete transactions more quickly than non-broadband service. Among the ASEAN countries, the number of broadband subscribers per 100 inhabitants (broadband penetration) varied from 0.02 in Burma to 21.7 in Singapore in 2008. Broadband penetration was far higher in Singapore than in any other ASEAN country, but increased in all members between 2004 and 2008 (figure 2.3).

A sound legal framework for e-commerce gives consumers and businesses confidence that transactions are secure and makes compliance with business laws and regulations possible without excessive paper-based documentation. ASEAN members have made progress in developing their legal frameworks for e-commerce in recent years. For example, in 2004, only Brunei, the Philippines, Singapore, and Thailand had adopted laws on electronic contracting; at present, all members have done so except Cambodia and Laos.[149] Vietnam was particularly active in developing its legal framework for e- commerce during this period. Between 2004 and 2008, it enacted a Law on Electronic Transactions, a Law on Information Technology, and 11 decrees and circulars to implement provisions of these laws.[150]

Nevertheless, getting the legal framework right is challenging for even the most active reformers. For example, despite Vietnam's numerous reforms, businesses there do not conduct transactions through electronic data interchange (EDI) systems.[151] Industry representatives reported that electronic payments are difficult to implement because financial and accounting regulations require companies to obtain many physical signatures (e.g., from a company's chief accountant and its director), and tax regulations require paper-based receipts. One multinational automobile manufacturer in Vietnam continues to use paper-based payments in Vietnam despite having implemented EDI at virtually every other location in the world.[152]

ICT access, facility with ICTs, and willingness to make electronic transactions are more difficult to measure. One proxy for the first two factors is the percentage of the population that regularly uses the Internet. In five ASEAN countries—Burma, Cambodia, Indonesia, Laos, and the Philippines—less than 10 percent of the population used the Internet in 2008 (figure 2.4). In contrast, more than 50 percent of the populations in Malaysia and Brunei and nearly three-quarters of the population in Singapore used the Internet that year. Growth of Internet usage in Vietnam has been particularly impressive. It increased from 7.6 percent in 2004 to 23.9 percent in 2008, moving Vietnam from a position well below the global mean to just above it (figure 2.4).

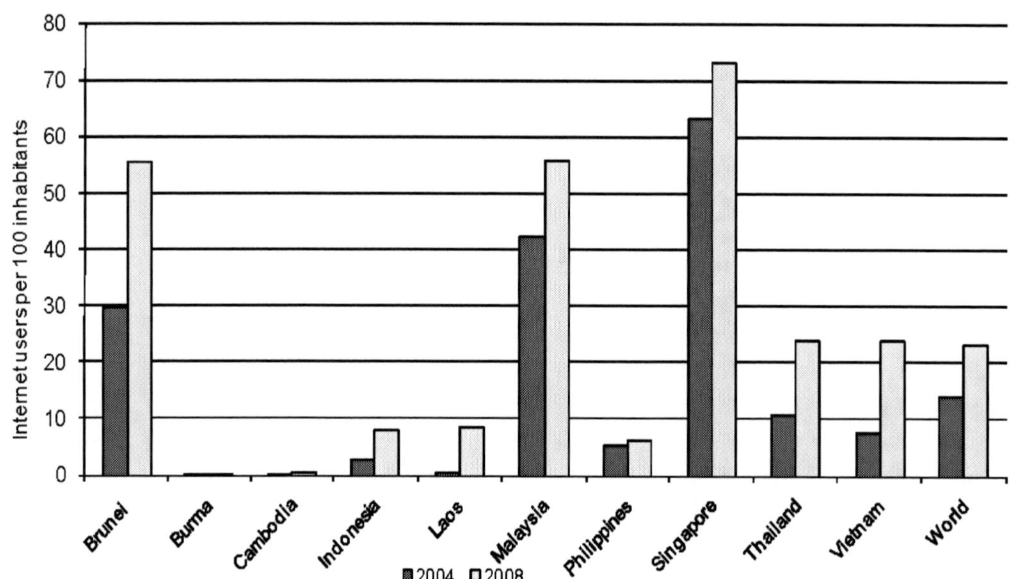

Source. International Telecommunications Union (ITU), ICT Eye database, accessed April 21, 2010.

Figure 2.4. Internet usage: The majority of people in Brunei, Malaysia, and Singapore used the Internet in 2008, while less than 10 percent did so in five other ASEAN countries.

Willingness to make electronic transactions is a function of the aforementioned factors and, more broadly, of the amount of trust one places in electronic transactions and one's openness to changing established business habits. In at least some ASEAN countries, trust may be one of the important factors slowing the growth of e-commerce. Returning to the example of Vietnam, consumers frequently make travel reservations over the Internet, but rarely purchase other goods or services online.[153] One source said that consumers mistrust online transactions because they fear that their financial information will be stolen.[154] Their reluctance may also be a function of established habits: officials at the Vietnam E-commerce and Information Technology Agency (VECITA) remarked that most Vietnamese Internet users are young people, yet executives of many companies are older. They do not often use the Internet and are accustomed to doing business offline.[155]

Promotion of e-commerce is embedded in ASEAN's economic integration agenda. The e-ASEAN Framework Agreement in 2000 enjoined members to "facilitate the growth of e-commerce in ASEAN" by enacting "laws and policies relating to electronic commerce transactions based on international norms."[156] The AEC Blueprint in 2007 called upon members to harmonize laws for electronic contracting and dispute resolution and reiterated the e-ASEAN Framework Agreement's call for members to facilitate mutual recognition of digital signatures.[157] ASEAN members have also committed to regional cooperation on ICT infrastructure development. The e-ASEAN Framework Agreement says that members will "work towards establishing ... an ASEAN Information Infrastructure Backbone."[158]

ASEAN's most significant contributions to the development of e-commerce appear to have been in the domain of e-commerce laws. Between 2004 and 2008, the Harmonisation of E-Commerce Legal Infrastructure in ASEAN project, sponsored by the Australian Agency for International Development (AusAID) under the ASEAN-Australia Development Cooperation Program, assisted members in developing e-commerce legislation in accordance with international best practices.[159] The project played a key role in helping Indonesia to prepare its draft law (adopted in 2007) and draft laws for Laos and Cambodia (awaiting ratification by those countries' National Assemblies). ASEAN Secretariat staff noted that ASEAN has not received donor funds dedicated to development of e-commerce laws since the aforementioned project concluded. This may limit the Secretariat's ability to continue meaningful work in this area.[160]

Regarding development of ICT infrastructure, ASEAN's work has mostly been limited to facilitating discussions among member countries on topics such as prioritization of ICT investments. More recently, the Secretariat has been seeking to promote dialogue with private sector partners that are potential sources of ICT investment, such as the World Bank's International Finance Corporation and the U.S.-ASEAN Business Council.[161] It is unclear whether this outreach has led to any new investments.

3. Electronics: Computer Components

Electronics Overview[162]

Electronics products are a major source of manufacturing in the ASEAN region.[163] ASEAN countries exported $267 billion in electronics products to the rest of the world in 2008 (table 2.4), making it the largest in terms of value among ASEAN's priority integration sectors. In addition, ASEAN countries serve as important production and research and development (R&D) locations for multinational electronics firms in virtually all industries within the electronics sector.

Taken as a whole, the ASEAN region is a major producer of electronics. However, ASEAN remains 10 distinct countries with differing policies and competitive strengths. In order to "develop, strengthen, and enhance the competitiveness of the ASEAN electronics sector and promote ASEAN as an integrated platform in which to do business in electronics," the ASEAN Secretariat, in conjunction with member countries, developed the Roadmap for Integration of Electronics Sector (Roadmap).[164] The Roadmap proposes 58 measures to improve regional integration in the electronics sector, of which 13 are sector-specific. The other 45 are measures that address common issues of importance across many sectors. An example of a sector-specific measure is the promotion of electronics trade exhibitions, while an example of a cross-sectoral measure is the development of harmonized ASEAN customs declaration forms.

Among the measures specific to the electronics sector, the most notable success has been the implementation of the ASEAN Electrical and Electronic Mutual Recognition Arrangement (MRA). The electrical and electronics MRA is the first ASEAN-wide MRA,[165] and it establishes ASEAN-wide acceptance of test certificates from approved laboratories that assess electric and electronic products for conformity with safety and other regulations. In addition to the MRA, the electronics sector also has the most active public-private dialogue group among the priority integration sectors: the ASEAN Electronics Forum (AEF).[166] The AEF meets regularly to prioritize integration measures and assist with implementation.

Although there have been some ASEAN-wide efforts to integrate the electronics sector, competition between ASEAN countries remains an obstacle to integration. ASEAN produces

10 percent of the world's electronics and consumes 5 percent.[167] Most electronics production is for export, and export-oriented production is primarily driven by multinational firms. The electronics industry in ASEAN "relies on medium and large . . . firms from its neighbor countries in North East Asia and the US and EU for its further expansion and integration."[168] Because the large multinational firms mainly produce electronics for worldwide export, intra-ASEAN trade in electronics only accounts for 20 percent of the region's trade in these products.[169] Countries worldwide, including ASEAN countries, compete for investment from multinational electronics firms, and ASEAN countries may grow at the expense of one another. Vietnam surpassed the Philippines and Indonesia in total foreign direct investment (FDI) inflows in the late 1990s[170] and has become an important recipient of electronics sector investment in the past several years. Several multinational electronics companies confirmed that they compare the benefits of investing in various ASEAN countries before deciding on a location, or that they have shifted their activities within ASEAN due to changes in countries' competitive advantages.[171] The competitive dynamic between countries for export- oriented investment makes substantial cooperation between ASEAN countries on integration efforts difficult.

Computer Components[172]

Key Findings

Within the electronics sector in ASEAN, the computer components industry is among the most important in terms of export competitiveness and inbound investment. Computer components account for nearly 25 percent of electronics exports from ASEAN to the rest of the world and approximately 22 percent of intra-ASEAN trade in electronics.[173] The ASEAN region occupies a globally competitive position in this industry as the world's second largest exporter of computer components. ASEAN countries have also been important locations for direct investment from major computer components producers for more than 40 years. At present, ASEAN countries specialize in various aspects of computer components production, including hosting regional headquarters, performing R&D, and assembling products. Despite the region's overall competitiveness in the computer components industry, it faces a challenge from China, which has increased its share of components production as it has become the single largest assembler of finished computers. Many components producers are increasing production in China, both in order to be closer to their customer (the computer assembler) and because China performs well on several measures of competitiveness in this industry, including labor mix, existence of competitive local suppliers, and transportation infrastructure. In response, ASEAN member countries have acknowledged the importance of regional integration in order to remain competitive. However, as seen throughout the broader electronics sector, regionwide efforts to integrate this industry are hampered by competition between ASEAN countries to supply components to the global computer supply chain. An important way in which countries become involved in export-oriented computer components production is through direct investment from multinational components firms, which several ASEAN countries compete heavily to attract.

Background

The following sections will describe ASEAN economic integration, export competitiveness, and inbound investment as they relate to the competitive factors impacting the computer components industry in the region. Among the diverse set of computer components products, key items produced in the ASEAN region include hard disk drives (HDDs), computer central processing units (CPUs), and assembled printed circuit boards. Inbound investment is particularly important to regional competitiveness in this industry, as a limited number of large multinational firms account for most of the production of many types of components (either directly or through their suppliers and contract manufacturers). Many countries worldwide, including several ASEAN countries, vie for investments from these firms as a source of skilled manufacturing and R&D jobs and integration into the highly globalized supply chain for computer products. This dynamic is complex given that many factors—availability of skilled labor, existence of supporting industries, political stability, ease of doing business, and financial incentive packages offered to investors—are important aspects of a country's ability to compete in this industry.

Within ASEAN, there is substantial variation between member countries in terms of the types of components they produce and the extent to which they participate in non-manufacturing activities, such as R&D. The three ASEAN countries most heavily involved in the computer components industry are Malaysia, Singapore, and Thailand. The Philippines plays a smaller role in exporting computer components, but it has been an investment destination for firms such as Intel (until recently) [174] and Integrated Microelectronics, Inc., and it also serves as country coordinator for the ASEAN electronics sector Roadmap. Over the past few years, Vietnam has emerged as an important location for the computer industry, attracting direct investment from multinationals such as Compal Electronics, Intel, and Jabil Circuit. Vietnam's exports of components are small by volume at present, but given its rising importance as an investment destination, Vietnam's role in the computer industry, and specifically in components production, is likely to continue to grow. Indonesia exports a relatively small volume of computer components, although it is more heavily involved in other parts of the electronics industry. Brunei,[175] Laos, Cambodia, and Burma do not produce computer components. The latter three of these countries lack a large enough pool of skilled labor for this industry at present, and their limited transportation infrastructure is also a barrier to investment.[176]

In the computer components industry, Singapore specializes in hosting regional headquarters of multinational firms, performing R&D, investing in other ASEAN countries, and certain skill-intensive manufacturing tasks such as that of producing specialized computer processors.[177] Due to high labor costs, manufacturing activity has gradually dwindled in Singapore since the 1980s.[178] Malaysia has also become increasingly involved in regional R&D functions while also maintaining an important role in high-technology manufacturing.[179] For example, while Malaysia's role in the HDD industry has shrunk as manufacturing has moved to lower cost locations, it remains a competitive location for the more skill-intensive CPU industry. Thailand currently plays a leading role in HDD production; between 2004 and 2008, its share of global computer storage unit exports grew from under 10 percent to over 23 percent.[180] Vietnam's role in the industry largely remains to be determined, but Intel, the world's largest producer of CPUs, will open a new assembly and test facility in 2010—its largest anywhere—near Ho Chi Minh City, an investment totaling about $1 billion.[181] The value of the products that Intel plans to export from this facility substantially exceeds that of Vietnam's total computer components exports in 2008, so it is likely that Intel's CPUs will dominate Vietnam's computer components exports for at least the next several years.[182] The competitive factors behind these production patterns are examined in greater detail later in this chapter.

Currently, China is regarded as presenting the greatest challenge to ASEAN's regional competitiveness in the computer components industry. Between 2004 and 2008, China's share of global exports of computer components grew from 27 to 35 percent, while the ASEAN region's share remained flat at 20–21 percent.[183] As China has come to dominate final assembly of computers over the past 10 years, it has also captured an increasing share of components production and investment.

ASEAN has responded to this challenge from China through its attempts to more fully integrate the electronics sector in the region. The goal of the ASEAN electronics Roadmap was to work with the private sector to "strengthen regional integration" in order to "promote ASEAN as an integrated platform in which to do business in electronics."[184] However, while the electronics sector is one of the more active among the priority integration sectors in discussing how to implement Roadmap measures, there are few examples of substantial cooperation between member countries regarding efforts to improve integration because, as noted, ASEAN member countries still view each other as rivals for inbound investment. Closer regional integration within ASEAN would likely benefit the region's overall competitiveness in the global computer components industry, but competition between member countries for direct investment from multinationals, such as through the use of tax breaks and other financial incentives, impedes further integration.

Regional Integration, Export Competitiveness, and Inbound Investment

Leading Competitive Factors

The ASEAN region has long been an attractive destination for the production of computer components and other electronics. Singapore began producing electronics for export in the 1960s, and Malaysia became a competitor shortly thereafter, attracting foreign investment from Intel, AMD, Hewlett-Packard, Hitachi, and others during the 1970s.[185] Beginning in the 1980s, components firms from Japan, Taiwan, Korea, and Singapore sought to lower their costs by moving more manufacturing offshore, largely to locations such as Malaysia, the Philippines, and Thailand.[186] As costs rose and labor became scarce in Malaysia in the 1990s, Thailand attracted more investment, particularly in the HDD industry.[187] Since the early 2000s, Vietnam has emerged as a new competitor in the industry. This section will examine the competitive factors that, in addition to government policies, determine the competitiveness of ASEAN countries within the global supply chain for computer components. These include diversity, cost, and productivity of the labor pool, development of supporting industries, transportation infrastructure, macroeconomic conditions, market size, and other factors.

Labor

One of the most important competitive factors in the computer components industry is the availability and relative cost[188] of both skilled professionals and assembly workers. Compared with other manufacturing sectors, the computer components industry is skill-intensive in that, due to the rapid pace of innovation, it relies more heavily on an adequate supply of well-trained engineers and specialized technicians.[189] Companies must be able to find engineers and technicians locally or easily bring them into a country in order for that country to become or remain competitive. In addition, in order to produce components competitively, a country must have an available pool of relatively unskilled, productive assembly workers that can be hired at a competitive wage. Thus, it is not only skill availability or labor cost that determines the competitiveness of a country's labor pool in this industry. Rather, it is the diversity of the labor pool—the availability of workers at a wide range of skill levels needed in the components industry and their corresponding cost—that gives a country a competitive advantage in attracting investment and jobs in the computer components industry.[190]

Each of the countries in ASEAN has its own unique labor challenges. In Singapore, it is difficult to find labor at most skill levels that is cost-competitive with that found in other components-producing countries. The manufacturing jobs that remain in Singapore tend to be very specialized technical positions. One of the multinational firms that continues to operate a manufacturing facility in Singapore has moved many of the jobs that used to be done there to China. The firm retains a few high-level manufacturing positions in Singapore, however, because it has found the skill set of engineers in Singapore well suited to its business, and it also appreciates the fact that its staff in Singapore can speak both Chinese and English.[191]

In Malaysia, as the labor force has developed the skills needed to participate in R&D and other professional activities, labor for assembly has become scarce.[192] Assembly workers are typically brought in from other Asian countries, some from ASEAN and some from outside the region. Indonesia is a major source of assembly workers for Malaysian components factories.[193] In addition to the shortage of assembly workers, one company noted that while Malaysia has skilled engineers, there is a great deal of competition for these workers because multiple industries in Malaysia rely on them.[194] The combination of the need to bring assembly workers in from other countries and the shortage of skilled engineers in Malaysia creates a potential constraint on the country's competitiveness in this industry.

In Vietnam, labor is cost-competitive, but there is a shortage of experienced managers and engineers. Companies generally state that they have little difficulty recruiting new college graduates for professional positions or securing an adequate supply of relatively unskilled assembly workers. Vietnamese engineers reportedly are cost-competitive, as they typically make salaries about half of those in China.[195] Electronics assembly workers are also highly cost-competitive in Vietnam; the minimum wage for this type of work is less than half the average manufacturing wage in prime manufacturing areas of China,[196] although foreign firms generally report paying more than the minimum wage. However, in 2006, the overall labor productivity level in Vietnam was only 53 percent of that of China.[197] While Vietnam's labor productivity is rapidly increasing, it is currently comparable with China's only for labor-intensive activities.[198] In addition to relatively low labor productivity in Vietnam, finding experienced professional managers and engineers was uniformly cited as a major challenge for components producers, and firms have responded by instituting extensive in-house training programs, bringing in experienced professionals from other countries, or both.[199]

Finally, Thailand is generally considered "competitive in labor costs, availability, and productivity."[200] However, there is a shortage of engineers, and companies report that a larger pool of electronics engineers would improve the country's competitive position.[201]

Supporting industries

Another important competitive factor in the computer components industry is the existence of a base of suppliers capable of working with multinational firms. When firms are located very close to a competitive base of suppliers, such as in a cluster, it "facilitates the sharing of information, speeds logistics, minimizes plant inventory, and ensures a ready supply of locally produced parts and components."[202] The degree to which ASEAN countries have adequate supplier bases to meet components firms' needs varies. Singapore has a number of indigenous firms focused on supplying a wide range of services and advanced technology components to multinationals. [203] A competitive network of suppliers also exists in the Penang region of Malaysia, which is an electronics cluster.[204] One HDD company operating in Thailand reported that they have been satisfied with the availability of competitive suppliers from Singapore and Malaysia, but that Thai firms generally lack the capital to compete with larger suppliers from other countries.[205] In Vietnam, an industry group noted that local firms' capacity to act as suppliers to multinationals is generally very low due to inadequate basic infrastructure and lack of experience producing to export standards. [206] Jabil Circuit's experience [207] in Vietnam confirms this. The company has found the supporting local electronics industry to be underdeveloped and, as a result, has had to import 80 percent of its inputs. [208] In the future, the company hopes to be able to source more of its inputs locally in order to speed production. To address this gap, governments in Thailand and Vietnam are both pursuing policy initiatives to develop the local electronics supplier base. At present, Vietnam has created a master plan to develop the supporting industry in electronics, but lacks funds to implement these measures.[209] Developing supplier networks is especially important to ASEAN countries' relative competitiveness in this industry because it is an area in which China excels.[210]

Other competitive factors

Beyond the availability of labor and existence of supporting industries, there are several other competitive factors that affect the computer components industry. While they were mentioned by industry representatives less often, they clearly affect the region's overall competitiveness in this industry. They include:

- *Transportation infrastructure.* Some computer components are shipped by air, while others are shipped by ocean. In either case, speed to market is important in this industry, so competitiveness depends on adequate infrastructure. Transportation infrastructure remains a major challenge in Vietnam, which just opened its first deepwater port in 2009 and lacks adequate roads in many areas.[211]
- *Size of the domestic market.* While most components production is for export to computer assemblers, a large domestic market for electronic products is an added incentive for some firms in the industry in choosing where to locate production.For example, producers in Vietnam noted that the growth potential of the domestic market for computers and other electronics had some impact on their location decision.[212]
- *Macroeconomic conditions.* In particular, inflation in Vietnam is a concern for many multinational firms.

- *Intellectual property protection.* Enforcement of intellectual property rights affects a country's competitiveness in the components industry. Counterfeit components are a major concern for firms, as virtually every part of a computer from CPUs to HDDs can be counterfeited. Addressing problems with components is costly because "a component that may be worth only $2 can cost $20 to replace if it is found to be counterfeit after it is mounted onto a circuit board."[213] Countries that effectively enforce intellectual property rights therefore have a competitive advantage over those where enforcement is weak. Within ASEAN, intellectual property enforcement varies widely. A more uniform, higher standard of enforcement would likely improve the region's overall competitiveness.

Regional integration

Given the sector's importance in the regional economy, the ASEAN Secretariat has initiated several efforts to improve integration in electronics production. The most industry-specific of these is the electronics sector Roadmap. Initiated in 2004, the Roadmap is designed to "develop, strengthen, and enhance the competitiveness of the ASEAN electronics sector and promote ASEAN as an integrated platform in which to do business in electronics."[214] Despite its ambitious agenda, attempts to implement the Roadmap have had somewhat mixed results. As noted, one of the key achievements under the electronics sector Roadmap was the implementation of the ASEAN Electrical and Electronics MRA. Computer components are not subject to the same consumer safety certification requirements as consumer electronics, however, so while the MRA was a step forward for regional integration in the broader electronics sector, it had little effect on computer components production. Meanwhile, some of the other Roadmap measures were never implemented.[215]

Under the Roadmap, the Philippines was designated as the country coordinator for implementation of Roadmap initiatives. In its role as country coordinator, the Philippines was instructed to convene representatives of the private sector from across ASEAN and obtain feedback on what improvements were needed in order to make the region a "truly integrated platform in which to do business in electronics." [216] The result of this consultation is the ASEAN Electronics Forum (AEF).

The AEF is unique within the region in that it has been the only active sectoral group in any of the priority integration sectors.[217] The purpose of the AEF is to promote dialogue between the public and private sectors and to help implement the measures contained in the Roadmap. The AEF has actively sought input from multinational firms as well as local companies, and it has had an ambitious agenda in moving toward implementation of many of the measures contained in the Roadmap. However, although the AEF played an important role in its first several years, its level of activity seems to have waned recently. The AEF convened 10 times between its inception in 2003 and October 2008, holding at least one meeting each year. However, the AEF did not meet in 2009 and, although a meeting was scheduled for April 2010, it did not take place.

In addition to the Roadmap, promoting investment in the electronics industry is also a focus of the ASEAN Secretariat. Initially, both the Secretariat and the AEF thought that electronics investment could be promoted by expanding the framework of the ASEAN Industrial Cooperation Scheme (AICO). AICO was intended to promote regional supply chains in manufacturing. However, the incentives offered under AICO were limited to duty waivers. For many electronics firms, especially computer components firms, this incentive was of little value, given that most of their products already enjoyed duty-free treatment. The intent was to expand the incentives offered under AICO at a later time.[218] However, despite recommendations from the AEF that AICO's scope be broadened,[219] incentives never expanded beyond duty waivers. As a result, AICO has had little impact on the computer components industry.

Beyond the official integration efforts led by the Secretariat, there is some evidence that ASEAN's role in integrating member countries into the regional economy can boost that country's competitiveness within the global supply chain for computer components. The primary evidence of this comes from Vietnam. Vietnam's integration into the ASEAN Free Trade Area accelerated its adoption of international norms on trade and investment policy[220] and thereby facilitated its entry into the WTO. WTO accession, in turn, has provided an enormous boost to Vietnam's ability to sell in international markets and attract investors.

Export Competitiveness

The ASEAN region is a globally competitive exporter of computer components. Tables 3.1 through 3.3 provide trade values for computer components for 2004–08. Overall, computer components exports from the ASEAN region totaled nearly $56 billion in 2008, confirming the importance of the industry to the regional economy and making the region the second largest source of computer components after China. On an individual country basis, Thailand and Malaysia were the fourth- and fifth-largest global exporters of computer components, respectively.[221]

Trade patterns during this period primarily reflect China's increasing importance as a trading partner for ASEAN countries due to its growing dominance of computer final assembly, a trend which is discussed in greater detail below. As seen in tables 3.1 and 3.2, ASEAN's exports of components to China grew 93 percent between 2004 and 2008, and imports of components from China grew 51 percent. These increases in ASEAN's components trade with China occurred despite the fact that the global recession led to significantly lower demand for computers in 2008 than in 2004, [222] indicating the strength of the overall trend. Meanwhile, ASEAN exports to the EU, Japan, and the United States fell between 2004 and 2008.[223] In 2008, this was primarily due to the effects of the recession on demand from those trading partners. Throughout the 2004–08 period, computer assembly activities also continued to shift away from these countries in favor of lower-cost locations such as China,[224] in turn reducing these countries' share of ASEAN's components exports.

Intra-ASEAN exports in this industry are only 20 percent of ASEAN's total exports because production of computer components is primarily for export to final computer assemblers, many of which are in China. However, there are a few computer assembly facilities in ASEAN countries,[225] and some components firms engage in intra-ASEAN sourcing, both of which contributed to the $11 billion in ASEAN exports to other ASEAN countries in 2008.

Trade barriers and nontariff measures (NTMs) are not major factors affecting the computer components industry, and ASEAN countries' elimination of tariffs on computer components and lack of restrictive NTMs are important aspects of the region's export competitiveness in this industry. Import duties on computer components have been virtually eliminated in intra-ASEAN trade, as shown in table 3.4.

Table 3.1. Computer Components: ASEAN Exports to Selected Markets, by Value, 2004–08

Country	2004	2005	2006	2007	2008	% change, 2004–08
	thousand $					
United States	16,959,870	16,699,224	17,877,001	12,841,911	11,282,201	(33)
China	7,740,109	10,211,999	11,533,966	12,156,735	14,919,132	93
EU-27	15,732,687	16,610,906	18,415,212	14,375,013	13,947,009	(11)
Japan	5,211,506	5,120,751	4,840,379	3,047,110	3,075,177	(41)
Subtotal	45,644,172	48,642,880	52,666,558	42,420,769	43,223,520	(5)
All other	14,550,024	16,118,365	17,062,641	12,993,593	12,508,378	(14)
Total	60,194,196	64,761,245	69,729,200	55,414,363	55,731,898	(7)

Source. WITS, Integrated Data Warehouse (accessed January 5, 2010).

Note. Data are based on ASEAN partner country imports because Brunei, Burma, Cambodia, Laos, Philippines, and Vietnam do not report trade data in all years. All other export value excludes intra-ASEAN trade.

Table 3.2. Computer Components: ASEAN Imports from Selected Markets, by Value, 2004–08

Country	2004	2005	2006	2007	2008	% change, 2004–08
	thousand $					
China	4,276,712	5,109,884	5,915,214	6,111,411	6,460,434	51
United States	2,656,017	3,328,901	2,946,240	2,728,021	2,945,541	11
EU-27	1,497,883	1,646,904	1,558,772	1,598,000	1,451,028	(3)
Japan	2,515,173	2,385,489	2,257,999	833,706	784,086	(69)
Subtotal	10,945,785	12,471,179	12,678,225	11,271,138	11,641,089	6
All other	5,759,733	5,706,861	5,508,312	4,907,841	4,267,909	(26)
Total	16,705,518	18,178,040	18,186,537	16,178,979	15,908,998	(5)

Source. WITS, Integrated Data Warehouse (accessed January 5, 2010).

Note. Data are based on ASEAN partner country exports because Brunei, Burma, Cambodia, Laos, Philippines, and Vietnam do not report trade data in all years. All other export value excludes intra-ASEAN trade.

As for components imported from non-ASEAN countries, all of the products discussed in this chapter are covered by the WTO's Information Technology Agreement (ITA).[226] This means that the ASEAN member countries that are signatories to the ITA must eliminate duties on computer components on an MFN basis.[227] Computer components generally are not subject to restrictive NTMs in the ASEAN region, either. A recent study reviewing these measures in ASEAN concluded that the only NTM affecting the computer industry in the region was the use of automatic import licensing in Indonesia.[228] Indonesia's licensing system became more restrictive in 2009 with the implementation of Decree 56, which designated additional import licensing requirements for electronics and has the potential to raise costs for importers.[229]

Given the established presence of the computer components industry in ASEAN countries and the general absence of tariffs and restrictive NTMs, the region is a leading exporter of these products and retains a competitive position globally. However, as noted, ASEAN countries are increasingly facing competition from China for components production. This is due to China's growing role as the leading location for final assembly of computers and the advantages of locating components production near customers. Between 2005 and 2008, China's production of desktop computers grew 82 percent, and production of notebook computers grew 142 percent.[230] During this period, China overtook the United States and the European Union to become the largest export market for computer components from ASEAN, as its imports of computer components from ASEAN countries grew by 83 percent. While this growth of computer assembly in China certainly offers a large and growing export market for ASEAN components, as China has captured an increasing share of final computer assembly, some components producers have moved to China or expanded production there as well in order to be closer to the customer.[231] Locating components production and finished goods assembly in close proximity to one another shortens production time, lowers inventory costs, and facilitates joint R&D.[232] These advantages of locating near the customer create a challenge for ASEAN's export competitiveness in computer components. This challenge is evident from the relative market shares of the two regions. Between 2004 and 2008, China's share of exports of computer components grew from 27 percent to 35 percent of the world total, with the share growing in each year during the five-year period. Meanwhile, ASEAN's share of global exports was flat throughout the period, remaining between 20 and 21 percent each year.

Table 3.3. Computer Components: ASEAN Member Exports to ASEAN Members, by Value, 2004–08

Country	2004	2005	2006	2007	2008	% change, 2004–08
			thousand $			
Malaysia	4,309,582	4,997,453	4,420,801	2,108,608	4,474,471	4
Thailand	2,870,551	3,065,174	2,896,029	1,662,688	2,404,821	(16)
Singapore	1,387,402	1,392,126	1,691,559	2,395,553	1,890,899	36
Indonesia	1,611,089	2,034,033	2,026,839	662,927	1,299,259	
Philippines	860,248	1,139,927	1,215,090	857,623	691,448	
Vietnam	38,706	157,049	357,646	408,826	451,937	(a)
Brunei	66	481	412	249	974	(a)
Cambodia	7	38	79	19	85	(a)
Burma	504	30	37	14	41	(92)
Laos	8	7		4	8	10
Total	11,078,163	12,786,316	12,608,492	8,096,512	11,213,944	1

Source. WITS, Integrated Data Warehouse (accessed January 5, 2010).

Note. Data are based on ASEAN partner country imports because Brunei, Burma, Cambodia, Laos, Philippines, and Vietnam do not report trade data in all years. Trade between these members may be understated. Totals may not add due to rounding.

[a] > 1,000 percent.

Table 3.4. ASEAN Country Tariffs for Computer Components, 2004, 2008, and 2010 (ad valorem Percent)

ASEAN members	2004		2008		2010[a]
	Imports from other ASEAN countries	Imports from non- ASEAN countries (MFN)	Imports from other ASEAN countries	Imports from non- ASEAN countries (MFN)	Imports from other ASEAN countries
Brunei	0	0	0	0	0
Burma	1.5	1.5	0–1.5[b]	1.5	0
Cambodia	15	15	0–5[c]	15	0–5[c]
Indonesia	0[d]	0[d]	0	0	0
Laos	0	5	0	5	0
Malaysia	0	0	0	0	0
Philippines	0	0	0	0	0
Singapore	0	0	0	0	0
Thailand	0	0	0	0	0
Vietnam[e]	0–5	5–10	0–5	0–8	0

Source: Agreement on the Common Effective Preferential Tariff (CEPT) Scheme for the ASEAN Free Trade Area, Consolidated Package of Tariff Reductions for 2008, http://www.aseansec.org/19802.htm; Royal Malaysia Customs Department, HS-Explorer (accessed April 2010); WTO, Tariff Analysis Online (accessed April 2010).

[a] 2010 MFN rates for ASEAN imports from non-ASEAN countries were not available to USITC staff for every ASEAN member country.

[b] Burma's 2008 duty rates on the computer components were mostly 1.5 percent in 2008, but a few products may be imported free of duty from ASEAN countries, including computer printers and assembled printed circuit boards.

[c] Cambodia's duty rates on computer components imported from other ASEAN countries were mostly 5 percent in 2008 and 2010, but a few products, including hard disk drives, may be imported free of duty.

[d] Indonesia's duty rates on all computer components covered in this chapter were zero in 2004, other than those on computer printers. Imports of computer printers from all countries carried a 5 percent rate of duty in 2004, including those imported from ASEAN countries.

[e] Vietnam's duty rates on the computer components covered in this chapter varied within the ranges shown in 2004 and 2008. In 2008, the majority of MFN duties on these products were 3 percent, and the majority of AFTA duties were zero. All remaining AFTA duties on these products were reduced to zero in 2010. In addition, Vietnam acceded to the WTO and the Information Technology Agreement (ITA) in 2007 and is currently phasing in zero MFN duties on all of these products as required by the ITA. The phase-in is scheduled to be completed by 2012.

In addition to competing with China's growing computer components industry, ASEAN exports are also affected by competition between countries within the region. For example, labor cost differences within the region were a major factor in the HDD industry's shift of production from Malaysia to Thailand.[233] Similarly, Vietnam's increasing export competitiveness has come, to some degree, at the expense of exports from other countries in the region. As one analyst has pointed out, within the broader electronics sector, "Indonesia experienced a contraction in exports and Philippines a slowdown in the 2000–2006 period following the emergence of Vietnam as a competitor."[234] This suggests that as individual

ASEAN countries strive to improve their export competitiveness, production may shift from one ASEAN country to another, and total exports from the region may not necessarily grow.

Inbound Investment

Several ASEAN countries provide attractive investment environments for computer components firms. Because, as noted, a country's competitiveness in the computer components industry depends in no small part on investment from multinational firms that provide links to global markets, countries actively compete to attract FDI and generally do not place restrictions on investment. Production of most types of computer components is dominated by a few large firms that exercise "control over critical resources and capabilities that facilitate innovation and . . . [possess the] capacity to coordinate transactions and the exchange of knowledge"[235] in the global computer supply chain. Location decisions by this relatively small set of multinational components firms drive production in the countries where they invest because these firms are the direct link to the major computer assemblers, which are the customers for components. In 2009, the top two HDD manufacturers together accounted for over 60 percent of the market.[236] In CPUs, production is even more concentrated: Intel accounted for nearly 82 percent of the microprocessor market in 2008, and its nearest competitor, AMD, accounted for over 17 percent, for a combined market share of over 99 percent.[237] Competition between countries for investment from these firms is therefore intense, especially given that many countries worldwide are equipped with the skilled labor, infrastructure, and other necessary factors to participate in electronics assembly. These firms choose production locations based on many competitive factors, and most of them have operations in more than one ASEAN country, as well as in China. Negotiating power regarding the terms of investment favors the firms holding the technology for computer components, not the countries where production takes place. As such, at present, none of the ASEAN countries places restrictions on foreign investment in computer components manufacturing beyond those that are applied either to all foreign investment or to all foreign investment in manufacturing.[238]

Competition between countries for investment from multinational components producers leads governments to offer generous incentive packages to firms to induce them to invest. Conversations with multinational firms in this industry confirmed that incentive packages offered by governments can act as a "tiebreaker" in their decisions about where to place new facilities.[239] In the incentives race, ASEAN countries must compete not only with countries outside the subregion, such as China, but also with each other in order to attract investment from major firms. Such competition exists for each of the various activities along the supply chain performed by firms in this industry: regional headquarters,[240] R&D, and assembly. Singapore, Malaysia, Thailand, and Vietnam were all cited by multinational firms as offering competitive incentive packages to investors. Among the financial incentives typically offered to investors in the computer components industry are multiyear tax deferments or exemptions, free or reduced-price land or buildings, and the ability to operate in free trade zones.[241] All of the components-producing countries in ASEAN offer R&D incentives[242] in an effort to capture more of the value added in the computer components industry. Competition for assembly plants can also be intense because of the jobs associated with these facilities. To this end, governments in the region often grant "pioneer" status[243] to computer components firms, especially if they are willing to conduct some R&D in the country, and this status carries particularly attractive tax exemptions.[244] This status is intended to designate an investment that will "create or expand a high value added industry"[245] and is commonly applied to firms in high- technology sectors such as electronics.

Between 2004 and 2008, Singapore was the source of the largest number of FDI activities in the computer industry in ASEAN, followed closely by the United States, Taiwan, the EU, and Japan, in that order. Singapore was also by far the leading destination for investment in the computer industry in ASEAN, followed by Thailand and Malaysia. Vietnam was the fourth most common destination, surpassing the Philippines and Indonesia.[246]

As a relative newcomer to the electronics industry, and especially to the computer components industry, Vietnam offers an example of the incentives that countries offer in order to compete for investment (box 3.1). In competing for Intel's newest assembly and test facility, the government of Vietnam worked directly with the company to meet Intel's needs and secure a commitment to invest. While the full details of the incentive package offered to Intel are not public, it is known that the government built a power substation to meet the plant's requirements and worked with Intel to develop a unique electronic processing system for its customs documents.[247] The government also signed an anticorruption memorandum of understanding with Intel.[248] Finally, Intel likely was accorded "pioneer" status, which provides "preferential tax treatment for perhaps as long as fifty years."[249] In general, Intel company representatives indicated that this government responsiveness was one of several important factors in its final investment decision, in conjunction with the other competitive factors described throughout this chapter.[250]

Political stability is also important to investors as they determine where to locate facilities. Among the components-producing countries in ASEAN, Singapore, Malaysia, and Vietnam score highly on World Bank indicators of political stability, while Thailand, the Philippines, and Indonesia score much lower.[251] One industry official confirmed that the lack of political stability in Thailand was a factor limiting its competitiveness for that firm's investment.[252] Political stability is an area where ASEAN countries must also compete with China, and several firms noted China's political stability and overall government effectiveness as a factor in its ability to attract investment from multinationals.[253]

Overall, ASEAN countries have been competitive in attracting inbound investment from multinational electronics firms. However, competition for these firms' investment is intense, and China continues to grow as a presence in the computer industry. Still, many firms retain a presence in ASEAN countries as well. One reason cited for doing so was to diversify the production base. For example, one firm stated that it has a policy of not exceeding a certain percentage of its worldwide production capacity in any one country and that ASEAN countries are important as an alternative location in enabling it to follow this strategy.[254] ASEAN countries' ability to retain this position as an attractive alternative to China depends on maintaining competitiveness and a strong overall business climate. To this end, ASEAN leaders plan to implement the ASEAN Comprehensive Investment Agreement (ACIA), which provides greater legal protection and transparency for investors, including ASEAN-based foreign investors. If the scope of the regionwide improvements to the investment climate was broadened to include all foreign investors, this would likely be well-received by multinational computer components firms given that the agreement improves transparency—firms cited improved transparency as a factor that would make the region more competitive.[255] At present, however, multinational firms in the region indicated that they generally deal solely with individual country governments in structuring their investments in ASEAN countries.[256]

Box 3.1. Computer Components: Profile of ASEAN Integration

FDI plays a major role in computer components production in ASEAN. Some of the most notable examples of integration among ASEAN member countries center on direct investment links between the countries. Between 2004 and 2008, Singapore was the leading global source of direct investment in the computer hardware industry within ASEAN (as measured by number of investments), with the majority of this investment going to Malaysia and Thailand.[a] Increasingly, Singapore is also investing in Vietnam and, through these investment linkages, is improving Vietnam's competitiveness and more fully integrating the region in computer components production. The Vietnam Singapore Industrial Park (VSIP), which is a joint venture between several Singaporean investors and a Vietnamese state-owned corporation, is an example of this bilateral investment link. As FDI in manufacturing in Vietnam has grown in recent years, industrial parks providing a wide range of services to these investors have become common. VSIP is one of the largest providers of these services, operating four large industrial parks in Vietnam. These parks have attracted approximately 400 investment projects with a combined total investment capital of $2.4 billion, and VSIP is currently expanding the parks to meet further demand.[b]

An official from a U.S.-based electronic components firm operating a plant in one of the VSIP sites explained why the park has been so attractive to his company. According to this source, the fact that the upper management of VSIP is Singaporean is an asset because they are very comfortable working with multinational companies and addressing their needs. Meanwhile, VSIP's Vietnamese management is a government entity, which, in this firm's experience, has meant that the park staff is highly skilled at walking investors through all of the required government paperwork and obtaining the proper licenses. This combined management structure is very valuable to the firm and was cited as a key factor in the firm's success in Vietnam.[c] VSIP's success with investors from outside the region is one example of intra-ASEAN linkages improving member countries' ability to compete globally in the computer components industry.

[a] VSIP Web site, http://www.vsip.com/vn/ (accessed April 14, 2010).

[b] Industry representative, interview by USITC staff, Binh Duong province, Vietnam, March 11, 2010.

[c] Ibid.

Trade Facilitation, Logistics Services, and E-Commerce

Trade Facilitation

In general, the computer components industry has expressed concerns about customs transparency and adherence to international norms—concerns that are shared by multinational firms across many industries. However, due to the nature of the industry and its status as a desired source of investment in many countries, firms in this industry mentioned fewer trade facilitation problems than firms in many other industries. Because neither duties nor nontariff measures are common in the computer components industry, the aspects of trade facilitation that are most important to this industry center on reducing the administrative burden of customs procedures. This has led to a competitive form of integration of customs procedures whereby ASEAN countries tend to adopt similar trade facilitation measures in order to remain attractive destinations for investors.

The major components-producing countries in ASEAN all use free trade zones (FTZs) as a means of simplifying trade-related administrative procedures for firms. These FTZs have contributed to what one industry official called "market-forced harmonization,"[257] that is, an environment in which countries' policies have converged because of investor demand for faster and simpler customs procedures rather than as a result of any centralized integration effort. FTZs simplify and speed up customs procedures for firms by placing customs officers onsite within the zone, with a focus on providing fast and efficient service to tenants. Despite its status as a world leader in trade facilitation and the fact that it currently offers most customs services online, Singapore continues to offer several FTZs in addition to its normal trade facilitation measures. Components firms operating in Malaysia and Thailand indicated that FTZs in those countries were important to them in attracting their investment and that operating in FTZs had helped them bypass time-consuming paperwork.[258] Similarly, in Vietnam, components firms tend to be located in industrial parks (some of which are FTZs) that compete with one another on the level of service provided to their tenants. These services typically include on-site customs facilities and assistance with obtaining preclearance for imports of materials.[259]

Despite the competitive environment fostered by ASEAN FTZs in terms of facilitating trade, industry officials expressed some concerns regarding customs procedures in the region. Most often mentioned was the issue of customs transparency and predictability in Thailand, especially in the customs post-audit process. More generally, each of the sources expressed the view that improved customs harmonization throughout the ASEAN region, such as through the ASEAN Single Window, would be somewhat helpful. In the meantime, however, they stated that there were few problems in working with each country individually.

Logistics Services

Officials of computer components firms in ASEAN countries that were interviewed did not express any specific concerns regarding logistics (beyond basic infrastructure needs in some areas). Computer components firms generally rely on third-party logistics providers for both imports of their inputs and exports of their products.[260] Some computer components are air-shipped, making logistics easier.

E-Commerce

Many large components firms prefer to use electronic systems to manage their supply chains and work with suppliers. However, suppliers' ability to adopt adequate supporting technology to implement these electronic systems varies, especially since the network of suppliers with experience working with multinational firms is not well-established in some countries. One of the tools that multinational components firms use to promote e- commerce is RosettaNet. RosettaNet is a consortium of electronics firms throughout the supply chain "working to create, implement, and promote open process standards for ebusiness."[261] The associated software allows for linkage of enterprise systems between suppliers, customers, and logistics providers.[262] In the broader computer industry, participants include AMD, Dell, Hitachi, and Intel.[263] These firms often encourage suppliers to implement RosettaNet in order to expedite transactions and improve the accuracy of communication between companies.[264] ASEAN member countries have also encouraged adoption of RosettaNet, and to date, Singapore, Malaysia, Thailand, and the Philippines have joined the United States, Australia, China, Japan, Korea, and Taiwan in establishing regional affiliate offices to support implementation of RosettaNet standards.[265] The Malaysian government has been particularly active in providing grants to small and medium-sized suppliers to enable them to adopt RosettaNet standards.[266] Regionwide, one of the measures listed in the electronics sector Roadmap is to "institutionalize" RosettaNet, but the purpose of this measure was to establish the software for application to customs procedures (exchange of trade documents) rather than promote its use between firms in the private sector.[267] RosettaNet is an example of private sector-led e-commerce efforts in ASEAN and underscores the importance of measures that integrate ASEAN producers into the global computer supply chain for improving competitiveness in the region.

4. Textiles and Apparel: Cotton Woven Apparel

Textiles and Apparel Overview[268]

The ASEAN region and certain ASEAN countries in particular, are notable suppliers of textiles and apparel to the world. For the entire textiles and apparel priority sector, ASEAN countries accounted for 7.2 percent of total world exports in 2008, up from 6.7 percent in 2004. The priority sector includes a much broader group of products than this chapter's profile industry, cotton woven apparel. ASEAN exports of all textiles and apparel totaled $38.5 billion in 2008, compared to $6.3 billion for cotton woven apparel (table 2.4). The broader priority sector covers other important segments, such as manmade fiber yarns, fabrics, and apparel, which are important to some ASEAN producers, particularly Indonesia.

At present, integration in the ASEAN textiles and apparel sector consists mainly of production of yarn and fabric in one ASEAN country to be shipped to another ASEAN country, where they are made into apparel. A fully developed and integrated textiles and apparel sector in the ASEAN region would likely encompass all elements of the supply chain across several ASEAN members (i.e., the production of fibers, spinning of yarns, knitting or weaving of fabrics, and cutting and sewing of finished apparel), as well as at least some regional integration into global supply chains.

One factor influencing the level of integration in ASEAN textile and apparel production is ASEAN's Roadmap for Integration of Textiles and Apparel Products Sector (Roadmap). The Roadmap encompasses the full textile and apparel supply chain, including raw fibers (e.g., raw cotton, wool, polyester staple fibers), yarns, fabrics, and finished apparel or textile articles.[269] The Roadmap provides for the elimination of tariffs on intra-ASEAN trade in all sector products, as well as other measures to encourage integration. Tariff elimination is almost complete in six ASEAN countries. The remaining four—Vietnam, Cambodia, Burma, and Laos—have reduced but not eliminated tariffs on cotton woven apparel and its textile inputs, such as cotton woven fabrics.[270]

The Malaysian government is the coordinator for the textile and apparel Roadmap. Reportedly, Malaysian officials have worked with the ASEAN Federation of Textile Industries (AFTEX), a private association made up of textile and apparel industry members from the 10 ASEAN countries, to increase integrated apparel production among the ASEAN countries.[271] In addition, AFTEX is working with the U.S. Agency for International Development (USAID) ASEAN Competitiveness Enhancement (ACE)

Project to stimulate intra-ASEAN production as a means to increase the ASEAN region's global competitiveness.[272]

The removal of global quotas has had a larger influence on the degree of production integration among the ASEAN countries than the tariff eliminations and reductions mandated in the Roadmap.[273] The expiration of these quotas under the World Trade Organization (WTO) Agreement on Textiles and Clothing (ATC) on January 1, 2005, resulted in significantly increased competition among global suppliers to the U.S. and EU markets, including among the individual ASEAN countries.[274] Large U.S. and EU retailers consolidated their supply bases once the quotas were removed.

Further, according to industry representatives, to the extent that there has been ASEAN integration along the supply chain in the textiles and apparel priority sector, it has largely been driven by variations in the individual countries' economic environment.[275] The reason is that availability and relative cost of labor are driving factors in the textile and especially apparel industries.[276] For example, there is a trend to locate the more labor- intensive apparel factories in the countries with relatively less developed economies, such as Vietnam,[277] Cambodia, and Laos, where labor costs are relatively low and there is a large pool of available labor.[278] The Philippines, with a relatively abundant supply of labor, has also traditionally been an apparel-producing country, although it has struggled to remain cost-competitive compared with some of the other ASEAN countries. [279] Production of the more capital-intensive textiles (yarns and fabrics) and higher-quality apparel is concentrated in the ASEAN countries with higher labor costs—Thailand, Malaysia, and to a lesser extent, Singapore.[280] Cross-border integration is fostered by investment from the fabric-producing countries in ASEAN into the lower-cost-apparel manufacturing countries of Vietnam and Cambodia.[281] Indonesia, which has a vertically integrated textile and apparel supply chain, also has been able to export its fabrics to ASEAN countries that are primarily apparel producers.[282]

The remainder of this chapter focuses on the cotton woven apparel industry. The manufacture of cotton woven apparel lends itself to regional integrated production, and this type of production is a growing trend in the industry. Among the ASEAN countries, the level of economic integration in this industry is relatively small, but slowly increasing.

Cotton Woven Apparel[283]

Key Findings

As noted in the overview, there is a small but growing amount of integrated production of cotton woven apparel among the ASEAN countries. Intra-ASEAN trade in cotton woven fabrics—the primary input into cotton woven apparel—increased by 13 percent in four years, rising from \$168.4 million in 2004 to \$190.8 million in 2008 (table 4.1). This trade, an important component of regional integration for cotton woven apparel, represented approximately 9 percent of the region's total cotton woven fabric imports, valued at $2.1 billion in 2008 (table 4.2). The reduction in duties among member countries, as outlined in the sector Roadmap, has helped to facilitate increased integration in the cotton woven apparel industry. The ASEAN-Japan FTA has also generated some integrated production between ASEAN member countries because exports to Japan of apparel made from ASEAN fabric are entitled to duty-free treatment under that agreement.

Table 4.1. Certain Cotton Woven Fabrics: ASEAN Member Imports from ASEAN Members, by Value (2004–08)

Country	2004	2005	2006	2007	2008	% change, 2004–08
	thousand $					
Indonesia	36,145	40,146	34,261	38,182	36,334	1
Vietnam	20,864	17,524	19,794	21,581	29,520	41
Singapore	14,573	18,403	26,445	42,130	29,462	102
Cambodia	22,900	26,333	37,384	22,663	22,570	(1)
Burma	11,810	15,413	15,341	19,681	21,800	85
Laos	18,251	17,966	18,216	18,612	21,100	16
Malaysia	15,305	20,197	16,152	11,308	12,518	(18)
Philippines	15,479	16,811	11,404	11,676	8,730	(44)
Thailand	10,294	10,889	4,785	3,930	7,845	(24)
Brunei	2,743	2,311	1,277	1,271	939	(66)
Total	168,363	185,992	185,060	191,033	190,818	13

Source: WITS, Integrated Data Warehouse (accessed January 5, 2010). Indonesia's exports from Global Trade Services (accessed April 15, 2010).

Note: Data are based on partner country exports. Brunei, Burma, and Laos do not report trade data, and the following countries did not report data for selected years: Cambodia (2005–08), Philippines (2004–06), and Vietnam (2008). Trade between these countries may therefore be undervalued.

Table 4.2. Certain Cotton Woven Fabrics: ASEAN Imports from Selected Markets, by Value, 2004–08

Country	2004	2005	2006	2007	2008	% change 2004–08
	thousand $					
China	572,808	761,537	891,594	1,033,143	1,138,947	99
Hong Kong	488,024	486,937	397,529	410,259	379,357	(22)
Japan	101,241	105,903	104,042	110,124	105,944	5
Taiwan	126,939	114,644	97,625	88,723	76,698	(40)
Korea	79,596	76,558	69,456	67,539	61,914	(22)
EU-27	60,122	57,832	65,474	69,987	52,926	(12)
United States	12,509	12,959	8,310	8,714	8,182	(35)
Subtotal	1,441,239	1,616,371	1,634,029	1,788,489	1,823,967	27
All other	183,996	184,388	204,779	241,737	253,551	38
Total	1,625,235	1,800,759	1,838,808	2,030,226	2,077,518	28

Source. WITS, Integrated Data Warehouse (accessed January 5, 2010).

Note. Data are based on ASEAN partner country exports. Brunei, Burma, and Laos do not report trade data, and the following countries did not report data for selected years: Cambodia (2005–08), Philippines (2004–06), and Vietnam (2008). "All other" import value excludes intra-ASEAN trade. Trade between these countries may therefore be undervalued.

Table 4.3. Cotton Woven Apparel: ASEAN Exports to Selected Markets, by Value, 2004–08

Country	2004	2005	2006	2007	2008	% change, 2004–08
	thousand $					
United States	3,360,705	3,757,580	3,913,203	3,776,907	3,694,324	10
EU-27	1,147,099	1,010,868	1,356,781	1,400,340	1,569,228	37
Japan	198,399	218,612	234,832	268,657	288,441	45
China	3,582	5,399	6,141	7,107	11,977	234
Subtotal	4,709,785	4,992,460	5,510,957	5,453,011	5,563,970	18
All other	451,704	517,860	582,261	691,541	774,612	71
Total	5,161,489	5,510,320	6,093,218	6,144,552	6,338,582	23

Source. WITS, Integrated Data Warehouse (accessed January 5, 2010).

Note: Data are based on ASEAN partner country imports.

Table 4.4. Cotton Woven Apparel: ASEAN Member Exports to ASEAN Members, by Value, 2004–08

Country	2004	2005	2006	2007	2008	% change, 2004–08
	thousand $					
Indonesia	39,637	52,601	77,537	57,587	29,653	(25)
Malaysia	28,410	19,757	18,748	19,549	24,782	(13)
Thailand	16,074	13,449	16,234	15,135	15,946	(1)
Philippines	5,504	6,856	6,273	6,778	5,284	(4)
Cambodia	7,551	5,381	10,109	4,788	4,121	(45)
Singapore	546	673	1,315	1,685	3,428	528
Vietnam	5,138	3,648	3,137	2,448	3,341	(35)
Laos	233	760	775	1,072	1,509	547
Burma	2,621	435	587	971	940	(64)
Brunei	5,400	1,391	1,002	1,004	270	(95)
Total	111,114	104,951	135,720	111,018	89,275	(20)

Source. WITS, Integrated Data Warehouse (accessed January 5, 2010).

Note. Data are based on partner country imports. Brunei, Burma, and Laos do not report trade data, and the following countries did not report data for selected years: Cambodia (2005–08), Philippines (2004–06), and Vietnam (2008). Trade between these countries may therefore be undervalued.

Background

The following sections will describe the industry, with particular attention to two important aspects—export competitiveness and inbound investment—as they relate to the competitive factors affecting integrated production of cotton woven apparel among the ASEAN countries. The cotton woven apparel industry encompasses an assortment of different apparel items, including the production of most cotton trousers (including jeans and khakis), as well as cotton woven blouses and shirts (such as men's dress shirts). Other products include cotton woven shorts, jackets, dresses, skirts, and underwear. Although the industry's exports account for a relatively small share of total ASEAN textile and apparel exports (16 percent in 2008), the cotton woven apparel supply chain provides opportunities for integration across the ASEAN region.[284] This supply chain includes the production of cotton yarns, the weaving of the yarns into cotton fabrics, and the cutting and sewing of the cotton fabrics into cotton woven apparel. In addition, before shipping, the finished apparel must be packaged according to the buyer's instructions. This analysis will focus primarily on the production of cotton woven fabric and cotton woven apparel.

The role that this part of the supply chain plays in ASEAN countries' economies varies greatly. Some ASEAN countries, such as Vietnam and Cambodia, have a competitive advantage in apparel production, but need to import most of their cotton woven fabrics. Other countries, such as Thailand and Indonesia, produce both cotton woven fabrics and apparel. Even though they consume a large share of their fabric production domestically, they also have an incentive to trade in cotton woven fabrics. The production of cotton woven fabrics is capital-intensive, necessitating high capacity utilization rates in order to cover fixed costs. Further, factories tend to specialize in specific types of cotton woven fabrics (such as denim or twill fabrics for pants, or oxford cloth for men's shirts). Apparel manufacturers, on the other hand, require a large variety of different fabric styles to meet the needs of their customers. Thus, apparel-producing countries import fabric inputs from many different countries.

Regional Integration, Export Competitiveness, and Inbound Investment

Leading Competitive Factors

The expiration of global quotas under the ATC on January 1, 2005, resulted in significantly increased competition among the individual ASEAN countries, as well as among global competitors who supplied the U.S. and EU markets. Large U.S. and EU retailers consolidated their supply bases because the availability of quota was no longer a significant factor in deciding where to have their goods produced.[285] Competition among the ASEAN countries also increased as Vietnam, an efficient and low-cost apparel producer, gained full access to the U.S. market. In December 2006, the United States granted Vietnam normal trade relations (NTR) status, which significantly reduced the duties on U.S. imports from Vietnam.[286] On January 11, 2007, Vietnam became a member of the WTO.[287]

Without the global quotas, the repercussions of increased competition among retailers emerged as a major factor in the global production of cotton woven apparel, influencing both export competitiveness and inbound investment. Because of this intense retail competition, retail buyers were and still are demanding lower costs and quick turnaround and delivery times. Retailers do not want to be left with inventory and take markdowns at the end of a season. As a result, ASEAN industry sources report that short lead times have become a major factor in choosing suppliers.[288] Increasingly, retailers are demanding a greater variety, but fewer garments, in various styles and colors. Key factors affecting costs and delivery include wage rates, productivity, and infrastructure and other logistics affecting access to raw materials, as well as shipping the finished products. U.S. retailers attribute Vietnam's rise to a global supplier of cotton woven apparel to its high production efficiency (about 85 percent of that of China); low production costs (lower than China's with similar delivery times); and a ready supply of highly skilled and low- cost labor. In addition, Vietnam's proximity to China, with China's large supply of cotton woven fabrics and other apparel inputs, allows short delivery times for raw materials.[289]

Regional Integration

Overview

Although the amount of ASEAN cross-border production using ASEAN inputs is small relative to production of cotton woven apparel using inputs imported from outside the ASEAN region, trade and industry representatives report increases in integration and cite government and private sector initiatives to accomplish this goal. Industry sources report that the individual ASEAN countries are not able to compete globally without ASEAN integration. Specific examples of successful regional integration are described in box 4.1. Increasing ASEAN integrated production will give the region a comparative advantage that each country alone does not have.[290]

The Roadmap is a primary government initiative to integrate apparel production, including cotton woven apparel production, among the ASEAN countries. A key private sector initiative is the Source ASEAN Full Service Alliance (SAFSA), sponsored by USAID's ACE project, working with the ASEAN Federation of Textile Industries (AFTEX). Both the government and private sector programs are actively encouraging and providing procedures and incentives to increase integrated apparel production among ASEAN members.

Box 4.1. Cotton Woven Apparel: Profile of ASEAN Integration

There are several examples of integrated production in the cotton woven apparel industry in the ASEAN region. As discussed above, Malaysia's apparel manufacturers, facing high labor costs and a shortage of labor, have moved up the value chain, producing higher-quality, higher-priced, and branded goods. Because of constant price pressure even from the high-level branded customers, Malaysian apparel producers have set up factories in the lower-cost ASEAN countries of Vietnam, Cambodia, and Indonesia. Privately owned by the Malaysian firms, these factories perform the cut-and-sew production work while such activities as designing, marketing, and merchandising are conducted in the apparel company offices in Malaysia.[a]

Vietnam and Cambodia's apparel industries, in turn, depend heavily on imports of garment inputs, including cotton woven fabrics. Some of these imports come from within ASEAN. For example, Indonesia exports heavier-weight cotton woven denim fabric to Cambodia for the production of jeans.[b] Thailand supplies lightweight cotton woven fabric to factories in Vietnam and Cambodia for use in apparel, largely men's and women's tops, the majority of which are exported.[c] Thailand's trade in cotton woven fabrics is bolstered by its reputation for producing reliable quality goods, as well as its existing trade relationships with the ASEAN markets, especially Vietnam and Cambodia.[d]

Vietnam produces little fabric, and most of its fabric imports come from outside ASEAN. Thailand's shipments of lightweight cotton woven fabrics to Vietnam nearly doubled during 2004–08.[e] Nevertheless, in 2008, Thailand still only supplied roughly 3 percent of Vietnam's imports of these goods.[f]

Like Vietnam, Cambodia has very little fabric production, and over 90 percent of inputs used in apparel production must be imported.[g] Again, the majority of these inputs are sourced from non-ASEAN countries. Thailand was Cambodia's third-largest source of lightweight cotton woven fabrics in 2008, after China and Hong Kong.[h] Of the 10 factories operating in Cambodia that import lightweight cotton fabrics, Thailand supplied 4, which used the Thai- sourced fabrics largely to produce shirts, blouses, and jackets.[i]

[a] Industry official, e-mail message to Commission staff, May 6, 2010.
[b] Industry representative, interview by Commission staff, Jakarta, Indonesia, March 3, 2010.
[c] BDLINK Cambodia and ACE Project Team, "Diagnostics Report, Textile and Apparel Supply Chain Corridors, Light-Weight Woven Cotton Fabric (HS 5208)," July 2009, 16.
[d] ACE Project Team, "Textile and Apparel Supply Chain Corridor Diagnostics, SWOT Analysis," May 2009, 24.
[e] Based on data obtained from WITS Integrated Data Warehouse (accessed January 5, 2010).
[f] Based on data obtained from WITS Integrated Data Warehouse (accessed January 5, 2010).
[g] BDLINK Cambodia and ACE Project Team, "Diagnostics Report, Textile and Apparel Supply Chain Corridors, Light-Weight Woven Cotton Fabric (HS 5208)," July 2009, 25.
[h] Based on data obtained from WITS Integrated Data Warehouse (accessed January 5, 2010).
[i] BDLINK Cambodia and ACE Project Team, "Diagnostics Report, Textile and Apparel Supply Chain Corridors, Light-Weight Woven Cotton Fabric (HS 5208)," July 2009, 31–32.

ASEAN Roadmap for Integration of Textiles and Apparel Products Sector

The Roadmap was developed to strengthen regional integration through the liberalization and facilitation of trade in goods, services, and investments in the textile and apparel sector. It was also designed to promote private sector participation. The Roadmap has produced mixed results. To date, its major accomplishments have been the reduction and elimination of tariffs. As of 2010, tariffs on cotton woven fabrics and apparel made from these fabrics have been eliminated by the ASEAN-6 and reduced and/or eliminated by the ASEAN-4 (Vietnam, Cambodia, Burma, and Laos) (tables 4.5 and 4.6). According to the Roadmap, the tariffs on these products are to be completely eliminated by the ASEAN-4 on January 1, 2012. However, there is no evidence thus far that any of the other provisions of the Roadmap have been met. For example, the ASEAN nontariff barrier (NTB) database is incomplete and outdated.[291] The database was established as part of the Roadmap to help ensure transparency in rules governing the trade of goods. A national government source stated that the Roadmap is no longer a critical factor influencing economic integration in the cotton woven apparel industry.[292]

Table 4.5. ASEAN Country Tariffs for Cotton Woven Apparel, 2004, 2008, and 2010 (Percent ad valorem)

ASEAN members	2004		2008		2010[a]
	Imports from other ASEAN countries	Imports from non-ASEAN countries(MFN)	Imports from other ASEAN countries	Imports from non-ASEAN countries (MFN)	Imports from other ASEAN countries
Brunei	0	0	0	0	0
Burma	5[b]	10[b]	5	10	5
Cambodia	7–35[c]	7–35 c	5	7–35 c	5
Indonesia	0–5	15	0–5[d]	15	0
Laos	3–8	10	0–5	10	0–5
Malaysia	5	15–20	0	0–20	0
Philippines	5	15	0	15	0
Singapore	0	0	0	0	0
Thailand[e]	0–5	30	0	10–60[f]	0
Vietnam	5–15	50	5	20	5

Source: Agreement on the Common Effective Preferential Tariff (CEPT) Scheme for the ASEAN Free Trade Area, Consolidated Package of Tariff Reductions for 2008, http://www.aseansec.org/ 19802.htm; Royal Malaysia Customs Department, HS-Explorer (accessed April 2010); WTO, Tariff Analysis Online (accessed April 2010).

[a] 2010 MFN rates for ASEAN imports from non-ASEAN countries are not available to USITC staff for every ASEAN member country.

[b] The majority of tariff rates are 5 and 10 percent ad valorem, depending on if the supplying country is an ASEAN member or not. The exception is other men's cotton woven garments classified in HTS subheading 6211.32.10, which has MFN and ASEAN rates of duty of 2 percent ad valorem.

[c] The majority of tariff rates range between 15 and 35 percent ad valorem. The exceptions are babies' garments classified under HTS subheadings 6209.20.90, 6209.30.90, and 6209.90.90 that have a rate of duty of 7 percent.

[d] Most imports are duty free. Only women's and girls' cotton woven skirts and shirts classified in HTS subheadings 6204.52 and 6206.30 have a rate of duty of 5 percent ad valorem.

[e] Thailand's 2008 CEPT package of tariff rates in 2008 is the same as Thailand's 2007 CEPT package.

[f] Thailand's MFN rate for most cotton woven apparel imports was 60 percent in 2008.

Table 4.6. ASEAN Country Tariffs for Cotton Woven Fabrics, 2004, 2008, and 2010 (Percent Ad Valorem).

ASEAN members	2004		2008		2010a
	Imports from other ASEAN countries	Imports from non-ASEAN countries (MFN)	Imports from other ASEAN countries	Imports from non-ASEAN countries (MFN)	Imports from other ASEAN countries
Brunei	0	0	0	0	0
Burma	1–5	1–7.5	4	4	4
Cambodia	7	7	5	7	5
Indonesia	0–5	10–15	0	10–15	0
Laos	5	10	0	10	0
Malaysia	5	10	0	10	0
Philippines	5	10	0	10	0
Singapore	0	0	0	0	0
Thailand	0	30	0	5	0
Vietnam	15	40	5	12	5

Source. Agreement on the Common Effective Preferential Tariff (CEPT) Scheme for the ASEAN Free Trade Area, Consolidated Package of Tariff Reductions for 2008, http://www.aseansec.org/ 19802.htm; Royal Malaysia Customs Department, HS-Explorer (accessed April 2010); WTO, Tariff Analysis Online (accessed April 2010).

[a] 2010 MFN rates for ASEAN imports from non-ASEAN countries are not available to USITC staff for every ASEAN member country.

ACE Project's SAFSA Initiative[293]

The ACE project funded by USAID is likely to have a greater effect than the Roadmap in increasing intra-ASEAN production.[294] It is designed to stimulate intra-ASEAN production in order to increase the ASEAN region's global competitiveness. To accomplish this goal, the ACE project is supporting the restructuring of the ASEAN's textile and apparel sector so it can compete in the growing global trend of "fast fashion." As described above, the increasing competition in the global retail apparel market has driven apparel buyers to seek one-stop sourcing and hold smaller inventories in order to reduce costs. This has led them to place smaller orders for a greater variety of garments, delivered in shorter lead times. Although countries that are vertically integrated are most likely to have the means to develop "fast fashion" production, the goal of SAFSA is to develop an ASEAN industry from which global buyers would be able to source fully designed apparel that is produced with ASEAN inputs in ASEAN countries.[295]

The use of virtual vertical factories (VVFs)[296] can enable the production of fast fashion by linking suppliers and garment manufacturers in different countries. VVFs have the potential to allow the various players in the production chain to exploit their comparative advantages and market a complete suite of services to buyers, including design—a function traditionally controlled by buyers.[297] In June 2010, there was a SAFSA Textile and Apparel Global Forum held in Singapore. Eight buyers from the United States and the EU were present, and 14 VVFs (textile mills and garment factories) attended.[298]

Role of FTAs on Integration

FTAs are an additional driving force behind ASEAN integrated production.[299] For example, the rules of origin under the ASEAN-Japan FTA require that cotton woven apparel be produced from cotton woven fabric made in an ASEAN country or Japan. Indonesia has doubled its fabric exports to other ASEAN countries since the ASEAN- Japan FTA took effect in December 2008.[300] Companies based outside of the ASEAN region may use ASEAN countries as a platform for duty-free exports to Japan under the ASEAN-Japan FTA, and Indonesian industry sources attribute this success to the FTA.[301]

To a lesser extent, the new ASEAN-China FTA will also permit non-ASEAN-based firms to use ASEAN countries as platforms for duty-free exports to China. However, the ASEAN-China FTA appears less likely to encourage integration in the textiles sector among ASEAN countries, at least initially. In the near term, this FTA will likely allow imports from China, the world's largest textile and apparel producer and exporter, to compete more effectively in local ASEAN markets.[302] However, to the extent that production costs increase in China, making it less competitive, it is possible that the China FTA could result in increased market opportunities for ASEAN producers, particularly in the labor-intensive apparel sector.

Export Competitiveness

Trade

ASEAN exports of cotton woven apparel to the world increased in value from $5.2 billion in 2004 to $6.3 billion in 2008; however, ASEAN's share of world trade in this sector declined slightly. In 2004, ASEAN exports of cotton woven apparel accounted for 4.2 percent of world trade in these products; in 2008, this share dropped to 3.8 percent. As noted above, Vietnam and Indonesia are the two largest exporting countries of cotton woven apparel among ASEAN members, together accounting for 61 percent of total ASEAN cotton woven apparel exports to the world in 2008 (table 4.7).

The United States is by far the largest market for ASEAN exports of cotton woven apparel (table 4.3). Nevertheless, the share of ASEAN exports going to the U.S. market declined from 65 percent in 2004 to 58 percent in 2008. The share of ASEAN exports going to the EU has remained relatively steady at 20 percent. As a whole, ASEAN countries have diversified their exports away from the United States and the EU, the traditional large markets that were largely controlled by import quotas until the end of 2004. As explained earlier, following the removal of quotas, U.S. and EU retailers and apparel firms consolidated their sourcing to fewer countries to reduce costs and strengthen supplier relationships. Many countries, including some of the ASEAN countries, looked to expand their exports to other markets in order to maintain production levels. ASEAN exports to Japan, for example, increased by 45 percent during 2004–08, and the share of ASEAN exports going to Japan increased from 3.8 percent in 2004 to 4.5 percent in 2008 (table 4.3). Recent growth in ASEAN exports to Japan is also partly attributed to the ASEAN-Japan FTA. U.S. firms that contract with ASEAN apparel companies indicated that the ASEAN-Japan FTA provides an incentive to produce cotton woven apparel in the ASEAN countries for the Japanese market.[303]

As discussed in box 4.1, apparel-producing ASEAN countries rely heavily on imports of cotton woven fabrics from non-ASEAN members (see table 4.2).[304] Nevertheless, some ASEAN countries have seen significant increases in their exports of cotton woven fabrics to other ASEAN members during 2004–08 (table 4.8). For example, Thailand's exports of cotton woven fabrics to other ASEAN countries increased by about 70 percent during 2004–08, fueled largely by increased exports to Vietnam, and, to a lesser extent, by increased exports of higher-valued fabrics to Singapore (table 4.8). Malaysia's exports of cotton woven fabrics also increased by 42 percent during 2004–08, albeit from a much smaller base.

Competitiveness

Using export data to the world (excluding intra-ASEAN trade) as an indicator, Vietnam has emerged as the most competitive producer of cotton woven apparel in the ASEAN region (table 4.7). Over the last five years, Vietnam has more than doubled its exports of cotton woven apparel to the world, from $1.0 billion in 2004 to $2.0 billion in 2008, and is now the top ASEAN exporter of such products to the world.

The Vietnamese cotton woven apparel industry, like the Vietnamese textile and apparel sector as a whole, has been heavily influenced by government policies. The Vietnamese government sets targets for growth in sales and exports in the textile and apparel sector, including the cotton woven apparel industry.[305] Its goal is to increase vertical integration in the industry, including the production of raw cotton and increased investment in dyeing and finishing.[306] In addition, the Vietnam Textile and Apparel Group (VINATEX), a state-owned conglomerate, has partial ownership in about 100 firms and a majority share in 27 of those firms.[307]

Industry sources indicated that Vietnam can be competitive with China because it has an abundant, skilled workforce, particularly in sewing; relatively high production efficiency; and competitive labor rates.[308] It also has a stable business environment and, along with

Bangladesh, is considered one of the prime alternatives to China for firms looking to diversify their sourcing.[309] Vietnam is also attractive because of its proximity to Asian fabric suppliers (particularly China, Korea, and Taiwan), which allows short delivery times for raw materials. One industry source also indicated that Vietnam is an attractive place to invest compared with many apparel-producing locations around the world because, although its infrastructure is relatively underdeveloped, the government is actively working to improve it.

Table 4.7. Cotton Woven Apparel: ASEAN Member Exports to the World, by Value, 2004–08

Country	2004	2005	2006	2007	2008	% change, 2004–08
	thousand $					
Vietnam	973,463	1,112,532	1,428,897	1,724,382	2,028,985	108
Indonesia	1,238,362	1,604,961	1,854,065	1,845,342	1,846,860	49
Cambodia	779,371	791,364	780,795	784,870	811,557	4
Thailand	728,336	763,382	722,846	654,808	617,421	(15)
Philippines	839,550	774,287	769,150	601,446	508,278	(39)
Malaysia	320,927	285,639	321,073	311,693	309,023	(4)
Burma	145,531	89,964	124,713	137,049	141,578	(3)
Laos	59,191	54,474	54,355	59,340	56,106	(5)
Singapore	70,644	29,954	35,647	24,167	16,878	(76)
Brunei	6,114	3,762	1,675	1,454	1,895	(69)23
Total	5,161,489	5,510,320	6,093,218	6,144,552	6,338,582	

Source. WITS, Integrated Data Warehouse (accessed January 5, 2010).

Note. Data are based on partner country imports. Brunei, Burma, and Laos do not report trade data, and the following countries did not report data for selected years: Cambodia (2005–08), Philippines (2004–06), and Vietnam (2008). Trade between these countries may therefore be undervalued.

Table 4.8. Certain Cotton Woven Fabrics: ASEAN Member Exports to ASEAN Members, by Value, 2004–08

Country	2004	2005	2006	2007	2008	% change, 2004–08
	thousand $					
Thailand	67,002	73,416	96,601	112,353	115,471	72
Singapore	50,136	45,803	29,330	29,602	29,727	(41)
Indonesia	36,799	51,975	41,639	25,333	28,626	(22)
Malaysia	11,738	12,537	14,313	17,963	16,677	42
Philippines	(a)	(a)	(a)	1,419	319	(a)
Cambodia	27	(a)	(a)	(a)	(a)	(a)
Vietnam	2,661	1,984	2,994	4,363	(a)	(a)
Total	168,363	185,714	184,876	191,033	190,818	13

Source. WITS, Integrated Data Warehouse (accessed January 5, 2010); GTIS, Global Trade Atlas Database (accessed April 15, 2010).

Note. Data are based on exports. Brunei, Burma, and Laos do not report trade data, and the following countries did not report data for selected years: Cambodia (2005–08), Philippines (2004–06), and Vietnam (2008). Trade between these countries may therefore be undervalued.

[a] Not available.

Improvements include industrial zones that provide basic manufacturing infrastructure, as well as a new deepwater port southeast of Ho Chi Minh City that provides short transit times of 15 to 17 days to the U.S. West Coast.

Indonesia also has a large apparel industry, and historically it has been among the more competitive producers of cotton woven apparel in the ASEAN region. After Vietnam, it was the second largest ASEAN exporter of cotton woven apparel to the world in 2008 (table 4.7). Indonesia's exports of cotton woven apparel jumped in 2005 and 2006 following the expiration of the ATC global quotas on January 1, 2005; however, such exports leveled off after 2006, and the Indonesian cotton woven apparel industry is now experiencing difficulties in terms of logistics and rising costs. It does have a large workforce, with average monthly wages for the apparel industry reported to be lower than in Vietnam ($100 per month compared to $120 per month in Vietnam). On the other hand, wages are higher in Jakarta, which has pushed apparel production further inland where wages are lower, but transportation costs are higher.[310] Moreover, Indonesia's worker productivity is reportedly less than in Vietnam.[311]

Unlike Vietnam, which has to import most of its apparel inputs, Indonesia also has a sizable industry producing cotton woven fabrics, which it uses in its production of cotton woven apparel. It also exports cotton woven fabrics to other countries, including other ASEAN members. However, Indonesia's textile industry's capacity utilization rate is just under 75 percent, due in part to problems with the power supply.[312] Because of comparatively high transportation costs and poor infrastructure, and the fact that Indonesia does not have a major port, some ASEAN industry sources conclude that Indonesia has limited potential as a global competitor in garment production.[313] In spite of these challenges, however, some U.S. apparel buyers consider Indonesia a competitive producer and an attractive sourcing alternative to China.[314]

Cambodia, Thailand, and the Philippines are the next-largest exporters of cotton woven apparel in ASEAN. Cambodia's exports of these goods remained relatively flat at about $0.8 billion during 2004–08, while those from Thailand and the Philippines declined during the same period. Of these three countries, Cambodia has the lowest labor costs and is the most competitive producer of cotton woven apparel in terms of opportunities for growth and integrated ASEAN production. Cambodia's labor costs for the apparel industry, at approximately $80 to $100 per month,[315] are among the lowest of the ASEAN producers.

Thailand has a fully integrated textile industry, including yarn spinning, fabric weaving, and printing, dyeing, and finishing facilities.[316] Approximately 60 percent of Thailand's fabric production is consumed domestically, and the remaining 40 percent is exported.[317] Thailand's fabric producers ship their cotton woven fabrics to other ASEAN countries, primarily Burma, Cambodia, Laos, and Vietnam, for apparel production, to benefit from those countries' low labor costs.[318] Thailand's woven fabric industry is known for its high- quality cotton woven fabrics sold at relatively high prices.[319] The industry now faces increased manufacturing costs, in part due to the use of outdated machinery, rising labor costs, and relatively low capacity utilization rates (approximately 75 percent).[320]

The cotton woven apparel industry in the Philippines contracted during the 2004–08 period.[321] Exports of cotton woven apparel from the Philippines declined annually, falling by approximately 40 percent during 2004–08 (table 4.7). Facing intense competition from such lower cost producers as Bangladesh, Vietnam, Cambodia, China, and Indonesia, the Philippines lost market share in the United States, its largest cotton woven apparel market.[322]

The Government of the Philippines attempted to assist the industry by developing incentives to attract FDI. A special zone was set up in a former U.S. military base. All textile machinery imported into this special zone is duty-free.[323] In addition, because the United States is the Philippines' largest apparel export market, the Government of the Philippines is seeking preferential duty-free treatment in the United States for its apparel exports that are made with U.S. fabric.[324]

The industry in Malaysia suffers from high production costs fueled by a shortage of labor.[325] Consequently, Malaysia's factories tend to be small and focused on the production of higher-value goods and branded apparel for export.[326] At the same time, Malaysian producers have expanded into lower-end apparel production in Vietnam and Cambodia.[327] The Malaysian industry is known for its quality products and timely delivery, both of which are important competitive factors, especially for branded apparel companies.

Little information is available on the cotton woven apparel sector in Laos. Its exports to the world remained relatively flat during 2004–08. However, its exports of cotton woven apparel to other ASEAN members increased in value, from $223,000 in 2004 to $1.5 million in 2008. Some industry analysts suggest that Laos' low-cost work force makes it an attractive location for potential future investment for the production of apparel.[328] However, Laos has not yet been targeted for significant investment in apparel production because of its poor logistics. The country is landlocked, with a significant portion inaccessible by roads. The road network that does exist is in poor condition, and Laos has no rail system.[329]

Brunei is not a competitive producer of apparel, as is evidenced by its small and declining share of ASEAN exports to the world and declining intra-ASEAN exports. Brunei's industry virtually disappeared after the removal of quotas at the end of 2004.[330] Information is not available on the state of the cotton woven apparel industry in Burma.

Inbound Investment

According to industry and trade sources in the ASEAN countries, there are few, if any, restrictions on investment in the cotton woven apparel industry or in the industry that makes its inputs—cotton woven fabrics.[331] Malaysia and Singapore have been two of the major sources of intra-ASEAN investment in Vietnam and Indonesia. As discussed, the costs of production in Malaysia and Singapore are generally higher than that for other ASEAN countries, and firms in these countries have sought lower-cost options for production outside of and among other ASEAN countries, particularly for apparel.

Vietnam is the primary recipient of investment among the ASEAN countries. According to an industry source in Vietnam, approximately 30 countries, including Singapore, Malaysia, Taiwan, Hong Kong, Sri Lanka, Bangladesh, the United States, France, and the United Kingdom, have invested in textile and apparel production in Vietnam, and a high percentage of that investment is likely to be in the cotton woven apparel industry.[332] Several other ASEAN countries have attracted or encouraged FDI in the sector. Approximately 60 percent of Indonesia's apparel production is based on foreign investment,[333] with much of the investment in the apparel sector originating from other ASEAN countries (Malaysia and Singapore), as well as from Taiwan, Korea, and Japan.[334] Cambodia's apparel industry, including the cotton woven apparel industry, is based almost entirely on FDI from countries such as China, Taiwan, and Korea.[335] FDI in Thailand's textiles and apparel sector also increased in 2007 and 2008 from prior levels.[336] In 2007, Thailand attracted $71.1 million in new FDI, following a net outflow of FDI in 2006 ($7.8 million).[337]

Given the intensity of global competition in the apparel industry, the availability, cost, and productivity of labor[338] are major driving factors affecting inbound investment in the cotton woven apparel industry. Although Vietnam's apparel producers get most of their inputs and fabrics from China, there is growing investment in Vietnam to increase capacity in the production of textile and apparel inputs. VINATEX, the state-owned textile conglomerate, indicated that it is encouraging Japan, Korea, Hong Kong, and the United States to increase their investment in fabric production in Vietnam.[339] Industry sources indicate there has also been strong support for the industry from the Government of Vietnam through the development of industrial parks, making it relatively easy to set up a factory.[340] As indicated, logistics and infrastructure problems affecting access to imported raw materials and exports of finished goods have been important factors inhibiting inbound investment, though the government has been working to improve both these factors.

Trade Facilitation, Logistics Services, and E-Commerce

Trade Facilitation and Logistics Services

Trade facilitation and logistics services, which affect quick turnaround and speed to market, are important competitive factors in both global and intra-ASEAN trade in cotton woven apparel and fabrics. Most industry representatives cited underdeveloped infrastructure as a challenge faced by ASEAN fabric and apparel producers, particularly in Indonesia, Cambodia, and to a lesser extent, Vietnam.

Industry sources noted that Indonesia has poor transportation infrastructure, including a costly and slow trucking system, as well as a rail system that is underdeveloped and even more costly than trucking.[341] Moreover, since Indonesia is located furthest among the ASEAN countries from its key input suppliers and key markets (i.e., China and the United States), its lack of a deepwater port further slows the movement of goods in and out of the country.[342] On the other hand, a new trans-Java highway that could reduce ground transportation bottlenecks is being built; it is scheduled for completion by 2013.

Cambodia is a country whose strengths in producing cotton woven apparel are undercut by poor infrastructure. Transportation between Thailand, Vietnam, and Cambodia is by road, which presents challenges because of poor road conditions.[343] Cambodia's apparel industry must ship its exports through the small port of Sihanoukville, which is also served by poor roads. Vietnam would like to serve as an export gateway for Cambodia's cotton woven apparel industry.[344]

Vietnam also is hampered by poor infrastructure. Nevertheless, industry sources indicated that the situation is improving in Vietnam. Already, its new deepwater port facilitates faster delivery of inputs into the country and improves the speed to market for delivery of cotton woven apparel.[345]

Another common trade facilitation problem among the ASEAN countries is customs clearance,[346] which also affects quick turnaround and speed to market. Customs clearance reportedly takes longer in Vietnam than in some of the other ASEAN countries; however, through government efforts, the time needed for clearance has been reduced from about six days to two days. By comparison, Malaysia's customs clearance time is six hours and Singapore's is one hour. The Vietnamese government is attempting to further reduce the time to import and export by 30 percent through its Project 30, which is a multiyear, cross-cutting red tape reduction program.[347] Improvements in customs clearance in Cambodia have been limited.[348]

E-Commerce

The use of e-commerce in the ASEAN cotton woven apparel industry is limited but increasing. As an example of electronic procurement, a factory in Penang, Malaysia, produces shirts for an upscale U.S. retailer and a sister factory produces pants for this retailer. Both factories are electronically linked to the retailers that sell these garments. Thus, when inventory drops to a certain level, the factory automatically ships more.[349] Electronic procurement is likely to become more widespread in the future, especially if companies are able to provide superior speed to market and become one-stop sources— operations that can offer customers both fully designed and fully produced apparel from a single source.

5. Wood-Based Products: Hardwood Plywood and Flooring

Wood-Based Products Overview[350]

The ASEAN wood-based products priority sector includes processed wood products, such as lumber, plywood, other panel products, flooring, and wooden furniture. Unprocessed wood, such as logs, poles, and posts, are specifically excluded from the sector, as are pulp and paper products. This chapter provides a brief discussion of the wood-based products sector and then the region's hardwood plywood and flooring industry, an important wood products segment that is representative of the competitive challenges and trends facing the region's wood products producers.

Historically, ASEAN countries have been significant contributors to global trade in wood-based products. The region is known for products made of tropical wood species that are desirable for their technical attributes as well as for their aesthetic qualities when used in furniture, cabinetry, flooring, and architectural woodwork. Plantation-grown woods such as rubberwood and teak are increasingly used in manufacturing wood-based products, as wood supply from the tropical natural forests becomes less available.[351]

Indonesia and Malaysia are major ASEAN players in the global trade of lumber, plywood, veneer, moldings, and flooring. Malaysia is the world's largest exporter of hardwood lumber (i.e., non-coniferous species) and the second largest exporter of hardwood plywood, behind China. Indonesia is the third largest exporter of hardwood plywood. Vietnam has emerged as the second largest exporter of wooden furniture, again behind China. Malaysia, Indonesia, and Thailand are also significant exporters of wooden furniture and furniture components. ASEAN exports of wood-based products increased from $14.9 billion in 2004 to $17.7 billion in 2008 (table 2.4), although ASEAN global market share remained about the same, slipping from 11.6 percent to 11.1 percent.[352]

ASEAN countries have been historically competitive in this sector largely because of their abundant and relatively low-cost wood supply. However, the region's international competitiveness is being challenged by Chinese expansion and by concerns in major markets about deforestation and illegal logging. Wood supply has also become less certain in recent years as natural tropical forests are more fully exploited and more restrictions are placed on their harvesting. As wood supply becomes more of an issue, the region is shifting its manufacturing to more highly processed wood products, such as flooring and furniture.

Because of the uncertainty of securing adequate wood supplies, relatively little foreign investment flows into wood-based products manufacturing in the ASEAN countries. Furniture manufacturing and forest plantations are the two areas where foreign investment activity has been more pronounced over the past several years, with much of the foreign capital coming from outside of the region. Investors from Taiwan, Hong Kong, Japan and, to a lesser extent, the United States and Europe have taken advantage of competitively low wood costs and labor rates to locate wood furniture plants in the region. Often, these facilities import temperate species of logs, lumber, and veneer from North America and the European Union (EU) to make secondary products (such as furniture and flooring) to be exported back to those markets.[353]

To augment wood supply from natural forests, forest plantations have attracted some, albeit limited, foreign investment in the form of joint ventures. Japan is the largest outside investor in plantations in the region, with joint ventures reported in Thailand, Vietnam, and Indonesia.[354] Chinese investment in the ASEAN wood-based sector has focused on logging (mainly in Burma) and forest plantations in other ASEAN countries to secure raw materials for factories in China. Intra-ASEAN investment in plantations is reportedly highest in Indonesia, with investments by Malaysian and Singaporean companies; some additional ASEAN cross-border plantation investment is reported in Vietnam. With few exceptions, however, government control of forest land in ASEAN countries complicates foreign investment and is a major barrier for foreign investors because of ownership controls and regulations.[355]

The ASEAN Roadmap for Integration of Wood-based Products Sector (Roadmap) shares many of the same objectives and programmatic elements as other priority sector Roadmaps, including tariff elimination, free flow of investment, and better trade facilitation. Indonesia is the lead country for the ASEAN Wood-based Products Sector Roadmap. Measures specific to wood products aim to, among other things, enhance cooperation in certification of wood-based products; combat illegal trade in forest products; ensure sustainability of forest resources; counter negative publicity about the tropical hardwood products trade; develop joint marketing efforts for ASEAN wood products; and promote investments in forest plantations. While ASEAN institutions have successfully served as forums for dialogue, progress has been slow in developing substantive region-wide approaches to many of these issues.

Hardwood Plywood and Flooring[356]

Key Findings

Although unprocessed wood (referred to as industrial roundwood) is not included in the ASEAN wood-based products priority sector, it is nevertheless a dominant factor influencing economic integration. In fact, the single most important competitive factor in the production and export of hardwood flooring and plywood is the availability of competitively priced raw material; the basic cost for delivered logs typically accounts for more than half of all production costs.

Without exception, all of the ASEAN countries have imposed prohibitions of one kind or another on the harvesting and/or export of unprocessed roundwood (i.e., logs) and, in some cases, on the export of sawnwood (i.e. lumber) as well, both to encourage manufacturing of more highly processed products and to protect diminishing forest resources.[357] These restrictions affect intraregional trade as well as the overall supply of raw materials available for the region. In general, a variable wood supply and competition for exports between the member countries tend to hinder economic integration in hardwood plywood and flooring.

Two other elements—the ability to satisfy international consumers' concerns about the legality and sustainability of raw materials, and the elimination of tariffs—are also critical to achieving regional economic integration in the ASEAN hardwood plywood and flooring industry.[358] Although considerable progress has been made in lowering tariffs within ASEAN, policies in major export markets compelling ASEAN producers to assure that wood products are from legal, if not demonstrably sustainable, sources remain a major obstacle to the competitiveness of ASEAN hardwood plywood and flooring. Initiatives to collaborate on issues relating to deforestation and illegal logging are emphasized in the Roadmap, but progress has been complicated by the fact that ASEAN members are involved in various other simultaneous national, bilateral, and international processes aimed at achieving similar objectives.[359]

The reduction in tariffs put into effect with the ASEAN Free Trade Area (AFTA), along with implementation of other ASEAN measures for economic integration, could potentially induce greater intra-ASEAN investment and trade at a time when wood supply in the region is tightening. Historically, tariffs on hardwood plywood and flooring in ASEAN countries were very high—as high as 40 percent in many cases. As of January 1, 2010, intra-ASEAN tariffs on hardwood plywood and flooring are mostly zero or less than 5 percent. Lower tariffs might allow more flexibility in sourcing materials and producing different varieties of hardwood plywood and flooring, but tariff reduction by itself has not yet triggered increased intra-ASEAN trade in this sector.

Economic integration and intra-ASEAN trade could benefit the hardwood plywood and flooring industry as the region becomes more dependent on plantation-grown trees and imported raw materials, and as the region develops its downstream manufacturing capacity for wood products such as flooring, furniture, and millwork. However, plywood production in the region is declining. Suitable raw material is becoming scarcer and China has become a formidable competitor taking away global market share. While plywood, engineered wood flooring, and wood laminate flooring offer opportunities for cross-border investment, there has been relatively little intra-ASEAN investment in this sector. ASEAN institutions have had some positive effects on the hardwood plywood and flooring industry in terms of lower tariffs and increased dialogue on sector-specific competitiveness issues, but the industry itself has not experienced measurable regional economic integration.

Background

Hardwood plywood and flooring are mainstay industries within the ASEAN wood-based products sector, and both are export oriented. ASEAN shipments of hardwood plywood and flooring represent an estimated 30 percent of all primary manufactured wood products (excluding roundwood) and 20 percent of wood-based products as defined in the ASEAN priority sector (which includes wood furniture).[360] However, in 2008, only an estimated 15 percent of ASEAN production of hardwood plywood was consumed within the region.[361] In that year, ASEAN exports of hardwood plywood and flooring accounted for $4.9 billion (table 5.1), or 27 percent of total world exports of these commodities.[362] ASEAN's largest market is Japan, which accounts for approximately 40 percent of the region's hardwood plywood and flooring exports.

Malaysia and Indonesia dominate production in the ASEAN region, accounting for over 90 percent of ASEAN production (an estimated 57 percent and 34 percent, respectively).[363] Malaysia and Indonesia are also the region's two largest exporters, together accounting for over 95 percent of ASEAN exports (an estimated 52 and 44 percent, respectively).[364] These figures mask some significant shifts that have occurred in the region over time. Indonesia's plywood capacity has declined significantly over the past five years, while Malaysian production has remained steady. The number of operating plywood plants in Indonesia declined from 128 in 2003 to only 25 in 2009.[365] The Indonesia plywood industry is contracting because of less government support, tightening timber supplies, and lost market share in industrialized countries concerned about the legality of wood sourcing.[366] Malaysia currently has 91 operating plywood plants.[367] Only a handful of plywood facilities operate in the other ASEAN countries, which collectively account for less than 6 percent of regional production.[368] Brunei Darussalam (Brunei) is the only ASEAN member without any plywood production capacity. The Philippines was once the largest global hardwood plywood producer, but it exhausted its supply of exploitable timber resources by the 1980s and now accounts for only 2 percent of the region's output. The country's forest management agency reported 39 operating plants in 2007.[369] Separate data on wood flooring manufacturing by ASEAN countries are not available, but Malaysia and Indonesia are also the region's largest producers and exporters of engineered and wood laminate flooring products.[370]

Construction is the major end use for hardwood plywood produced in the ASEAN countries, but other uses for high-quality tropical hardwood plywood include furniture, door skins, architectural woodwork, and flooring.[371] ASEAN plywood and flooring producers typically use tropical wood species such as meranti, kapur, and keruing, among many others. For flooring, they also utilize locally grown bamboo, as well as imported temperate species such as oak, cherry, maple, and walnut.

Three general types of hardwood flooring are manufactured in the ASEAN region—solid, engineered, and laminated. The latter two types have grown in popularity and are increasingly manufactured in both China and the ASEAN region.[372] The local market for wood flooring is small; ceramic tile is the most common type of flooring used in the region.[373]

Regional Integration, Export Competitiveness, and Inbound Investment

Leading Competitive Factors

The factors affecting integration, export competitiveness, and inbound investment in the ASEAN hardwood plywood and wood flooring industry are dominated by issues related to raw materials—specifically, stable access to wood supplies and international concerns regarding legal sourcing and forest governance. Other factors that affect the region's integration and global performance are the emergence of China as a major global competitor and product standards that ASEAN producers must meet to sell into the industrialized country markets. Of lesser importance to the competitiveness of ASEAN plywood and flooring producers (and not discussed at length in this profile) are labor costs, infrastructure, and availability of investment capital.

Table 5.1. Hardwood Plywood and Flooring: ASEAN Exports to Selected Markets, by Value, 2004–08

Country	2004	2005	2006	2007	2008	% change, 2004–08
			thousand $			
Japan	2,160,069	1,957,806	2,424,880	2,163,012	1,901,615	(12)
EU-27	779,370	804,845	914,017	944,344	903,538	16
Korea	316,986	345,374	358,585	401,350	419,753	32
United States	695,708	642,676	659,525	600,886	383,402	(45)
Taiwan	302,804	296,123	299,615	305,685	302,921	0
Australia	125,760	139,430	147,582	208,490	195,167	55
China	400,583	312,797	214,665	181,324	167,854	(58)
Subtotal	4,781,281	4,499,051	5,018,869	4,805,091	4,274,250	(11)
All other	430,320	619,364	525,985	740,248	646,847	50
Total	5,211,601	5,118,415	5,544,854	5,545,339	4,921,096	(6)

Source. WITS, Integrated Data Warehouse (accessed January 5, 2010).

Note. Data are based on ASEAN partner country imports because Brunei, Burma, Cambodia, Laos, Phillipines, and Vietnam do not report trade data in all years. All other export value excludes intra-ASEAN trade.

Wood costs remain the single most important factor in relative competitiveness for hardwood plywood and flooring. Delivered wood costs to a plywood plant are typically upwards of 60 percent of the total manufacturing costs.[374] As measured by the export market, tropical log prices in the ASEAN region are generally lower than in other world regions.[375] However, hardwood plywood manufacturers are facing increasing scarcity in raw materials.[376] This is particularly the case for the large-diameter logs that are most often utilized for making plywood.[377]

ASEAN countries regulate the volume of timber harvesting under concession systems to private and/or state enterprises, and for environmental reasons, some have imposed complete bans on harvesting in natural forests where large logs typically originate. With a few exceptions, most ASEAN countries restrict or ban exports of unprocessed or minimally processed wood material (table 5.2). With raw materials being so critical and not easily secured, opportunities for investment in the hardwood plywood and flooring industry are limited, particularly in the import-dependent countries.[378] Thailand and Vietnam have complete bans on harvesting in natural forests, so manufactured wood products such as hardwood plywood and flooring are made from plantation-grown resources or imported raw material. Plywood manufacturers in Thailand rely principally on plantation-grown rubberwood. Malaysia is also expanding the use of rubberwood for core plywood material.[379]

Major industrialized market countries are placing greater emphasis on assuring that imported hardwood plywood and flooring products are sourced legally and obtained from sustainably managed forests.[380] By one estimate, as much as 25 to 40 percent of demand for tropical timber products in major markets could be subject to verification that they meet legality and/or sustainability requirements. These procurement-related policies are often government-mandated, but in some cases they are also privately imposed.[381] In Japan and some European countries, procurement policies dictate that wood products must be produced according to specified guidelines for legality and/or sustainability.[382] In some situations, wood products must be certified according to an approved certification scheme—a requirement viewed as a nontariff barrier by some producers in ASEAN countries.[383] In the United States, the Lacey Act prohibits trading in plant-derived products taken illegally. Furthermore, it requires U.S. importers of most plant-derived products to file a declaration that includes the species name and country of harvest of the plant material contained in each imported shipment.[384] The EU is also considering a regulation requiring importers to employ a due diligence system that verifies legality of timber products sold in the European market. Given these burgeoning trade-influencing measures, ASEAN producers of hardwood plywood and flooring recognize that, to remain competitive in the major industrialized markets, they must meet requirements for ensuring legality, if not also the sustainable sourcing, of their products.[385]

Table 5.2. Harvesting and/or Export Bans Imposed by ASEAN Countries

Country	Year enacted	Description
Brunei	1989	Has a moratorium on new logging permits and a complete ban on log exports.
Cambodia	1996	Has a complete ban on exports of logs and rough lumber.
Indonesia	1980–1992 1992–1998 2001: logs 2004: lumber	Log exports banned from 1980 to 1992; ban replaced by high export tax from 1992 to 1998. Since 2004, exports of both logs and rough lumber from natural forests are banned; also prohibits harvesting certain species such as ramin (*Gonystylus* spp.).
Laos	1999	Exports of logs, lumber and "semi-finished" products from natural forests are banned, but government-issued exemptions are common.
Malaysia	1985 (P. Malaysia) 1993–1996 (Sabah) 1992 (Sarawak)	Peninsula Malaysia has complete log export ban; Sabah imposed export ban from 1993 to 1996; Sarawak imposes an export quota.
Burma		No specific export ban noted.
Philippines	1986 (export ban) 1992 (harvesting ban)	Has a complete ban on exports of logs from natural forests; has a harvesting ban on steep slopes and elevations over 1,000 meters.
Singapore		Has no productive forests, so imposes no restrictions on log trade.
Thailand	1989	Has a harvesting ban in natural forests; log exports are plantation based or re-exports.
Vietnam	1992	Prohibits exports of both logs and rough lumber from natural forests; has a harvesting ban in natural forests.

Sources. Resosudarmo and Yusuf, table 1, 2; FAQ, ASIA and the Pacific National Forestry Programmes Update No. 34, 35, 41; Bibi and Berudin, 19; Forest Trends, 4; Boungnakeo, 5; Macek, 1; Tachibana, 58–59.

Hardwood plywood manufactured from tropical species also faces global competition from hardwood plywood made of birch and poplar in Russia and China.[386] ASEAN countries mainly compete against China, which has become the world's largest producer of hardwood plywood and the largest exporter of both hardwood plywood and wood flooring.[387] From 2004 to 2008, ASEAN's global market share declined from 36 percent to 27 percent, while China's share increased from 14 percent to 26 percent.[388] To produce these goods, China is importing raw materials from ASEAN (principally from Sabah and Sarawak in Malaysia and from Burma), other Southeast Asian countries, Russia, North America, and Africa. Despite a high dependence on imported raw material, China is a low-cost world producer. Some observers assert that the Chinese industry is becoming more focused on its own large and growing domestic market, which could actually restore ASEAN competitiveness elsewhere and/or provide new opportunities for ASEAN hardwood plywood in China.[389] Another analyst suggests that, by driving down wood flooring prices, the success of Chinese wood flooring exports has also made a wider range of flooring choices, including tropical wood flooring made in ASEAN countries, available to global consumers.[390]

The ability to meet product standards is important to the international competitiveness of ASEAN hardwood plywood and flooring producers, but there are no ASEAN-wide standards, and little progress to date has been made on harmonizing standards in the region.[391] Instead, product standards are imposed on hardwood plywood and flooring products in a number of major ASEAN export markets, where they are focused on performance or appearance (depending upon end use). Plywood used for concrete forming or structural applications must meet specific standards in Japan, Canada, the United States, and the EU. Hardwood plywood is often graded according to grading standards maintained by the associations in the importing countries. Compliance is verified by accredited independent testing agencies, and for some standards, testing labs have been approved in the ASEAN region. However, testing and certification requirements reportedly create delays and bottlenecks for ASEAN exports.[392]

Hardwood plywood, engineered flooring, and laminated wood flooring products must meet strict formaldehyde emission standards in Japan, Europe, and, most recently, the United States. In 2007, the state of California implemented restrictive formaldehyde emission standards for plywood, reconstituted wood products (e.g., particleboard, medium-density fiberboard, hardboard), and for products made from those materials. ASEAN producers, like all producers selling into the California market, must meet the state's formaldehyde emissions standards and be certified by a California-approved certification agency.[393] One Indonesian certifier, the only one approved by the state of California in ASEAN, has certified 25 producers in Indonesia and 10 producers in Malaysia. An additional 9 facilities in Malaysia and 2 in Vietnam have also been certified by U.S.- and Hong Kong-based agencies. On July 7, 2010, President Obama signed into law the Formaldehyde Standards for Composite Wood Act that establishes national emission standards similar to those in California and will require all plywood and composite wood products sold in the United States to be certified as compliant with the new national standards.[394]

Regional Integration

As of January 1, 2010, intra-ASEAN tariffs on hardwood plywood and flooring were scheduled to become zero in all but the four least developed countries (Burma, Cambodia, Laos, and Vietnam). It is too early to assess the effects of tariff reductions on intra-ASEAN trade in hardwood plywood and flooring, but other equally or more important factors continue to hamper regional integration involving these products. Clearly, government controls over wood supply (through land ownership and concession policies) and restrictions on the harvesting and trade of raw materials (unprocessed wood) have encouraged discrete, country-specific hardwood plywood and flooring industries. Although export restrictions on raw materials may foster the development of downstream hardwood plywood and flooring industries within a particular ASEAN country, they tend to form a barrier to the development of an integrated regional industry that could otherwise draw upon wood materials at various stages of processing sourced throughout the ASEAN region.

Another factor influencing regional integration in ASEAN plywood and flooring is the pressure that is mounting in external markets for assurances that wood products are derived from legal and sustainable sources. Illegal harvesting and trade are alleged to be significant in the ASEAN region, and cases have been documented by a number of NGOs and governments.[395] The EU, Japan, United States, Australia and other external markets for ASEAN hardwood plywood and flooring have instituted various policies to address concerns about illegal logging and tropical deforestation.[396] In response, ASEAN industries and governments are attempting to collaborate on addressing these issues. The Roadmap contains measures to address legality issues, sustainability and certification, protection for endangered species, outreach activities, and public relations efforts to counter negative publicity on tropical hardwood products.[397] ASEAN collaboration in responding to these concerns may be helping the industry to integrate in some respects, but ASEAN countries are also working independently in other forums to protect their global competitive interests.

In addressing legal trade and sustainability issues, ASEAN reports that progress has been made on a number of the Roadmap objectives.[398] Several separate but related ASEAN- wide work plans and agreements have been signed for the purpose of strengthening enforcement of forest and wildlife laws, curbing transboundary pollution, and promoting sustainable practices.[399] Separately, Vietnam and Laos have entered into a Cooperative Action Plan targeting illegal hunting and forest products trade.[400] The overarching ASEAN Economic Blueprint includes a goal of developing a phased approach to forest certification by 2015.[401]

An Ad-Hoc Working Group to develop a Pan-ASEAN Timber Certification Initiative is working to develop criteria and indicators that can be applied in each ASEAN country, and agreement has been reached on core elements of principles for an ASEAN legality standard.[402] Malaysia has made the most progress on forest certification, having independently developed its own national forest certification scheme endorsed by the Programme for the Endorsement of Forest Certification (PEFC).[403] As of 2009, 4.4 million hectares, representing 21 percent of the country's forests, have been certified by the Malaysia Timber Certification Council (MTCC). Another internationally recognized forest certification scheme, the Forest Stewardship Council (FSC), and an Indonesian certification scheme, LEI (Lembaga Ekolabel Indonesia), have collectively certified nearly 3 million hectares in ASEAN countries, mostly in Indonesia (the FSC has certified 1.1. million hectares in Indonesia and approximately 70,000 hectares in Vietnam, Thailand, and Laos).[404]

Table 5.3. Major Intergovernmental Processes to Address Legal and Sustainable Forest Products Production and Trade

Process	Sponsor	When Initiated	Countries	Summary
Asia-Pacific Regional Dialogue to Promote Trade in Legally Harvested Forest Products	U.S. and Indonesia	2009	Indonesia, U.S., Australia, Brunei, Malaysia, Papua New Guinea, Singapore, Solomon Islands, and Vietnam	Convened by the US and Indonesia in September 2009 pursuant to the U.S.- Indonesia MOU (see separate entry). The dialogue is intended to discuss issues related to trade in forest products among the invited countries. Both ministries of trade as well as ministries of forestry/environment participate. The second meeting is being planned.
Memorandum of Understanding (MOU) on Cooperation to Combat Illegal Logging and Associated Trade	U.S. and Indonesia	2006	U.S., Indonesia	The MOU was developed under a Trade and Investment Framework Agreement concluded between Indonesia's Minister of Trade and Minister of Forestry and the United States Trade Representative. It created a Working Group that meets two or three times per year and provides a mechanism for consultation and cooperation to combat trade associated with illegal logging and promote sustainable management of Indonesia's forests.
East Asia Forest Law Enforcement and Governance Initiative (EAFLEG)	World Bank, governments, others	2001	Countries in East Asia, EU, U.S.	Launched as a coalition of governments and stakeholders to devise ways of addressing illegal logging and forest governance issues.
Forest Law Enforcement, Governance and Trade (FLEGT) voluntary partnership agreements (VPA)	European Commission	2007	Malaysia, Indonesia, EU	Malaysia and Indonesia are in negotiations with the European Commission to form FLEGT VPAs whereby the EU would recognize a legality assurance system (LAS) involving the issuance of licenses to exporters that would ensure legal sourcing. The incentive offered is that a VPA would provide more secure access to the European market and give those countries potentially greater access to European foreign aid. Vietnam has started pre- negotiations toward a VPA.
Pan-ASEAN Timber Certification Initiative	ASEAN and International Tropical Timber Organization (ITTO)	2002	ASEAN countries	ASEAN forestry officials established a working group to develop a pan-ASEAN timber certification scheme with the goal of adopting an initiative by 2006. A progress report was presented in 2005 at an ITTO workshop in Berne, Switzerland. No additional information is available

Process	Sponsor	When Initiated	Countries	Summary
Forest Law Enforcement, Governance and Trade (FLEGT) due diligence regulation	European Commission	Proposed	EU members	Would require importers and European producers to implement due diligence system for ensuring legal sourcing.
Roadmap for Integration of Wood-based Products Sector measures	ASEAN	2005	ASEAN countries	Several measures to address legality and sustainability are included in ASEAN's Integration Roadmap for Wood-Based Products. In addition to the Pan-ASEAN Timber Certification Initiative, measures include developing a cooperation program among relevant authorities to combat illegal trade in forest products and to exchange wood import information (through ASEAN Customs Director Generals and the ASEAN Wood Based Products Working Group). They also include a program to promote regional cooperation in addressing issues related to the Convention on International Trade in Endangered Species of Wild Fauna and Flora (CITES) through the ASEAN Experts Group on CITES and the ASEAN Wildlife Law Enforcement Network.
Asia Forest Partnership	Consultative Group on International Agricultural Research (CGIAR) – an organization funded by the World Bank, FAO, United Nations Development Program (UNDP), Internat-ional Fund for Agricultural Development (IFAD), and individual countries	2002	20 countries in Asia (includes Cambodia, Indonesia, Malaysia, Philippines, Thailand, and Vietnam), the EU, and the U.S.	Organizes multi-stakeholder forums to support cooperation and information exchange on sustainable forest management issues in Asia and the Pacific.

Sources. USTR, "MOU between United States and Indonesia," November 2006; USTR, "USTR Announces New Asia-Pacific Initiative," September 2, 2009; World Bank, "Overview" (accessed January 8, 2010); EC, "Forest Law Enforcement, Governance and Trade (FLEGT)" (accessed January 8, 2010); ASEAN, "Forest and Timber Certification" (accessed March 8, 2010); "EU FLEGT: Due Diligence," www.illegal-logging.info (accessed January 16, 2010); ASEAN, "Roadmap for Integration of Wood-Based Products Sector," 2007; Asia Forest Partnership, "AFP in Brief," July 28, 2008.

Beyond specific ASEAN measures and certification programs, other multilateral and bilateral processes with participation by ASEAN member countries have been undertaken to address various aspects of legal and sustainable sourcing and trade of wood products (table 5.3). Malaysia and Indonesia are in the process of negotiating with the European Commission to enter into voluntary partnership agreements (VPAs) under the EU's Forest Law Enforcement, Governance and Trade action plan to enable easier exports to the EU.[405] The VPAs would create a legality licensing system that verifies the legal origin of exports to the EU. The United States has a memorandum of understanding (MOU) with Indonesia on legal trade in forest products. One outcome of the MOU has been the Asia-Pacific Regional Dialogue to Promote Trade in Legally Harvested Forest Products, which currently includes four other ASEAN members in addition to Indonesia, as well as several other trading partners in East Asia.[406] These processes have resulted in a number of initiatives endorsed at very high levels of government and industry, but they have also produced an assortment of criteria and requirements that, as a practical matter, are complicated and confusing to producers, including ASEAN hardwood plywood and flooring exporters.[407]

Export Competitiveness

In 2008, ASEAN exports of hardwood plywood and flooring products totaled an estimated $4.9 billion (table 5.1).[408] External trade accounted for over 95 percent of the total for the region—approximately $4.7 billion in 2008. Exports to Japan accounted for nearly 40 percent of total ASEAN hardwood plywood and flooring exports; the EU, 18 percent; South Korea, 9 percent; and the United States, 8 percent. Malaysian exports of hardwood plywood and flooring alone were an estimated $2.4 billion in 2008, or nearly half of total ASEAN exports of these products (table 5.4). While Indonesian exports totaled approximately $2.0 billion in 2008, or 40 percent of the export total, this represented a decline of 30 percent from 2004 (table 5.4). The decline reflects a loss in industry capacity due to tighter wood supplies and lost sales because of concerns over the legality of Indonesian supply in major markets.[409] In contrast, Malaysian exports increased by 23 percent in 2008, as compared to 2004 (table 5.4), possibly because of its more advanced certification program and aggressive marketing of its products. The other promising producer is Vietnam, which has been exporting bamboo plywood and flooring for secondary manufacturing to Taiwan as well as to end-use flooring markets in the United States.[410] While modest compared to Malaysia and Indonesia, Vietnam's exports of hardwood plywood and flooring have more than tripled since 2004, totaling an estimated $73 million in 2008, principally to South Korea, the EU, Japan, and the United States (table 5.4).[411]

While intra-ASEAN trade has increased over the past five years, it remains relatively small, accounting for no more than 5–10 percent of total exports. Indonesia and Malaysia supplied 75 percent of total intra-ASEAN hardwood plywood and flooring exports.[412] Vietnam is the largest market for intra-ASEAN hardwood plywood and flooring exports, followed by Singapore, Malaysia, Indonesia, and Thailand (table 5.5). Hardwood plywood is used in Vietnam's burgeoning wood furniture industry. Imports of hardwood plywood and flooring from outside of the region, most notably from China, exceed intraASEAN trade in these products and increased more than intra-ASEAN trade during 2004–08. Imports from China increased by over two and half times (table 5.6), indicative of that country's strengthening global position. The majority of ASEAN imports of these products are plywood, as opposed to flooring; the plywood is used primarily for concrete forms in local construction and in the manufacturing of furniture for export.[413]

Table 5.4. Hardwood Plywood and Flooring: ASEAN Exports to the world, by Value, 2004–08

Country	2004	2005	2006	2007	2008	% change, 2004–08
	thousand $					
Malaysia	1,947,973	2,003,652	2,474,831	2,518,147	2,405,281	23
Indonesia	2,793,577	2,589,393	2,468,830	2,351,073	1,950,498	(30)
Philippines	181,246	189,201	238,905	255,440	218,114	20
Thailand	166,888	179,797	193,467	229,084	206,862	24
Vietnam	22,353	37,425	41,329	62,674	73,060	227
Singapore	30,604	37,975	43,917	40,570	46,045	50
Burma	34,678	37,896	32,131	33,181	13,890	(60)
Laos	6,506	5,876	8,922	7,062	7,309	12
Brunei	7	100	64	518	36	400
Cambodia	4,460	2,948	354	218	1	(100)
Total	5,188,293	5,084,263	5,502,750	5,497,997	4,921,096	(5)

Source. WITS, Integrated Data Warehouse (accessed January 5, 2010).

Note. Data are based on ASEAN partner country imports because Brunei, Burma, Cambodia, Laos, Phillipines, and Vietnam do not report trade data in all years. Data excludes intra-ASEAN trade.

Table 5.5. Hardwood Plywood and Flooring: ASEAN Member Imports from ASEAN Members, by Value, 2004–08

Country	2004	2005	2006	2007	2008	% change, 2004–08
	thousand $					
Vietnam	24,564	36,860	57,333	75,113	48,185	96
Singapore	20,471	23,867	29,423	36,033	34,538	69
Malaysia	18,739	14,443	20,199	29,434	29,150	56
Indonesia	11,141	13,694	21,104	32,591	27,076	143
Thailand	21,508	23,021	23,531	28,300	22,193	3
Philippines	27,003	30,866	41,543	40,459	11,208	(58)
Brunei	3,297	2,408	3,310	3,492	1,044	(68)
Burma	971	715	750	1,050	939	
Laos	584	448	586	1,196	560	
Cambodia	662	526	522	671	257	(61)
Total	128,940	146,845	198,301	248,339	175,150	36

Source. WITS, Integrated Data Warehouse (accessed January 5, 2010).

Note. Data are based on ASEAN partner country exports because Brunei, Burma, Cambodia, Laos, Phillipines, and Vietnam do not report trade data in all years. Trade between these members may therefore be understated.

Table 5.6. Hardwood Plywood and Flooring: ASEAN Imports from Selected Markets, by Value, 2004–08

Country	2004	2005	2006	2007	2008	% change, 2004–08
	thousand $					
China	39,902	37,018	55,471	112,157	106,349	167
United States	22,556	33,073	43,354	54,346	52,487	133
EU-27	39,340	40,464	60,930	47,150	35,530	(10)
Japan	4,473	3,985	3,250	8,115	17,222	285
Subtotal	106,271	114,540	163,005	221,769	211,588	99
All other	144,232	156,677	187,046	230,848	275,822	91
Total	250,502	271,217	350,051	452,618	487,410	95

Source. WITS, Integrated Data Warehouse (accessed January 5, 2010).

Note. Data are based on ASEAN partner country exports because Brunei, Burma, Cambodia, Laos, Phillipines, and Vietnam do not report trade data in all years. All other export value excludes intra-ASEAN trade.

Tariff reductions within ASEAN have made its member countries more competitive in each other's markets relative to non-ASEAN imports. Before the implementation of the Common Effective Preferential Tariff (CEPT) scheme, plywood and flooring tariffs in the ASEAN countries were as high as 40 percent—among the highest for any wood products in the region. While MFN tariff rates applied to external trade remain high, the CEPT scheme has effectively reduced tariffs on plywood and flooring for intra-ASEAN trade to 5 percent or less, effective January 1, 2010. ASEAN has not yet harmonized MFN tariff rates on non-ASEAN imports. Applied MFN tariffs for non-ASEAN countries for most hardwood plywood and flooring categories range from zero in the case of Singapore to a high of 35–40 percent for Malaysia and Laos (tables 5.7 and 5.8).

ASEAN has entered into free trade agreements (FTAs) with China, Japan, the Republic of Korea, Australia-New Zealand, and India that either have or will effectively lower tariff barriers in those markets, although hardwood plywood and flooring products are typically listed as "sensitive" sectors and are thus subject to gradual staging of tariff reductions. Under the ASEAN-China FTA, for example, final reductions for tariff lines on the "highly sensitive" lists, which include China's plywood and flooring imports, will not occur until 2018.[414]

The full effects of the CEPT tariff reductions and the various FTAs on plywood and flooring trade patterns both within ASEAN and between ASEAN countries and other FTA partners are uncertain.[415] In general, Malaysia and Singapore are expected to be most advantaged by the ASEAN-China FTA because of their English-Chinese language facility and already well-established trading relationships.[416] In contrast, the ASEAN- China FTA sparked protests in Indonesia earlier this year from groups fearful of the impact from Chinese imports.[417]

Table 5.7. ASEAN Country Tariffs for Hardwood Plywood, 2004, 2008, and 2010 (Ad Valorem Percent)

ASEAN members	2004			2008	2010[a]
	Imports from other ASEAN countries	Imports from non-ASEAN countries (MFN)	Imports from other ASEAN countries	Imports from non-ASEAN countries (MFN)	Imports from other ASEAN countries
Brunei	5	20	0–5	20	0
Burma	10	15	5	15	5
Cambodia	15	15	5	15	5
Indonesia	0	10	0	10	0
Laos	20	40	5	40	5
Malaysia	5	25–40	5	25–40	0
Philippines	5	15	0–5	15	0–5
Singapore	0	0	0	0	0
Thailand	5	12.5	5	5	0
Vietnam	5	10	5	8	5

Source. Agreement on the Common Effective Preferential Tariff (CEPT) Scheme for the ASEAN Free Trade Area, Consolidated Package of Tariff Reductions for 2008; Royal Malaysia Customs Department, HS-Explorer (accessed April 10, 2010); World Trade Organization, Tariff Analysis Online (accessed April 2010).

[a] 2010 MFN rates for ASEAN imports from non-ASEAN countries are not available to USITC staff for every ASEAN member country.

Table 5.8. ASEAN Country Tariffs for Engineered and Laminate Wood Flooring, 2004, 2008, and 2010 (Ad Valorem Percent)

ASEAN members	2004		2008		2010[a]
	Imports from other ASEAN countries	Imports from non- ASEAN countries (MFN)	Imports from other ASEAN countries	Imports from non-ASEAN countries (MFN)	Imports from other ASEAN countries
Brunei	0–5	0–20	0	0–20	0
Burma	3–5	3–5	3–5	3–5	3–5
Cambodia	(b)	(b)	10	15	5
Indonesia	0	5–10	0	5–10	0
Laos	10	20	1–5	20–30	1–5
Malaysia	5	20	0	20	0
Philippines	0–5	5–15	0	5–15	0
Singapore	0	0	0	0	0
Thailand	5	12.5–20	0–5	5–30	0
Vietnam	5	5–10	0	3–8	0

Source. Agreement on the Common Effective Preferential Tariff (CEPT) Scheme for the ASEAN Free Trade Area, Consolidated Package of Tariff Reductions for 2008; Royal Malaysia Customs Department, HS-Explorer (accessed April 10, 2010); World Trade Organization, Tariff Analysis Online (accessed April 2010).

[a] 2010 MFN rates for ASEAN imports from non-ASEAN countries are not available to USITC staff for every ASEAN member country.

[b] Not available.

Inbound Investment

Data specific to foreign direct investment (FDI) in hardwood plywood and flooring are not available, but industry sources indicate that there is not much FDI in hardwood plywood and flooring manufacturing in ASEAN countries.[418] Some specific examples of FDI in hardwood plywood and flooring include one U.S. company known to have a flooring plant in Malaysia[419] and one Japanese company that recently announced plans to construct a major new $250 million plywood mill in Vietnam.[420] Taiwan is also often cited as a major investor in Vietnam, mostly in furniture, but also reportedly in enterprises that produce bamboo for flooring manufacturing.[421]

Many ASEAN countries restrict foreign investors from majority ownership and/or do not offer them national treatment. For example, countries may impose additional licensing, registration or other regulations on foreign enterprises not required of domestic investors. When fully implemented, the ASEAN Comprehensive Investment Agreement (ACIA) will provide additional investor protections, particularly to ASEAN-based investors, although it is too early to assess the effects of the ACIA on flows into the hardwood plywood and flooring sector.[422]

Hardwood plywood and flooring differ from other manufacturing sectors because of their close link to production in forests, where foreign investment is typically tightly controlled. With few exceptions, ASEAN countries prohibit foreigners from owning forest land outright, and government concessions are generally restricted to joint ventures that have majority local ownership.[423] Restrictions currently apply equally to ASEAN investors as well as investors from outside of the region.

Consistent with measures in the Roadmap, all ASEAN countries encourage new forest plantations, but limited land availability and cumbersome regulatory requirements tend to discourage foreign investment. For example, the Malaysian government provides favorable loan terms to investors for establishing and managing forest plantations, but loans are available only to wholly owned Malaysian non-publicly listed companies or to publicly listed companies with majority Malaysian equity ownership.[424] Nevertheless, some foreign investment in plantations has been permitted from enterprises in non- ASEAN countries seeking to secure wood supplies for their own (or joint venture) industries. Korean and Chinese investors have undertaken forest-related investments in Vietnam, Cambodia, Laos, Indonesia, and Thailand. Singaporean and Malaysian companies have invested in Indonesia.[425] Overall, outside investments are not as extensive as those from local investors.

The Roadmap recognizes that investments in securing stable wood supplies are necessary to foster integration in the hardwood plywood and flooring industry. To that end, the Roadmap includes measures to promote plantations as a way of augmenting the region's wood supplies. Many ASEAN members have incentive programs and other policies in place to promote forest plantations. The vast majority of plantations have been established to supply the pulp and paper industry, but plantations of teak, rubberwood, khaya, and other species intended for making plywood, furniture, and other products are being established. In the case of rubberwood, the latex is tapped for a number of years, and then the trees are harvested for wood and veneer to be used for lumber, plywood, engineered flooring, and other products. Similarly, investments in oil palm are envisioned to supplement wood supply for plywood and other wood products.[426] The plywood industry is slowly adapting to smaller diameter rubberwood and oil palm trees that can be used after their productivity for latex and oil wanes (see box 5.1). The continuing addition of rubberwood and other forest plantations is expected to add to the region's wood supplies and therefore contribute to ASEAN competitiveness.

Box 5.1. Hardwood Plywood and Flooring: Profile of ASEAN Integration

By most measures of cross-border investment and cross-border links in manufacturing, the hardwood plywood and flooring industry has not experienced significant regional integration. The industry is highly parochial because of its long-established history within the major producing countries and close linkages to locally available forest resources. Nevertheless, there are instances of plywood and flooring companies investing and/or operating in more than one ASEAN country. The most successful cross-border investments involve vertically integrated companies. For example, one Malaysian company is in both the oil palm and the wood flooring business.[a] It manufactures flooring in Malaysia and maintains oil palm plantations in both Malaysia and Indonesia. Although the technology is not yet fully commercialized, oil palm wood is beginning to be used for core layers in some plywood manufacture. The potential advantage of using oil palm wood in plywood production is that it would reduce reliance on wood from natural forests. Some regional integration is also occurring by virtue of FDI from outside of ASEAN. For example, Taiwanese companies are reportedly sourcing lumber and veneer of coconut palm from Indonesia and sugar palm from Thailand for use in furniture and flooring products manufactured both in Vietnam and in Taiwan.[b]

ASEAN efforts to reduce and eliminate intra-ASEAN tariffs on hardwood plywood and flooring may provide additional opportunities for regional industry integration. If tariff elimination makes it more economically attractive, core plywood material could be imported from another ASEAN country and finished with a locally manufactured decorative outer ply or "wear" layer. Similarly, wood laminate flooring uses medium-density fiberboard (MDF), which could be imported and finished with a locally printed paper and melamine overlay, assuming there are cost-savings or other advantages for doing so.

[a] The company referenced is TSH Resources Berhad, which has a wood flooring subsidiary called Ekowood.

[b] Industry representative, interview by USITC staff, Washington, DC, March 30, 2010.

Trade facilitation, logistics services, and E-commerce

Trade Facilitation

ASEAN trade facilitation efforts have led to some improvements—streamlining procedures at ports of entry and harmonizing documentation requirements—that have benefited trade in hardwood plywood and flooring, along with trade in the region more generally.[427] Further progress is anticipated over the next several years through the development of national and ASEAN single windows.[428] Considerable effort has also been made to harmonize intra-ASEAN tariffs and trade by implementing a common 8-digit tariff nomenclature based on the Harmonized System (HS), although not all ASEAN countries are reporting both trade data and applied tariffs consistently yet.[429]

Trade across land routes has been and continues to be a significant issue for wood products, particularly along remote borders. Relatively little plywood and flooring is transported by rail, and transport by truck across borders can pose bureaucratic difficulties.[430] At border crossings in the Greater Mekong Subregion (GMS), goods being trucked from one country to another are often unloaded, inspected by officials of both countries, and then reloaded onto separate trucks for the balance of the journey. Thailand requires permits and collects fees for transit of goods from Laos.[431] The Cross Border Transit Agreement signed in 1998 by Vietnam, Laos, and Thailand and later joined by Cambodia, Burma, and China was designed to streamline the land route border crossings.[432] However, not all of the annexes and protocols have been ratified by all signatories, and differences in laws and regulations continue to inhibit full implementation. That Thailand, Malaysia, and Singapore use the British left-side road convention while other ASEAN countries use a right-side driving system also presents difficulties for transport and at border crossings. Nonetheless, government officials report that progress has been made towards making border crossings smoother.[433]

Export taxes and cess fees (i.e., additional excise or product taxes) are assessed by some ASEAN members and add costs to exported plywood and flooring. Malaysia applies a cess rate of US$1.55 per cubic meter for plywood, and much higher rates for wood products made of rubberwood.[434] In 2008, Vietnam began imposing a 10 percent export tax on wood products, including plywood and flooring, made from imported wood. The tax was modified in 2009 to provide an exemption for wood boards less than 100 mm wide and 30 mm thick, which would include many solid wood flooring products.[435] In addition to excise and export taxes, unofficial payments to government officials at various levels in order to process exported goods are reportedly fairly common, particularly in the lesser-developed ASEAN countries, and can complicate export clearing procedures.[436]

Logistics services and E-commerce

Logistics services are not often used in producing and distributing plywood and flooring. These products are bulk items, most often shipped in large volumes by truck and ocean freight. Similarly, e-commerce is not widely used in the hardwood plywood and flooring business in the ASEAN countries. Businesses, particularly those in the flooring industry, use Web sites as the principal advertising medium for their products, and e-mail has clearly both expedited placement of orders and reduced the lead time for fulfillment.[437] However, the use of business-to business (B2B) electronic interchange for ordering and delivery scheduling does not appear to be significant in the industry.

6. Healthcare: Healthcare Services

Healthcare Overview[438]

The ASEAN Roadmap for Integration of the Healthcare Sector (Roadmap) covers five industries, four of which involve healthcare products. However, increased trade and industry development are making healthcare services an increasingly important sector of the ASEAN healthcare sector. For this reason, this chapter begins with a brief survey of the healthcare products covered by the Roadmap, but focuses primarily on healthcare services. Comparable data on trade and investment in healthcare services do not exist in the same way as for healthcare goods; hence, the trade and other data presented in this overview pertain only to healthcare goods, unless otherwise noted. The discussion of trade in healthcare services has been approached using the number of foreign patients that have received treatment as a proxy for cross-border trade, and data on individual investment deals have been used in the absence of aggregate investment statistics.

The healthcare priority sector,[439] as defined by the Roadmap, has grown in the past five years due to growing demand for healthcare goods and services, both inside and outside the ASEAN region. In 2008, trade within ASEAN of healthcare goods covered by the Roadmap[440] reached an estimated $2 billion (table 2.4).[441] In that same year, the world exported $9.3 billion of such goods to ASEAN and imported $10.2 billion of healthcare goods from ASEAN countries. Intra-ASEAN exports of these goods increased at an average annual rate of approximately 23 percent per year, compared to 30 percent annual growth in world imports of such goods from ASEAN.

ASEAN members have made limited progress towards regional integration of the healthcare goods sectors, as measured against the Roadmap. Measures set forth in the Roadmap that have thus far been achieved through the integration effort, which is led by Singapore, include progress in developing regional regulatory standards for healthcare goods (a primary focus of the Roadmap) and lowering barriers to trade.[442] When fully integrated, the region will have common standards for each of the four healthcare goods subsectors; presently, development of ASEAN-wide standards and regulatory harmonization is limited to the cosmetics industry.[443] However, members have reiterated their commitment to achieving harmonization in the remaining three healthcare goods subsectors, the markets for which have remained fragmented due to wide variation in national regulations.[444] Relatively more progress has been made in lowering trade barriers in the region. Most of the region's Common Effective Preferential Tariff (CEPT) rates were significantly lowered by 2003, and as of 2008, CEPT rates for most priority healthcare goods were at zero.[445]

Extra-ASEAN exports of healthcare goods listed in the Roadmap have remained competitive as governments have leveraged advantages in related, technologically intensive industries—specifically electronics and chemicals—to become an attractive production location for multinationals from the United States and Europe.[446] Although Asian manufacturers have traditionally focused on low-end medical devices and supplies, in recent years, governments such as Malaysia and Singapore have promoted the production of more technologically advanced products,[447] and extra-ASEAN exports have increasingly been higher-value products; pharmaceuticals and medical devices had the largest export growth from 2004 through 2008, as measured by world imports from ASEAN.

Healthcare goods covered by the Roadmap have become competitive within ASEAN as well, largely due to tariff reductions. In contrast with ASEAN's extraregional trade, the bulk of intraregional trade in healthcare goods—and growth in such trade[448]—has remained in lower-value products, primarily cosmetics or personal care items such as shampoos, soap, and deodorants. Expansion and privatization within the regional healthcare services industry, and lower prices due to the reduction of the CEPT rates have made regional products more competitive.

In recent years, the traditional investment destinations among the ASEAN countries—Malaysia, Singapore, and Thailand—have continued to attract investors from the United States, Europe, and Japan.[449] These countries' governments have promoted foreign investment in healthcare by offering tax exemptions and other incentives to foreign manufacturers that enter their healthcare industries.[450] However, as ASEAN has become more competitive globally, investment in the production of healthcare goods by other Asian countries (China, India, and South Korea) has increased and has targeted new destinations, such as Indonesia and Vietnam. These countries have attracted investment due to their rapidly growing pharmaceutical and medical device markets, as well as lower production costs, particularly as such costs rise in China.[451] For example, Vietnam's liberalization of investment restrictions upon its accession to the World Trade Organization (WTO) in 2007, combined with rapid economic growth and the government's investment in healthcare services infrastructure, led to increased investment in both its medical device and pharmaceutical markets—58 foreign-owned pharmaceutical businesses were established in 2007 alone.[452]

The healthcare services industry is the fifth segment of the healthcare priority sector. Healthcare services encompass a wide range of services and treatments provided by medical professionals and healthcare facilities such as hospitals, clinics, or laboratories. Services include inpatient services (services provided by hospitals to in-house patients), outpatient medical treatments provided by clinics, and other related medical services, such as pathology services provided by laboratories. Trade in the ASEAN healthcare services industry is estimated to be smaller than trade in the four goods industries included in ASEAN's healthcare priority sector. For example, in 2007, the world (including ASEAN countries) imported US$7.1 billion of priority goods from Singapore; in comparison, Singapore's healthcare industry provided services to 348,000 patients, valued at S$1 .7 billion (US$750 million).[453] However, the designation of healthcare services in ASEAN's healthcare priority sector reflects the industry's economic potential. Healthcare services generate export revenue,[454] attract foreign investment, and can support development in other industries by providing the necessary support services to attract highly skilled, educated employees.

Healthcare Services[455]

Key Findings

The initial stages of integration in healthcare services in the ASEAN region have largely been driven by increased trade and investment in the healthcare services industry.[456] In the ASEAN region, much of the investment in, and exports of, healthcare services have been undertaken by facilities belonging to healthcare groups, which are networks of hospitals and other healthcare facilities owned by one entity that share resources and coordinate care throughout the region. Private healthcare providers[457] in Malaysia, Singapore, and Thailand have been able to capitalize on the comparative advantages in their sectors, such as the availability of low-cost labor and highly skilled medical practitioners, and the liberalization of investment barriers to provide competitively priced healthcare services to foreign patients and meet increasing demand in the regional market. In the last 5 to 10 years, demand for quality healthcare services in Cambodia, Indonesia, and Vietnam has continued to grow, encouraging providers in Malaysia, Singapore, and Thailand to invest in these markets in order to better serve the local population. As with healthcare goods, the opportunities presented by economic growth, liberalization of investment regulations, and government incentives for investment in the healthcare industry have attracted a growing volume of international investment in regional healthcare groups.

As more patients have traveled across borders within the ASEAN region to receive healthcare, and investment in healthcare services has grown, governments have acted individually—and through ASEAN, collectively—to develop programs and policies to support industry growth and integration. Certain national governments, such as Brunei and Singapore, have entered into bilateral agreements regarding healthcare services, intended to further strengthen trade and investment relationships. They have also implemented programs simplifying procedures for the movement of foreign patients across borders. ASEAN has facilitated additional regional integration in the healthcare services industry by developing the Roadmap to coordinate country efforts, establishing the ASEAN Framework Agreement on Services (AFAS) to secure commitments on healthcare services, and initiating the effort to harmonize licensing standards and qualifications for healthcare services workers across countries.

Background

Private sector[458] participation across all ASEAN healthcare markets has increased in recent years, although healthcare systems in the region vary in terms of their private- public mix (table 6.1).[459] In Brunei, Malaysia, the Philippines, and Thailand—the largest healthcare markets in the ASEAN region as measured by per capita healthcare expenditure—the private healthcare sector plays a larger role; for example, in Singapore, 7 of its 13 hospitals are private facilities.[460] Vietnam, which opened its healthcare market to private facilities in 2003, had 20 private hospitals in 2008, two of which were foreignowned.[461] In contrast, Burma and Laos only recently opened their markets to private providers, and so continue to have largely public healthcare systems.[462] Healthcare systems in Cambodia and Indonesia are in transition as well. Both of these countries, due either to lack of infrastructure or to patient preference, meet excess demand by importing significant volumes of healthcare services from the more developed markets in the region.[463]

Only a few countries export large volumes of healthcare services, although all ASEAN countries trade at least some small amount of medical services.[464] Many ASEAN governments are promoting their healthcare exports (i.e., their medical travel industries). Governments in Malaysia, the Philippines, Singapore, Thailand, and Vietnam have all announced initiatives to promote their healthcare industries to foreign patients.[465] However, of the ASEAN countries, Singapore, and Thailand are the most competitive in the global healthcare services market, as measured by numbers of foreign patients they receive, although Malaysia's industry has exhibited rapid growth in recent years. Singapore and Thailand's largest competitors are largely in Southeast Asia, as the global market is fragmented regionally and consumers prefer local care.[466] Outside of the region, India is Asia's largest exporter of healthcare services, based on the number of foreign patients treated, followed by South Korea. China and Taiwan, too, are working to increase their exports of healthcare services.[467]

Table 6.1. Healthcare Services: Healthcare Systems of ASEAN Countries, Latest Year Available

			No. of physicians		No. of hospitals		
Indicator	Health expenditure per capita, PPP (constant 2005 s$), 2007	Out-of-pocket spending (percent of total healthcare spending), 2007[a]	Public	Private	Public	Private	Hospital beds per 1,000 people
Brunei	1149	99	506 (2008)	58 (2008)	4 (2008)	2 (2008)	2.8 (2006)
Burma	26	95	7,976 (2007)	13,823[b] (2007)	839 (2007)	(c)	0.6 (2006)
Cambodia	108	85	(c)	(c)	(c)	(c)	0.1 (2004)
Indonesia	81	66	(c)	(c)	(c)	(c)	0.2 (2002)
Laos	84	76	(c)	(c)	145 (2005)	0 (2008)	1.2 (2005)
Malaysia	604	73	15,096 (2008)	10,006 (2008)	137 (2008)	209 (2008)	1.8 (2007)
Philippines	130	84	3,047 (2007)	(c)	701 (2007)	1,080 (2007)	1.1 (2006)
Singapore	1643	94	4,297 (2008)	3,051 (2008)	6 (2008)	7 (2008)	3.2 (2007)
Thailand	286	72	15,343 (2004)	3,575 (2004)	927 (2004)	292 (2004)	2.2 (2002)
Vietnam	183	90	(c)	(c)	974 (2008)	20 (2008)	2.6 (2008)

Sources: World Bank, WDI database (accessed May 3, 2010); Ministry of Health Brunei, *Health Information Booklet 2008;* Ministry of Health Burma, *Hospital Statistics Report 2007;* World Health Organization, "Lao People's Democratic Republic," 2009; Ministry of Health Malaysia, *Health Facts 2008* ; Philippines Department of Health, "Health Facilities and Government Health Manpower 1995–2006"; Ministry of Health Singapore, "Health Facts Singapore"; Thailand National Statistical Office, *Statistical Yearbook Thailand 2007;* General Statistics Office of Vietnam, "Statistical Handbook of Vietnam 2009"; *International Medical Travel Journal* , "Indochina," November 20, 2008.

[a] Out-of-pocket expenditure is any direct outlay by households, including gratuities and in-kind payments, to health practitioners and suppliers of other healthcare goods and services. It is a part of private health expenditure. World Bank, Data Indicators, n.d. (accessed June 24, 2010).

[b] Data include cooperative doctors.

[c] Data not available.

While it is difficult to compare the competitiveness of ASEAN healthcare providers with extra-ASEAN providers,[468] anecdotal evidence and accreditation statistics suggest that ASEAN providers, particularly Singapore and Thailand, are making strides in becoming globally competitive. Accreditation statistics reported by the Joint Commission International

(JCI) (table 6.2), which list the hospitals that have voluntarily pursued an international standard of hospital quality, indicate that as of 2010, 31 hospitals in the ASEAN region have received JCI accreditation, including 13 in Singapore and 9 in Thailand.[469] These statistics also illustrate the rapid growth of healthcare services in other developing regions; for example, the Middle East increased its number of JCI-accredited hospitals from 7 in 2005 to 76 in 2010. However, industry representatives indicate that although the Middle East has vastly improved its healthcare infrastructure, Middle Eastern patients continue to travel to ASEAN (largely Thailand) for treatment, due to the high cost of quality care at home.[470]

Regional Integration, Export Competitiveness, and Inbound Investment

Leading Competitive Factors

The competitive landscape of the healthcare services industry in the ASEAN region has undergone substantial change in recent years, reflecting industry growth driven by increased demand and competition in the export market as well as new opportunities in developing markets. The primary driver of three related phenomena—regional integration, export competitiveness, and inbound investment—has been the growth of private healthcare groups. Following the 1997 Asian financial crisis, Thailand, followed by Singapore and Malaysia, promoted healthcare service exports (provision of services to foreign patients) to fill excess capacity in the market caused by a decline in domestic demand, as well as to capitalize on cost advantages stemming from the devaluation in regional currencies.[471] In the past five years, the healthcare export market has become increasingly competitive as new providers have entered the market. Competition in the market is based on price and quality, with firms seeking to differentiate perception of their services through marketing. Other factors which have contributed to gains in export competitiveness include government policies and programs, and the Mutual Recognition Agreements (MRAs) negotiated through ASEAN.

Private healthcare providers have invested aggressively both to expand beyond their domestic markets and to enter neighboring markets with growing demand for high- quality healthcare services. Regional expansion has drawn investors from outside the ASEAN region, as these indigenous healthcare groups offer stable investment opportunities through which investors can access the ASEAN market. The rising wealth, growing populations, and increasing demand for healthcare services in these markets provide attractive opportunities for investors.

Private healthcare groups have played an important role in the rise of regional integration throughout the region. New entrants to the market have driven providers to offer higher-quality services, increasing export competitiveness. Similarly, intra-ASEAN investment has risen as private healthcare groups look to regional expansion for new revenue streams. The development of the marketplace for healthcare services exports has made patients more likely to travel throughout the region to receive treatment. Other factors which have contributed to the deepening integration are shared economic growth; improved transportation and telecommunication linkages; bilateral agreements on healthcare between governments; and the liberalization of regional investment rules under the AFAS.[472]

Table 6.2. Healthcare Services: Number of JCI-Accredited Hospitals, 2005 and 2010

Region	2005	2010
Asia	15	65
ASEAN	11	31
Other Asia	4	34
Europe	38	91
Middle East	7	76
The Americas[a]	4	26

Source. Joint Commission International, "Joint Commission International (JCI) Accredited Organizations," n.d.

Note. 2005 statistics are based on accredited hospitals as of December 2005; 2010 statistics are accurate as of April 2010.

[a] The Americas region does not include the United States.

Regional Integration

As defined by the Roadmap, regional integration encompasses free movement of patients across borders as well as liberalized investment policies.[473] As mentioned above, much of the integration achieved has been driven by the development and expansion of regional private healthcare groups, but governments and the ASEAN Secretariat have established the necessary policies to support progress achieved by individual firms. Anecdotal evidence suggests that few regulatory barriers prevent patients from seeking healthcare services abroad within the region; indeed, under the AFAS, the seven members that made commitments on hospital services have completely liberalized consumption abroad.[474] Improved transportation and telecommunication linkages and revised visa procedures help patients move across borders in growing numbers. With respect to investment, new commitments to liberalization in the AFAS, implemented in 2009, have been augmented by unilateral and bilateral initiatives.

Strong economic growth in member economies, particularly in skilled or technical sectors such as electronics, has attracted populations of higher-income employees and has increased demand for high-quality healthcare services in the region, along with more cross-border travel by patients.[475] Additionally, increased per capita income[476] has changed the types of services demanded. Demand for higher-quality preventative and elective healthcare has increased, as has treatment for noncontagious afflictions, such as diabetes and hypertension.[477] ASEAN healthcare providers have addressed this growing demand by increasing their regional presence, through both expansion and marketing. For example, Malaysia's KPJ Healthcare Berhad has established a network of facilities throughout Malaysia and Indonesia, and has advertised throughout the region, targeting markets such as China and Vietnam.[478]

Intra-ASEAN travel by these growing populations of medical travelers has been facilitated by increased capacity in interregional air transportation, and may benefit in the future from more efficient visa procedures. In 2005, the ASEAN transport ministers initiated a plan—the ASEAN Open Skies agreement—to accelerate the liberalization of air travel. This agreement allowed regional budget airlines to enter the market and offer more flights between ASEAN destinations, such as Kuala Lumpur and Singapore or Kuala Lumpur and Rangoon (Burma).[479] Additionally, one of the initiatives set forth in the Roadmap specified reduced visa requirements to simplify the cross-border movement of patients within the ASEAN region,[480] although it is unclear if advances have been made on this issue.

By providing the organizational framework (through the AFAS) for regional cooperation and liberalization, ASEAN appears to have made progress in facilitating both increases in overall healthcare investment and further regional integration. Historically, governments have tended to maintain barriers to foreign healthcare establishments because they are often perceived as infringing on national sovereignty.[481] However, the seventh package of commitments in services in the AFAS, which was implemented in 2009, substantially liberalized commercial presence restrictions, largely removing equity limitations.[482] For instance, some countries such as Indonesia have liberalized regulations to permit foreign firms to hold majority shares; in Indonesia, foreign equity up to 51 percent is permitted. Vietnam also liberalized its regulations to allow 100 percent foreign ownership, although there is a minimum capital requirement of $20 million for hospitals.[483] Given the relatively recent implementation of these commitments, it is likely that the full impact of the rule changes on regional investment flows have not yet been captured in ASEAN member data.

In addition, member governments have begun efforts to support this process by participating in bilateral agreements on healthcare to encourage regional trade and investment. For example, Brunei signed memorandums of understanding (MOUs)[484] on healthcare training and healthcare services with Singapore in 2004 and 2007, respectively, and a MOU on healthcare services with Thailand in 2010.[485] Singapore also has a healthcare MOU with Cambodia.[486] These MOUs are intended to increase the cooperation in healthcare services between the respective countries across a number of areas, including training, research, investment, and the spread of infectious diseases.

Export Competitiveness

Thailand, Singapore, and Malaysia are the ASEAN region's leading exporters of healthcare services. In 2008, Thailand reported the largest volume of medical travelers among ASEAN member countries (table 6.3),[487] with the majority coming from outside the ASEAN region. Thailand's largest export market is reported to be the Middle East, accounting for over half of its foreign patients, although it also receives patients from developed nations such as Japan and the United States (5 percent and 3 percent, respectively).[488] Neighboring Burma is reported to be Thailand's largest source of patients in the ASEAN market, accounting for 5 percent of medical travelers.[489] In contrast, Malaysia and Singapore largely export healthcare services to ASEAN neighbors.

Singapore's largest export markets are Indonesia and Malaysia,[490] although Singapore has entered into bilateral agreements with certain Middle Eastern countries, such as the United Arab Emirates and Bahrain, to diversify its export base.[491] The ASEAN market accounts for 85 to 90 percent of Malaysia's exports of healthcare services, as patients from Indonesia, Vietnam, and Cambodia travel to Malaysia for care, while the Middle East and developed countries account for the remainder.[492]

The market for healthcare services exports within the ASEAN region has become more competitive in recent years; regional healthcare groups have sought to differentiate their services by adopting marketing strategies based on cost and perceived quality.[493] Malaysia, Singapore, and Thailand all have low labor costs relative to developed countries, which allow them to specialize in healthcare services exports.[494] Each has found a way to differentiate its services within the ASEAN market based on a combination of cost and quality. Thailand has positioned its healthcare exports as low-cost, high-quality care in a luxury setting, targeted at the extra-ASEAN market,[495] while Malaysia has focused largely on price-sensitive patients seeking quality care within the ASEAN region.[496] By contrast, Singapore has focused on exporting specialty care, such as cancer and heart treatment; this focus allows Singapore's medical professionals to maintain expertise in specific specialties and address excess capacity in Singapore's healthcare infrastructure, such as available hospital beds and underutilization of advanced medical technology.[497] Additionally, by creating a niche market, Singapore's exports command premium prices, as the country's industry has positioned itself to compete more on quality than on price, avoiding direct price competition with other regional providers.[498] Malaysia's rise as a destination for low-cost healthcare exports has provided the largest challenge to Singapore, particularly as the two countries are neighbors.[499] Patients from Indonesia, Vietnam, and Cambodia travel to Malaysia seeking more developed healthcare services infrastructure, while Singaporeans travel to Malaysia seeking lower prices.[500]

ASEAN member governments have aided their domestic exporters of healthcare services by implementing marketing campaigns and other initiatives to promote the industry (box 6.1). For example, the governments of Singapore, Malaysia, Thailand, and the Philippines have established programs, most often in conjunction with tourism authorities, which market directly to potential patients through online marketing campaigns.[501] Additionally, some governments have begun to offer incentives to encourage healthcare providers to export. For example, the government of Malaysia has established an institutional framework to facilitate exports, comprising nationwide pricing guidelines, an accreditation system, tax incentives, and promotional activities.[502]

Table 6.3. Healthcare Services: Exports of Healthcare Services, by Estimated Numbers of Foreign Patients, 2004–08

Country	2004	2005	2006	2007	2008
Malaysia	174,189	232,161	296,687	341,288	374,063
Singapore	320,000	374,000	410,000	348,000	(b)
Thailand	1,103,905	1,249,984	1,450,000 (a)	(b)	1,300,000

Sources: Association of Private Hospitals Malaysia, cited in Malaysia-German Chamber of Commerce and Industry, *Market Watch 2010* , 2010, 7; *Hospitals.sg,* "Singapore's Medical Tourism Figures Revealed by Health Minister," January 28, 2009; Yap, "Medical Tourism and Singapore," 2006/2007, 025; Tourism Authority of Thailand, "Thailand Projects Around 2 Million Visitors for Medical Services," n.d.; *International Medical Travel Journal* , "Thailand: Thailand to Boost Medical Tourism in 2010," January 20, 2010.

Notes. Singapore and Thailand do not report annual statistics on the estimated number of foreign patients treated. Instead, estimates for Singapore in 2004 and 2005 are from Yap, "Medical Tourism in Singapore;" 2006 and 2007 are from the Health Minister, as reported by *Hospitals*.sg. Estimates for Thailand are from two reports, both citing Tourism Authority of Thailand estimates. Data are not available for Singapore and Thailand for the years 2008 and 2007, respectively.

[a] Data are likely underestimated, as specific statistics are not provided. Available data cite over 1,450,000 foreign patients treated in Thailand in 2006.

[b] Not available.

Box 6.1. Healthcare Services: Profile of ASEAN Integration

In the ASEAN region, Singapore and Malaysia have moved furthest in terms of integrating healthcare markets. Patients move freely between the two countries, and the industries have strong investment ties. Geographic proximity and a shared history of British colonization have promoted integration, as has the complementary specialization of the two markets.[a] Singapore specializes in higher-cost, complex medical procedures, whereas Malaysia focuses on more routine procedures at a competitive price.[b] Consequently, Singaporeans in need of routine medical procedures increasingly travel to Malaysia for care, whereas Malaysians in need of more complex medical procedures travel to Singapore. The governments in these countries have supported increased trade in healthcare services by implementing policies facilitating medical travel. For example, Malaysia has reportedly implemented a "green lane" system to facilitate customs clearance for medical travelers at major entry points, including the Singapore-Johor causeway.[c]

Singapore and Malaysia also enjoy a strong investment relationship.[d] Parkway Holdings, one of the largest healthcare providers in Singapore, owns majority shares in two hospitals in Malaysia and is a minority shareholder in Malaysia's Pantai Hospital Group through a joint venture with the Malaysian government.[e] The Singaporean and Malaysian governments make efforts to support these bilateral investment flows, such as the Malaysian Minister of International Trade and Industry's trade mission to Singapore in 2008.[f]

Singapore's recent initiative permitting portability of healthcare financing is expected to deepen integration of the two markets.[g] In 2010, the Singapore Ministry of Health announced that beginning on March 1, Singaporeans would be able to use their Medisave[h] funds to pay for medical treatment abroad at approved facilities.[i] Industry representatives indicated this is expected to provide a large benefit to Malaysian healthcare providers.[j] The expansion of the Medisave program also allows the estimated 100,000 Singaporean workers who live in Malaysia to use their Medisave funds in their country of residence.[k]

[a] Industry representative, interview with USITC staff, Bangkok, Thailand, March 17, 2010.

[b] Industry representatives, interviews with USITC staff, Singapore, March 8 and 9, 2010.

[c] Industry representative, interview with USITC staff, Singapore, March 8, 2010; Association of Private Hospitals Malaysia, "Health Tourism in Malaysia," October 2008.

[d] Data on bilateral investment in healthcare between Malaysia and Singapore do not exist; assertions of the investment relationship are based on anecdotal information.

[e] ParkwayHealth Web site, "Our Hospitals," http://www.parkwayhealth.com/hospitals/ index.asp.

[f] International Enterprise Singapore, "Annual Report 2008/2009," 23.

[g] Industry representatives, interviews with Commission staff, Singapore, March 8 and 9, 2010.

[h] The Medisave program is a mandatory health savings account that Singaporean workers contribute to for healthcare expenses. The Medisave program, in conjunction with a catastrophic health insurance plan, are the Singaporean equivalent of national healthcare. Singapore government, Central Provident Fund Board, "Health Financing Framework in Singapore," n.d.

[i] Straits Times, "Use of Medisave Overseas," February 10, 2010.

[j] Straits Times, "Use of Medisave Overseas," February 10, 2010; industry representative, interview with USITC staff, March 9, 2010.

[k] Industry representatives, interviews with USITC staff, Singapore, March 9, 2010.

Few initiatives have been adopted by ASEAN to liberalize exports of healthcare services because, as mentioned above, few barriers to cross-border trade exist. However, under the Roadmap, MRAs have been negotiated to coordinate and harmonize licensing qualifications and standards for medical professionals among the 10 countries and to facilitate movement of professionals throughout the region.[503] MRAs for nursing, medical, and dental professionals are being implemented, and coordinating committees are working to develop a framework for core competencies and standards, relevant laws and regulations, entry procedures, and other related issues for each profession.[504] The MRAs are expected to increase the competitiveness of ASEAN healthcare service providers by expanding the pool of skilled labor.[505]

Inbound Investment

Both international and intra-ASEAN investment in healthcare in ASEAN have reportedly increased in recent years, although limited data exist on regional investment flows.[506] Anecdotal evidence suggests that intra-ASEAN investment has grown as regional healthcare groups have established operations throughout the region; at the same time, most extra-ASEAN investment has entailed private equity investors taking stakes in ASEAN private healthcare groups and facilities.[507] Leading ASEAN regional healthcare groups, largely from Malaysia, Singapore, and Thailand, have expanded into their developing-country neighbors, such as Cambcodia, Indonesia, and Vietnam, often by acquiring minority shares in existing facilities.[508] For example, KPJ Healthcare (Malaysia) and ParkwayHealth (Singapore) have both invested in Indonesia, and Thailand's Bangkok Hospital has established regional affiliates in Cambodia, Burma, and Vietnam.[509] At the same time, many of these leading ASEAN healthcare providers have attracted investment from outside of ASEAN. For example, in 2005, Newbridge Capital (United States) acquired a minority share of Singapore's Parkway Holdings, the parent company of the Parkway Hospital groups, for S$3 11 million (US$131 million).[510] Similarly, in 2008, U.S.-based Lombard Investments acquired a minority stake in the Philippine's Professional Services Inc., the owner of a private hospital.[511]

As mentioned earlier, growth in intra-regional investment has been driven by expansion strategies adopted by local healthcare providers, intended to maintain relationships with their patient populations and continue revenue growth.[512] In recent years, traditional medical travel destinations such as Singapore and Thailand have seen increased competition for foreign patients in the region[513] as more ASEAN countries export services and develop the infrastructure necessary to meet local demand.[514] Additionally, providers in Singapore and Thailand have watched as growing populations and economic growth have fueled demand for healthcare in neighboring markets. ASEAN healthcare groups have responded to the changing business environment by expanding their regional presence. Many have invested in hospitals outside their home market,[515] but more frequently they have established clinics or representative offices throughout the region. For example, while ParkwayHealth owns hospitals in four ASEAN countries (including Singapore), it has established patient assistance centers (which act as referral centers) in 19 countries, including in eight of the ASEAN members (except Laos and Thailand).[516] These patient assistance centers, such as the one in Vietnam, serve to expand Parkway's brand in a fast-growing new market, while avoiding issues such as infrastructure investment or the shortage of skilled workers by referring new patients back to Parkway's hospitals.[517] Similarly, Thailand's Bumrungrad Hospital has established 28 representative offices around the world, which promote Bumrungrad's services and refer patients to the Thai facility.[518] Adopting such investment strategies has allowed ASEAN healthcare providers to establish foreign commercial presences at lower cost, while also supporting hospital facilities in their home markets through patient referrals.[519]

While the ASEAN market has attracted growing investment in recent years, regional investors have reported that some measures continue to impede investment. Remaining equity limitations continue to reduce the region's attractiveness, because providers prefer to enter a market with full ownership, or at a minimum, majority-interest ownership, to exert more control over operations.[520] Although many countries have committed to few or no equity limitations in healthcare under the AFAS, other restrictions may prevent 100 percent foreign ownership. For example, many countries maintain restrictions on foreign land ownership. Healthcare firms initially used Real Estate Investment Trusts (REITs) to avoid the restrictions and maintain full foreign-ownership of the facilities; however, certain ASEAN member countries are reportedly implementing requirements that compel hospitals to own the land on which they operate, effectively forcing them to work with local partners.[521]

Trade Facilitation, Logistics Services, and E-Commerce

Trade Facilitation and Logistics Services

Review of current literature and interviews with industry representatives in the ASEAN region indicate that healthcare services firms have encountered few or no trade facilitation or logistics issues, and improvements in these two areas are unlikely to substantially increase the competitiveness of healthcare services exports.[522] Trade facilitation and logistics services are more frequently associated with trade in healthcare goods, rather than provision of healthcare services.

E-Commerce

Rising Internet usage in the region has expanded use of e-commerce in the healthcare services industry, which in turn has bolstered the competitiveness of healthcare services exports in the ASEAN region by increasing contact with patients and developing a more skilled workforce.[523] Many healthcare facilities have used the Internet to develop connections with patients before performing procedures or treatments, by answering questions and making advance arrangements.[524] Additionally, in countries where the governments are promoting healthcare exports, governments have established central Internet portals, generally linked to the country's tourism Web site, intended to attract future patients and direct them to participating private facilities. An online presence is vital for both private facilities and governments promoting a national healthcare services industry.[525]

Increased Internet access has also increased regional competitiveness by developing a more skilled healthcare workforce. Many healthcare facilities in ASEAN have encountered workforce shortages, as well as variation in levels of training and education. The availability of tele-education,[526] or training provided remotely over the Internet, has helped to improve healthcare workers' skills, allowing them to offer higher-quality, competitively priced healthcare services.[527] For example, in Thailand, nursing schools and public health colleges use teleconference and tele-education technologies to connect students with real-time lectures, conferences, and other events.[528] Tele-education has also offered healthcare groups in the region another avenue to expand their business and use their industry knowledge to generate revenues beyond their borders.[529] For example, a Singaporean provider provides training to students in Kazakhstan, Korea, and Vietnam using e-modules and webcams.[530]

To date, no ASEAN regional measures have been implemented to improve e-commerce specifically for healthcare services; however, initiatives underway from the ASEAN Secretariat include the MRAs for medical, nursing, and dental professionals. One aspect of the professional MRAs, which will harmonize and coordinate standards and licensing requirements across countries, is an electronic database of healthcare professionals. ASEAN member states are currently working to populate the database with information on foreign healthcare professionals operating in their markets and will eventually upload the database to the Internet to facilitate the movement of professionals throughout the region.[531]

7. Automotive: Motor Vehicle Parts

Automotive Overview[532]

The automotive industry, including the motor vehicle parts industry, is highly desired by many countries as a driver of economic growth, job creation, and technology development. The countries of the ASEAN region are no exception. They have succeeded at developing individual automotive industries over the past decades in part through the use of local-content requirements, high tariffs, investment incentives, and tax policies designed to promote and protect their respective industries. Moreover, the region's large population of 550 million has attracted investment from foreign automakers and their suppliers because of its potential market size.

To further integrate the region's automotive industries, the ASEAN Secretariat developed the Roadmap for Integration of the Automotive Products Sector (Roadmap) [533] to "strengthen regional integration efforts through liberalisation, facilitation and promotion measures to ensure full integration of the automotive sector by 2010" and to promote private sector participation. To achieve these goals, the Roadmap focuses on three groups of measures that are dedicated to expanding intra-ASEAN trade and investment, increasing the ASEAN automotive industry's technological capabilities, and improving human resources capability. Within these three groups are 60 measures largely related to tariff elimination, elimination of non-tariff measures (NTMs), customs cooperation, effective implementation of the ASEAN Industrial Cooperation Scheme (AICO)[534] and Common Effective Preferential Tariff (CEPT) schemes, improvement of the rules of origin, harmonization of standards and conformance, promotion of future investment, improvement of logistics services, enhancement of ASEAN car manufacturing capability, and establishment of a training and skill certification system. Although this Roadmap[535] is quite comprehensive in terms of initiatives and goals, no sectoral group reportedly exists to help move the Roadmap forward.[536] Some industry officials have indicated that the Roadmap is less significant now that ASEAN-6 duties have been eliminated, whereas other sources continue to see value in the Roadmap goals.

Within the broad priority sector, the automotive industry (in particular motor vehicles) is by far the most significant driver of regional trends in integration, export competitiveness, and inbound investment. The sector has achieved a considerable level of integration through the use of the AICO preferential duty system, which has allowed automakers and their parts suppliers to rationalize production throughout the region. The recent elimination of duties among the ASEAN-6 countries and reduction in NTMs are major achievements of the Roadmap that should help to further integrate the sector. However, the motor vehicle industry is still subject to numerous policies and taxes within the region, such as those imposed by the Malaysian government through its National Automotive Policy, that protect local industry and that channel investment and production to targeted motor vehicle products.

ASEAN automotive sector exports to the rest of the world more than doubled during 2004–08 to $27.1 billion (table 2.4). The four leading markets—the European Union (EU), Japan, Australia, and the United States—accounted for 65 percent of these exports. ASEAN automotive sector imports from the rest of the world grew at a slower pace (59 percent) during the period, reaching $33.3 billion in 2008, with Japan accounting for nearly one-half of these imports.

Motor Vehicle Parts[537]

Key Findings

Within the automotive priority sector, ASEAN's motor vehicle parts industry plays a pivotal role. This sector supports the region's large vehicle manufacturing industry and is a leading source of intra-ASEAN trade. In particular, the motor vehicle parts that are the subject of this analysis exhibit high intra-ASEAN trade intensity, accounting for 24 percent of regional priority sector exports in 2008. Moreover, AESAN exports of these motor vehicle parts to the rest of the world represented roughly 27 percent of comparable priority sector ASEAN exports in that same year.

The elimination of duties and the lowering of NTMs are by far the most significant achievements to date of the ASEAN automotive Roadmap in furthering integration, promoting export competitiveness, and attracting investment in the motor vehicle parts industry.[538] The ASEAN region has also been successful at implementing the AICO system,[539] a critical Roadmap target. The use of the AICO system by auto and parts manufacturers in particular allowed the industry to develop an integrated production scheme prior to the elimination of duties. Increased competition is expected within the region, which may help firms improve their export competitiveness by increasing their productivity, better managing their costs, and adding to their skill set.[540] At the same time, the more open ASEAN market may offer greater internal export opportunities for local firms.

Despite these achievements, the regional automotive industry and market have yet to fully integrate. This lack of cohesiveness is in part a legacy of national government intervention in support of local automotive industries through such measures as national car policies and local-content requirements that served to shield local markets and encourage development of small-country markets, as well as the limited progress currently being made in harmonizing safety and environmental standards and improving customs cooperation and coordination.

Further integration is likely as the region pursues Roadmap goals to reduce barriers in order to create a more seamless market. For some industry observers, however, the accelerated effort to integrate the region is considered to be too little, too late, given the extensive investment and development that is occurring in the industries of its regional rivals,

China[541] and India.[542] Others believe that those countries with well-established motor vehicle and parts industries, such as Thailand and Malaysia, have the most to gain from regional integration and product specialization.

BOX 7.1. AUTOMOTIVE INDUSTRY BACKGROUND

The motor vehicle parts industry serves two markets: original equipment manufacturers (OEMs, or automakers) and the aftermarket (e.g., dealers, retail parts outlets, repair facilities). Industry production ranges from labor intensive commodity products, such as brake rotors, to more sophisticated, capital- and R&D-intensive goods, such as wiring harnesses. Components for the OEM and aftermarket are often produced by different firms (or plants within the same firm) because of the more stringent manufacturing and material specifications required of the OEMs.

The automotive industry typically pursues a regional approach,[a] whereby automakers produce in their target markets to be responsive to local customers, reduce foreign exchange rate risk, and meet local vehicle pricing levels, for example.[b] In the initial stages of local production, automakers often export knocked down[c] kits of components and parts to the local market for assembly into motor vehicles. With greater production volumes, automakers typically shift to local manufacture and parts sourcing. Because of automakers' specific product,

According to a group within the ASEAN-Australa Development Cooperation Program when material, and quality requirements, they often purchase components from established producers with the necessary certifications, reputation, and experience to meet their quality, cost, and delivery (QCD) capability standards. Automakers' preference for sourcing many components locally to further benefit from just-in-time delivery and supplier support often means that their global suppliers will invest in greenfield operations or form joint ventures with local manufacturers.

Automakers and global suppliers also seek to develop a local supply base to help increase local content and lower manufacturing costs, and will help develop local producers' manufacturing and technical skills. Parts producers will also expand to foreign markets to diversify their customer base and tap expanding markets. In addition, local development of automotive parts allows automakers to produce and update vehicle models that are best suited to the local community. Automakers may continue to rely on imports of more high-tech components or signature systems (e.g., engines and transmissions) unless (or until) a reliable local supply base is established.

[a] Industry official, interview by USITC staff, Singapore, March 9, 2010.

[b] One parts industry representative indicated that its suppliers followed it to the ASEAN region. The firm must localize production to minimize its costs and reduce exchange rate risk. Industry official, interview by USITC staff, Singapore, March 9, 2010.

[c] The term "knocked down" essentially refers to an unassembled vehicle; with kits, all or some of the components and systems necessary to build a vehicle are packaged together and shipped by the automakers to the target market for assembly into a vehicle. Depending on the content of the kit, additional components may be necessary to procure for assembly of a complete vehicle.

By virtue of their already developed industries and large markets, these countries will attract greater investment because of the scale economies and manufacturing capabilities they offer.[543] Furthermore, complementary production of auto parts is expected to accelerate as a result of the AFTA duty eliminations, allowing scale production that would improve the regional industry's export competitiveness.[544]

Background

The ASEAN motor vehicle parts industry is concentrated in the four countries—Thailand, Malaysia, Indonesia, and the Philippines—where vehicle production is largely located.[545] The industry in Vietnam is growing but highly protected, and has yet to achieve the scale of the leading ASEAN countries. Singapore is also a major source of electronic automotive components, despite the lack of a motor vehicle industry, largely because of its historic role as a production base for many global electronics companies. The other ASEAN countries are not known to have significant motor vehicle parts industries in large part because of their lack of a large vehicle manufacturing base, technological expertise, manufacturing skills, and reliable infrastructure.

Japanese automakers and their parts suppliers have traditionally been the dominant players in the ASEAN automotive industry since their investments in the 1950s. In Thailand, for example, approximately 80 percent of auto assembling capacity belongs to

Japanese automakers. Their suppliers, many of which are located in the region to support their automotive customers, have a similarly prominent role in many of the region's industries, helping to develop the competencies of the local manufacturing industry (box 7.1).

Most of Thailand's nearly 700 Tier One[546] parts producers are members of Japanese *keiretsu* groups supplying their own customer base, whereas the majority of the domestic component manufacturers, numbering around 1,100 firms, are considered to be second- and third-tier suppliers.[547] In Malaysia, more than 350 firms manufacture automotive components, including local companies and subsidiaries of multinational parts producers.[548] Many of Malaysia's suppliers are believed to largely supply the country's leading national automakers, Proton and Perodua.[549]

The Indonesian industry totals at least 350 auto parts producers, with a focus on the domestic market.[550] Many of these firms lack state-of-the-art technology and managerial skills, and have limited financing resources.[551] The Philippine industry totals about 256 companies, one-half of which are Tier One companies with multinational connections. The remaining firms are largely Filipino-owned small and medium sized enterprises (SMEs) with limited technological capabilities and product quality issues.[552]

Regional Integration, Export Competitiveness, and Inbound Investment

Leading Competitive Factors

According to the funding component of the ASEAN-Australia Development Cooperation Program, when comparing ASEAN member countries, none reveals a comparative advantage in the production of automotive products.[553] Globally competitive automotive industry manufacturers typically possess significant financial resources, extensive R&D capability, technological expertise, and state-of-the art manufacturing skills, for example, that are not as widely available from ASEAN industry producers. However, the parts industries in Thailand

and Malaysia are probably the most advanced in terms of manufacturing competency and R&D ability.[554]

Economies of scale

Motor vehicle market size and the resulting economies of scale are particularly important for the auto parts industry, which relies on production volume to lower costs and supply numerous auto assembly plants from a small number of focused production facilities. The large combined market for auto parts is potentially a distinct advantage for the ASEAN region, creating the economies of scale necessary to manufacture components cost-effectively. Several sources cited the large regional market offered by the ASEAN countries as its leading competitive advantage,[555] noting that the ASEAN countries individually do not have large enough markets to support automakers and their suppliers.[556]

Quality, cost, and delivery (QCD) capability

Leading motor vehicle parts manufacturers generally employ a continuous improvement program for product quality, cost, and delivery that enhances competitiveness and improves margins. To qualify as a supplier to Japanese car manufacturers, for example, suppliers must be able to meet QCD standard manufacturing processes developed by Japan Automotive Standards Organization (JASO) or other national standard-making bodies in the United States or Europe.[557] Elements used to measure progress under a QCD program may include product return rate, employee productivity, stock turnover, meeting delivery schedules, equipment effectiveness, value added per person, and floor space utilization, for example.[558]

ISO certification

Motor vehicle parts suppliers must meet certain manufacturing and quality standards or certification requirements that are specified by their automaker customers. These may include International Organization for Standardization (ISO) standards such as ISO 9001 for quality management systems.[559] OEM suppliers are subject to first-party audits (self- assessments) and/or inspections by OEM personnel or independent auditors to verify compliance with OEM customer standards requirements (e.g., statistical process control).

R&D and design capability

Automakers have increasingly turned to their component suppliers for R&D and design capability to help reduce automaker costs and resource requirements, as well as improve vehicle quality and innovation. Tier One suppliers are encouraged to support cutting-edge product development and introduce new manufacturing techniques and technologies to maintain a long-term working relationship with the automakers.[560] This capability requires financial and resource commitments that are generally available only from large global suppliers.

Regional competitiveness

As previously noted, Thailand is generally considered to be the most competitive of the ASEAN countries producing motor vehicle parts. Because of its relatively open investment environment and large vehicle production base, the local industry has attracted numerous global component manufacturers. The presence of these producers in Thailand has helped to raise the overall performance level of the local industry, as has working with transplant automakers that require high manufacturing and product standards of their suppliers.

One global automaker supported this assessment, noting that auto parts makers are most competitive in Thailand, having progressed from supplying the domestic market to now supplying parts internationally. Because of its large auto assembly base, Thailand offers the level of scale economies and market size necessary to attract auto parts manufacturers and contribute to lower costs.[561] Another source indicated that Thailand is the "winner" in ASEAN, in part because it is the region's largest auto market. Automakers' main procurement activities also reportedly occur in Thailand, so it is important for component manufacturers to locate there to be in direct contact with their customers. According to this same industry source, Thailand is the future of the regional auto parts industry, as more products will be made there.[562]

Despite being the only ASEAN country to establish its own automakers and support a supply base, Malaysia's industry reportedly manufactures automotive components to the "Proton level"[563] rather than to the rigorous specifications (e.g., standards and quality) of global automakers, keeping many of its parts producers out of the global market. In fact, to enhance their export competitiveness, some Malaysian auto parts makers reportedly moved to Thailand so that they could compete globally.[564]

Many of the other regional domestic auto parts makers are believed to be smaller firms that largely lack the necessary skills and capabilities to be internationally competitive. Many have limited R&D and design capabilities and offer a narrow range of products and limited technological added value, all of which are important to supplying global automakers. Many of these parts producers are second- and third-tier suppliers, which by their nature are typically not active worldwide. A Toyota representative noted that the regional auto parts industry may have limitations on the availability of capable second- and third-tier suppliers.[565] Moreover, many of the region's small-scale parts producers are believed to supply the aftermarket, where product and manufacturing standards are not as strict.[566]

Regional integration

The existing regional integration of the auto parts sector has been achieved in part through the efforts of Japanese and foreign automakers and their suppliers within the framework of the AICO system to establish a manufacturing model that rationalizes production, thereby creating economies of scale and improved production efficiencies (box 7.2). ASEAN's success at eliminating duties and liberalizing investment in this sector has furthered this effort. With the elimination of duties on ASEAN components, further integration will likely be expected as auto parts producers will no longer be bound by the government approval process and reciprocal trading required of the AICO system. According to one U.S. trade group, the full implementation of the AFTA will result in scales of production that provide cost efficiencies, without having to establish a plant in each country.[567]

As a result, some industry sources acknowledge that industry integration will produce winners and losers in the auto parts industry.[568] One source indicated that although rationalization was part of the "grand plan" of the industry Roadmap, redistribution of plants and manufacturing would be difficult to accomplish. Obtaining agreement from the many industry players would likely pose a problem, particularly as the different levels of industrial development among the ASEAN countries make it hard to create a common integration plan.[569]

Box 7.2. Automotive: Profile of ASEAN Integration

The existing production integration in the ASEAN region has largely been driven by the use of the AICO system by foreign, particularly Japanese, automakers and parts producers. The AICO system, which was implemented on November 1, 1996, was included as an element of the automotive Roadmap to promote resource sharing among ASEAN members in order to increase industrial growth and investment, improve manufacturing scale economies, and widen the scope of ASEAN-based industries, and thus achieve greater ASEAN integration.[a]

AICO appears to have been somewhat successful in achieving its goals, at least in the automotive sector. According to the Philippine Institute for Development Studies, "The AICO Scheme has illustrated a successful integration initiative of the ASEAN as it encouraged multinational enterprises particularly those in the electronics and automotive industry to adopt efficient production networks regionwide."[b] Within this program, Japanese automakers have been able to consolidate and rationalize production throughout the region. The automotive industry has accounted for the majority of the AICO arrangements, and Thailand, Malaysia, Indonesia, and the Philippines have been the leading beneficiaries of this system, with Toyota being one of the primary movers and beneficiaries of this effort.

As an example of how the AICO system functions, Toyota was able to establish a production system whereby components from Thailand, Malaysia, and Indonesia were exported to the Philippines for assembly of certain motor vehicles (e.g., the Toyota Camry and Corolla). In return, Thailand, Malaysia, and Indonesia were obliged to import certain motor vehicle components from the Philippines for the assembly of designated motor vehicles (e.g., the Kijang and Soluna) in their respective countries.[c] In this way, Toyota was able to concentrate the production of components in individual countries and achieve the economies of scale necessary to produce cost-effectively. Mitsubishi, Honda, and Nissan used a similar approach in the ASEAN region to concentrate production of certain components in specified countries, which were then shipped to an ASEAN region vehicle assembly site. Denso (Japan), one of the world's leading motor vehicle parts manufacturers, was also able to use this system for production of a wide range of components, with shipments of parts largely traded among the same four countries.

The program came with certain limitations, however. Governments were concerned with balancing trade within the industry, which contributed to the obligations to purchase parts. Moreover, any revisions to the production and trading arrangements required approval by the parties to the agreement.[d] Additionally, each ASEAN member country applied its own methodology to the approval process, with no standardization across the region.[e]

Industry representatives generally indicated that the AICO system is no longer as important, since duties among the ASEAN-6 were eliminated as of January 1, 2010. As a result, the industry is freed from the paperwork and government approval process that were required under AICO to benefit from the duty preferences. One U.S. automaker indicated that under AICO the firm benefited from a 5 percent preferential duty rate for its automotive products traded within ASEAN, but it no longer participates in the program since duties have been eliminated.[f] Similarly, a trade group stated that the AICO is now meaningless because it only provided duty exemptions, and duties are now very low.[g]

New applications for AICO arrangements are reportedly on hold, although the program is still in effect for the ASEAN- 4 countries that have yet to fully eliminate their duties.[h] Some automakers still use the program, however, as it can be especially helpful in Vietnam.[i]

[a] Thailand in the 2000's, "Industry on the Move," n.d., 287.
[b] Aldaba and Yap, Investment and Capital Flows, January 2009, 31.
[c] Republic of the Philippines, Tariff Commission, Executive Order No. 215, February 15, 2000.
[d] Industry official, interview by USITC staff, Bangkok, March 16, 2010.
[e] Lau, "Distinguishing Fiction from Reality," 2006, 471.
[f] Industry official, interview by USITC staff, Bangkok, March 15, 2010.
[g] Association official, interview by USITC staff, Singapore, March 4, 2010.
[h] Association official, interview by USITC staff, Kuala Lumpur, March 12, 2010.
[i] Government official, interview by USITC staff, Jakarta, March 1, 2010.

Several other challenges to full integration also remain. Two difficult issues are the harmonization of automotive standards and regulations and the elimination of government policies that protect local industry, both of which are goals of the Roadmap. Moreover, although the ASEAN integration movement is creating conditions to foster greater cooperation among members, robust intraregional competition continues. As a result, the automotive industry is a particularly difficult sector in which to build consensus because individual countries (e.g., Malaysia, Indonesia, and Thailand) want to be the regional hub for the industry.[570] One association representative indicated that the region is trying "its level best" to integrate the industry, but that this process will take time.[571]

Standards/conformance

The lack of common automotive standards in the region weakens the industry's ability to integrate and compete effectively and is one of the Roadmap goals still under discussion. Because parts and vehicles must be manufactured to specific country standards rather than for a larger regional market, parts suppliers lose the benefits of economies of scale to lower costs. The harmonization of automotive environmental and safety standards throughout the region is reportedly moving slowly. Each ASEAN member country currently has its own standards, but the 10 countries are working on adopting a common set of standards for the automotive industry, using the UNECE 52 (1958) as a baseline[572] and Euro 4 standards for emissions by 2012.

The ASEAN Coordinating Committee on Standards and Quality (ACCSQ), which is instrumental to the regional effort to harmonize automotive standards, reportedly talks regularly with the private sector about automotive standards. Within ACCSQ, the Automotive Product Working Group (APWG) meets three times a year to discuss harmonization efforts, the last time in January 2010.[573] For autos and electronics, the general strategy is to first conclude a mutual recognition agreement (MRA),[574] and then to work on regulatory harmonization. With respect to motor vehicles and parts, ASEAN leaders have agreed to adopt an MRA based on the 1958 UNECE agreement, which could be signed by 2011. There is reportedly some international pressure (especially from the United States) to adopt the 1998 UNECE agreement which includes the option of self-declaration,[575] instead of the 1958 standard. This difference in approach has contributed to delays in progress toward harmonization in the automotive sector.

Thailand is considered to be the most active in the harmonization movement, likely because establishing a globally accepted standard in the region will be key to its ability to export internationally.[576] Japanese automakers are very involved in the standards discussions occurring in the ASEAN region, providing their expertise and conducting briefings (e.g., on EU standards).[577] These automakers often head the technical committees that are determining the standards regime for the region.[578] However, there reportedly is some resistance to adopting the EU standards because of the inference that local standards are somehow not up to international benchmarks.[579]

Although one group was unaware of any major obstacles to implementation of the MRA for the automotive industry, it stated that harmonizing the regulatory regime will be much more difficult because of the region's many different standards.[580] Another group noted that the costs associated with adopting a new standards regime are holding back progress.[581] Moreover, ASEAN reportedly lacks critical testing capability, which could take a number of years to establish, as well as the financial support to enforce these regulations.[582]

Government policies

The ASEAN countries have enacted a variety of measures to promote the growth and reach of their local automotive industries over the years, many of which have shielded these industries from regional and global competition and resulted in policy-led automotive production plans that limited the development of a regionally integrated industry (box 7.3).[583] According to one global parts maker, every ASEAN country is interested in developing an automotive industry,[584] largely because of their accretive impact on the economy and job creation. As a result, the industry is often considered to be a sensitive sector warranting protection, particularly for local producers.[585]

Box 7.3. Government Automotive Policies That Have been Eliminated

Numerous protectionist automotive policies were maintained by ASEAN member governments before 2005. The Philippines, for example, imposed requirements for local content and balancing of foreign exchange under the Car Development Programme (CDP) and the Commercial Vehicle Development Programme (CVDP) from 1995 through June 30, 2003. Thailand abolished local-content requirements in the automotive sector as of January 2000. Indonesia eliminated its extensive tariff and tax incentives for local content in the automotive industry in 1999, when the sector was deregulated and liberalized.

Sources: WTO, Report by the Secretariat, "Philippines Trade Policy Review," June 7, 2005, 53; WTO, Report by the Secretariat, "Thailand Trade Policy Review," October 15, 2003, 61; WTO, Report by the Secretariat, "Indonesia Trade Policy Review," May 28, 2003, 33, and "Indonesia Trade Policy Review," May 23, 2007, 82.

Many of these formal policies have been eliminated as the ASEAN member countries have come into compliance with their AFTA and WTO commitments, although some existing measures continue to limit the integration potential of the regional industry. U.S. automakers, for example, have indicated that they are hampered in the region by domestic policies that protect locally made products. These policies include domestic content-related benefits and tax structures that favor locally made products. The leading example of this protection is in Malaysia,[586] which arguably maintains the most restrictive automotive environment of the four leading ASEAN members with its National Automotive Policy (NAP).

Malaysia's NAP "aims to build a competitive automotive sector in Malaysia and promote exports of automotive products."[587] In light of its WTO and AFTA obligations, however, the Malaysian government has taken steps to liberalize its automotive industry and market, despite some internal pressure to maintain the industry's protected status.[588] In November 2009, the government issued a revised NAP aimed at liberalizing and deregulating the automotive sector in recognition of "the required transformation and optimal integration of the local automotive industry into regional and global industry networks within the increasingly liberalised and competitive global environment."[589] Although much of the NAP is focused on the motor vehicle industry, the policy identifies certain measures specific to the component sector. According to one industry source, the government wants to assist its parts producers through the NAP to make sure they survive to supply the national car brands.[590] These measures include –

- Phasing out imported used automotive products by June 2011;
- Introducing mandatory standards for parts and components;
- Promoting the production of critical and high-value-added parts and components;[591]
- Promoting development of hybrid and electric vehicles;[592]
- Enhancing the competitiveness of parts/components manufacturers through the continuation of the Automotive Development Fund and Industrial Adjustment Fund; and
- Increasing local content and enhancing the development of the Bumiputera[593] vendor program in any partnership of national automaker Proton and a globally established OEM.

Despite the continued implementation of a national auto policy, many observers expect that Malaysia's industry will continue to liberalize and improve its competitiveness. Some industry watchers see the revised NAP as significant progress towards liberalizing the Malaysian market while also acknowledging its limitations, whereas others have questioned its deregulatory impact. One industry association in the region, for example, noted that the Malaysian automotive industry is now more open, but still has doubts about the actual application of the policy and the scope of its liberalization.[594] Another group doubted the significance of the changes to the NAP.[595] One automaker indicated that it had pulled out of Malaysia, noting the NAP did not allow it to participate in the market. Moreover, its small production volume was not sufficient to allow local parts manufacturers to supply the firm.[596]

Tax incentives and excise taxes are other methods used to boost local content and shield domestic parts industries. Malaysia maintains an excise tax on vehicles[597] that favors higher local content. The value of the local content is deducted from the vehicle's taxable base before the tax is applied. Therefore, the net tax amount paid by the purchaser is lower for vehicles with higher local content, although the tax rate is the same. Such tax policies influence the direction of vehicle demand and create market distortions that filter down into the auto parts industry.[598]

Thailand is often accused of protecting its automotive industry with more informal measures. Its use of excise tax differentials to direct demand to pickup trucks, in which Thailand leads the region in terms of production, is frequently noted as a barrier to trade.[599] This approach has led foreign auto parts manufacturers to focus their investments on components for pickup trucks.[600]

Protectionist policies may help to develop component industries to support national carmakers, but they often are not sufficient to create an internationally competitive parts industry that can supply components of the quality, service, and delivery demanded by global automakers. In the case of Indonesia, for example, government protection was reportedly effective at expanding its auto parts industry, but largely in basic components rather than in the higher-technology goods.[601]

Export Competitiveness

As previously noted, the requirements to be a globally competitive supplier to the world's leading automakers are quite demanding, necessitating QCD capability and certification to automakers' manufacturing and materials specifications.[602] These qualities often require significant financial resources, extensive R&D capability, technological know-how, and state-of-the-art manufacturing skills that are not as widely available from local producers in the ASEAN region as in other leading automotive countries. However, the presence of a large number of multinational Tier One parts suppliers in the region has contributed to the competitiveness of the ASEAN auto parts industry in the OEM market. In any case, such requirements are less onerous in the aftermarket, where components do not have to meet global OEMs' stringent technical specifications and where local manufacturers are reportedly more active.

With the gradual elimination of duties by the ASEAN-6 (table 7.1) as a result of the AFTA/CEPT scheme[603] and the phase-out of many NTMs, the ASEAN region experienced greater trade both within the region and with external partners. ASEAN imports of auto parts from outside the region totaled $8.4 billion in 2008 and were dominated by Japan, which was the source for over one-half of such imports (table 7.2). Japan's leading role in ASEAN automotive production largely explains its prominent trading position. The EU and China are secondary import sources, although China's growth outpaced that of other sources over the period, more than tripling to $1.0 billion. Despite its role as a leading export market, the United States is only a small supplier to the ASEAN auto parts market, accounting for less than 4 percent of total imports in 2008.

Intra-ASEAN exports of auto parts doubled during the period, reaching $1.9 billion in 2008. Thailand was the leading intra-ASEAN export source during the period, supplying 37 percent of those exports in 2008, followed by Indonesia and the Philippines (table 7.3). ASEAN exports outside of the region more than doubled during 2004–08 to $7.4 billion (table 7.4).

Table 7.1. ASEAN Country Tariffs for Motor Vehicle Parts, 2004, 2008, and 2010 (Ad Valorem Percent)

ASEAN members	2004			2008	2010[a]
	Imports from other ASEAN countries	Imports from non- ASEAN countries (MFN)	Imports from other ASEAN countries	Imports from non- ASEAN countries (MFN)	Imports from other ASEAN countries
Brunei	0–15	0–20	0–5	0–20	0
Burma	1–5	1–5	1–5	1–5	1–5
Cambodia	(b)	15–50	5–20	15–35	5
Indonesia	0–5	0–15	0–5	0–15	0
Laos	5–8	5–10	0–2	5–10	0–2
Malaysia	0–5	0–30	0	0–30	0
Philippines	0–5	3–15	0–5	3–15	0–5
Singapore	0	0	0	0	0
Thailand	0–5	10–30	0–5	10–30	0
Vietnam	0–20	0–30	0–5	0–29	0–5

Source. Agreement on the Common Effective Preferential Tariff (CEPT) Scheme for the ASEAN Free Trade Area, Consolidated Package of Tariff Reductions for 2008, http://www.aseansec.org/19802.htm; Consolidated 2003 CEPT Package by Country, Cambodia, http://www.aseansec.org/19119.htm (accessed April 28, 2010); Royal Malaysia Customs Department, HS-Explorer (accessed April 2010); World Trade Organization, Tariff Analysis Online (accessed April 2010).

[a] 2010 MFN rates for ASEAN imports from non-ASEAN countries are not available to USITC staff for every ASEAN member country.

[b] Not available.

Table 7.2. Certain Motor Vehicle Parts: ASEAN Imports from Selected Markets, by Value, 2004–08

Country	2004	2005	2006	2007	2008	% change, 2004–08
	thousand $					
Japan	3,410,129	3,612,193	3,230,342	3,776,271	4,743,763	39
EU-27	608,011	676,844	739,148	964,049	1,093,308	80
China	311,027	426,218	505,696	776,027	1,029,389	231
United States	245,797	282,361	308,887	324,060	308,347	25
Subtotal	4,574,964	4,997,617	4,784,072	5,840,407	7,174,807	57
All other	698,584	926,597	763,444	971,528	1,206,138	73
Total	5,273,547	5,924,214	5,547,517	6,811,935	8,380,945	59

Source. WITS, Integrated Data Warehouse (accessed January 5, 2010).

Note. Data are based on ASEAN partner country exports because Brunei, Burma, Cambodia, Laos, the Philippines, and Vietnam do not report trade data in all years. "All other" export value excludes intra-ASEAN trade.

Japan and the United States were the leading export markets for ASEAN countries, accounting for 63 percent of total exports in 2008. ASEAN sector exports of $9.3 billion (tables 7.3 and 7.4) represented about 27 percent of total ASEAN automotive priority sector exports of $35.1 billion in 2008.

Table 7.3. Certain Motor Vehicle Parts: ASEAN Member Exports to ASEAN Members, by Value, 2004–08

Country	2004	2005	2006	2007	2008	% change, 2004–08
			thousand $			
Thailand	298,150	529,502	612,650	774,805	697,565	134
Indonesia	181,173	357,613	417,431	472,292	494,434	173
Philippines	246,975	310,007	304,631	311,342	385,526	56
Malaysia	143,796	159,875	170,217	224,045	211,781	47
Singapore	16,265	23,502	26,086	55,824	45,030	177
Vietnam	2,758	11,972	17,142	25,528	33,230	1,105
Laos	1,323	287	99	5,288	9,411	612
Brunei	3	40	133	64	74	2,061
Cambodia	26	0	2	21	46	77
Burma	0	12	7	8	30	12,418
Total	890,469	1,392,811	1,548,397	1,869,216	1,877,128	111

Source. WITS, Integrated Data Warehouse (accessed January 5, 2010).

Note. Data are based on ASEAN partner country imports because Brunei, Burma, Cambodia, Laos, the Philippines, and Vietnam do not report trade data in all years. Trade between these members may therefore be understated.

Although no direct evidence points to the implementation of new NTMs that could restrict trade, several industry sources have expressed some concern that as duties fall, new NTMs may be imposed to protect local manufacturing from the effects of production specialization.[604] In addition, as certain countries protect their industries, others may try to follow with similar protectionist policies.[605] One global automaker was aware of proposals for new NTMs, but acknowledged that none had been implemented. This firm expressed concern that ASEAN countries may enact measures that are not compliant with ASEAN's integration efforts, in part because the ASEAN countries do not seem to widely support the AEC 2015 agenda, which calls for progressively removing NTMs and a halt to new measures.[606] The U.S. Chamber of Commerce also noted that some governments have considered enacting domestic tax exemptions or reductions for locally manufactured vehicles or niche products over imported vehicles or parts.[607]

Ongoing efforts in the region to improve product quality and establish standards and certification systems highlight their importance to ASEAN's export competitiveness. The government of Malaysia, for example, recently announced its interest in increasing the number of export-oriented Tier One parts producers in the industry. The government plans on encouraging the local auto parts industry to export, but recognizes that these firms need to have the quality required for export markets.[608] In Indonesia, the industry is considering product and process standardization and the development of evaluation and certification systems to help local parts makers become more competitive and improve their access to international markets.[609]

Inbound Investment

The investment environment in the ASEAN region has been significantly liberalized within the ASEAN-6 countries,[610] as identified in the Roadmap, and most ASEAN member countries are actively seeking foreign direct investment (FDI) in their auto parts industries. The investment environment for the auto parts industry is, for the most part, open and unrestricted, with 100 percent foreign ownership permitted. According to the U.S. Chamber of Commerce, the creation of a single market would likely stimulate greater industry investment because of the opportunities it would offer to reach a large common market and to manufacture with scale economies.[611]

Through their investment policies, most governments are seeking ways to make it easier for foreign investors to enter and leave the market.[612] Thailand, for example, offers corporate income tax holidays, reduction or exemption of import duties on machinery and raw materials, land ownership rights for foreign investors, and other privileges.[613] Both Thailand and Malaysia are offering incentives to automakers and parts producers to jump-start their green vehicle initiatives.

Table 7.4. Certain Motor Vehicle Parts: ASEAN Exports to Selected Markets, by Value, 2004–08

Country	2004	2005	2006	2007	2008	% change, 2004–08
	thousand $					
Japan	1,559,410	1,825,210	2,352,678	2,797,953	3,265,143	109
United States	991,669	994,112	1,236,321	1,479,077	1,399,823	41
EU-27	314,467	368,167	464,996	671,189	740,260	135
China	97,225	100,495	170,942	171,656	134,683	39
Subtotal	2,962,771	3,287,984	4,224,938	5,119,874	5,539,909	87
All other	641,588	967,723	1,201,655	1,550,397	1,898,432	196
Total	3,604,359	4,255,707	5,426,593	6,670,271	7,438,342	106

Source. WITS, Integrated Data Warehouse (accessed January 5, 2010).

Note. Data are based on ASEAN partner country imports because Brunei, Burma, Cambodia, Laos, the Philippines, and Vietnam do not report trade data in all years. "All other" export value excludes intra-ASEAN trade.

ASEAN investment trends by source and destination continue to reflect the production patterns already established in the region. Because Japanese automakers and parts producers are ingrained in this region, Japan will likely continue as the leading investor in the ASEAN auto parts sector. Toyota, for example, reportedly considers the ASEAN region to be its "second mother country."[614] Japan accounted for 62 percent of the total investment in the ASEAN auto components industry, with investments of nearly $1.6 billion since 2003 in 123 projects.[615] Of Japan's 1.1 trillion yen ($10.6 billion) investment in the transportation sector worldwide in 2008 (on a balance of payments basis), ASEAN accounted for 10 percent (112 billion yen or $1.1 billion), which exceeded Japan's investments in China (nearly 102 billion yen, or $987 million).[616]

U.S. investment in the region totaled $139 million, with most of that investment occurring in the last two years in Thailand and Vietnam. Investment by Germany totaled $450 million and was more diverse in its destinations, covering all four major ASEAN auto parts-producing countries and, to a lesser extent, Cambodia, Singapore, and Vietnam. Only one project was reported by China, a $14 million project in Malaysia implemented by the Shanghai Automotive Industry Corporation.[617] In terms of completed mergers and acquisitions as reported by Zephyr,[618] investment was concentrated in Vietnam, Malaysia, Thailand, and Singapore.

Although inbound investment data specific to this industry segment are not available, investment statistics for broader aggregations within the transport industry suggest that Thailand is the leading recipient of transport-related FDI in the ASEAN region. Thailand is not only an attractive investment site because of its large established automotive industry, technological and R&D capabilities, support network, and business-friendly atmosphere, but it also serves as a doorway to other markets via its trade agreements with Japan, Australia, New Zealand, and India.[619] Of total identified investments of $2.6 billion in the ASEAN automotive components sector since 2003, investment in Thailand amounted to $1.6 billion, nearly 62 percent of the total with 114 projects. Vietnam came in second at $468 million with 27 projects. The total number of projects identified in the ASEAN region reached 189.[620]

Malaysia's focus on its national car brands has made it a somewhat less attractive investment site for global auto parts manufacturers. It is too early to tell whether the recent changes to its automotive policy will encourage the type of investment Malaysia seeks in order to raise its competitiveness to the level of its neighbor, Thailand.[621] For example, Malaysia has been seeking an investment partner for its automaker Proton but it has not been successful, in part because Proton intends to keep its current suppliers and is reportedly not open to any new suppliers.[622]

Indonesia is considered by many to be a growth opportunity for the automotive industry, given that it has the region's largest population (over 240 million)[623] and a relatively low rate of vehicle ownership.[624] The operating environment in the Philippines is open, but its industry is relatively small and lacks the economies of scale and range of capabilities that would attract significant levels of investment.

By comparison, FDI from all sources in China's transport equipment manufacturing sector amounted to $2.6 billion in 2007 alone, from a total of 1,242 projects.[625] Although not directly comparable, these FDI data clearly highlight the preeminent role played by China's automotive sector in attracting FDI relative to that of the ASEAN region. Its dominance has likely "sent a clear message that ASEAN needs to act like a single country, providing economies of scale to compete for those investment dollars."[626]

Trade Facilitation, Logistics Services, and E-Commerce

Improved trade facilitation, logistics, and e-commerce are all important to the greater competitiveness of the ASEAN auto parts industry. Advancement in these areas, however, does not appear to be moving rapidly and remains problematic in certain countries, according to some sources.

Trade Facilitation

The timely movement of automotive parts through customs is essential to reducing costs and meeting the just-in-time delivery system of the automotive industry. Customs delays and administrative complexities impose additional burdens that hamper industry competitiveness, which explains the importance of trade facilitation to the ASEAN automotive Roadmap.

Although ASEAN members have developed an itemized list of programs to improve customs procedures and trade facilitation, including the ASEAN Single Window (ASW), many industry representatives have the impression that ASEAN is making slow progress on these tasks and has not established firm concrete steps to achieve its goal.[627]

The ASW is one of the more high-profile efforts within the region to create a relatively seamless market. One multinational automaker noted that the ASW will be a big help with customs clearance issues,[628] while a leading global parts manufacturer thinks it will speed the customs clearance process for imported components. Currently the clearance time for its imported goods is two days, and its goal is one-day clearance.[629]

According to an industry official, Thailand's post-audit customs process creates an uncertain business environment.[630] Under this formal system, Thai customs authorities have up to 10 years to review a product's tariff classification and reclassify the product under the appropriate category. The violating company must then pay penalties that have accrued since the good's clearance and release. The Customs agents then split a share of the penalty award collected, which creates a conflict of interest that, despite being quite transparent, may encourage reclassification of goods. This system appears to be under review, and much progress has reportedly been made with the Thai government to modify this program.[631]

An industry official also pointed out that Thai Customs officials had a tendency to classify goods at higher duty rates than the importer, who would enter goods in classifications with lower rates when that option was available. Now that multiple-tier duties have been eliminated for intra-ASEAN trade, the problem of classification at different tariff levels is no longer experienced.[632]

Logistics Services

Logistics services are another critical element in the automotive production scheme and a key element of the Roadmap. The extensive reliance of the industry on just-in-time delivery places great importance on the ability to move components quickly to the auto assembler, whether from the local parts producer or from the port where components are imported. In some cases, logistics firms also provide sequencing to the automakers, in which parts are provided in the desired sequence to the automaker for the vehicle assembly.[633] This just-in-time delivery system often requires suppliers to locate near the automaker, thereby reducing logistics costs.[634]

In most cases, logistics issues were not cited as particularly burdensome or difficult, despite lingering infrastructure issues throughout the region. According to one automotive association, ASEAN countries recognize that regional infrastructure must be improved, but infrastructure priorities must be identified (e.g., rail, road) and then must be financed, both of which present problems.[635]

E-Commerce

It is less clear how far along the use of e-commerce is in the auto parts sector in the ASEAN region and its impact on competitiveness and integration. Although e-commerce has reportedly been introduced by nearly every ASEAN country as they build out infrastructure, the ASEAN countries operate at different levels of sophistication.[636] Many local suppliers are believed to have Web sites that advertise their products and provide basic company information, but business-to-business e-commerce may not be broadly used.[637] Multinational companies located in the region, however, generally employ e- commerce in their business operations. Several sources noted that, although e-commerce has its place, the conventional approach of personal meetings with potential parts purchasers and suppliers continues to be an important business tool.

A leading global automaker provided an example of its use of e-commerce in Vietnam. It employs only a basic form of electronic supply chain management, and this system is mostly used with suppliers located outside the country. The firm can place orders electronically, but payment/invoicing rules require them to use a hard copy. For example, for its dealerships in Vietnam, local rules require that a record of each individual transaction be provided, whereas in every other country where it operates, the company supplies a single monthly settlement with its dealers. However, the government is becoming friendlier to e-commerce. Vietnam has initiated Project 30, the government- wide effort to eliminate 30 percent of the administrative burden in transactions with the government, acknowledging that a paper-based system is not in line with globally competitive procedures.[638]

8. Agro-based Products: Palm Oil[639]

Agro-based Products Overview

The agricultural products covered by the ASEAN Roadmap for Integration of Agro-based Products (Roadmap) are limited both in number and by value, particularly when compared to the breadth of agricultural products traded within ASEAN and between ASEAN and its global trading partners.[640] The Roadmap, which primarily covers peas and beans, certain seeds, tomatoes and related products, and vegetable oils, represented only 12 percent of intra-ASEAN trade in agricultural products in 2004, declining to 10 percent in 2008.[641] Nonetheless, intra-ASEAN trade in those products, by value, increased 105 percent during 2004–08, while ASEAN external trade in those products increased 135 percent, indicating that demand for these products among ASEAN's external trading partners is growing faster than demand within the ASEAN region (table 2.4).

Among the products covered in the Roadmap, palm oil was identified as the industry that has experienced the greatest changes in terms of regional economic integration, export competitiveness, and inbound investment in recent years. The crucial factor in selecting palm oil for the profile industry was its considerable share of the total value in intraASEAN and ASEAN external agricultural trade. Even with the "negative list" exceptions on palm oil taken in whole or in part by the Philippines and Thailand (which slowed those countries' liberalization of intra-ASEAN palm oil trade), palm oil remained the dominant component of the agro-based products sector. In 2008, trade in palm oil represented almost 80 percent of intra-ASEAN trade and more than 70 percent of ASEAN external trade in the agro-based products sector defined in the Roadmap. It also represented more than 25 percent of ASEAN external agricultural products exports overall. As the world's leading producers of palm oil, Indonesia and Malaysia are the most competitive global producers and the most attractive markets for intra-ASEAN and external foreign investment in palm oil.

Although the Roadmap remains the clearest indicator of ASEAN priorities for agro-based trade integration, the ASEAN Economic Community Blueprint (Blueprint) also addresses recent developments in regional integration of agro-based products. The focus of the

Blueprint includes (1) standards and safety harmonization for agricultural products and (2) the development of ASEAN cooperation and approaches toward international bodies and agreements, such as the International Plant Protection Convention and the Codex Alimentarius Commission.[642] But the Blueprint sets targets for lowering tariffs for agricultural products only for imports from non-ASEAN countries, indicating that ASEAN countries may still view their fellow members more as competitors than as regional partners in the agro-based products sector.

Palm Oil

Key Findings

Joint decisions among ASEAN members to further regional integration, export competitiveness, and inbound investment in the palm oil industry, particularly through the Roadmap, appear to have had little impact on the structure of the industry. Instead, two groups of influential actors have had the primary effect on developments over the last five years: multinational corporations, which produce and refine palm oil and process it into finished and semifinished goods, and international groups, such as the Roundtable on Sustainable Palm Oil (RSPO), through which the corporations, among other actors, seek to influence developments in the palm oil market.

The oil palm tree can be cultivated only in a few areas of the world, centered within 10 degrees latitude north and south of the equator. In addition, oil palm fruit must be milled into crude oil quickly after harvest to maximize the quality of the oil. Consequently, only two ASEAN countries—Indonesia and Malaysia—possess the necessary climate for commercially viable palm oil growing and milling, and no significant competitors to Indonesian and Malaysian palm oil have emerged inside or outside ASEAN. The growing conditions and infrastructure development in these countries are so conducive to palm oil production that they supply most of the global demand for the product. Because of these advantages, harvesting and milling of oil palm fruit takes place largely within ASEAN. The natural limitations inherent in growing and milling oil palm fruit severely restrict most investment opportunities in ASEAN to Indonesia and Malaysia. Investments in other ASEAN countries would largely be in downstream products, such as refined palm oil or chemical derivatives, that do not require processing in close proximity to where the oil palm fruit is grown.

Most competition for palm oil customers outside ASEAN comes from other vegetable oils, such as soybean oil, but that competition is muted in South and East Asian markets because of palm oil's comparatively low price. Demand for this low-cost oil and the long experience of Indonesian and Malaysian companies with palm oil cultivation serve to drive intra-ASEAN and external investment in the palm oil value chain for both countries.

Background

Palm oil is produced from the pulp of the fruit of the oil palm tree, which is cultivated best in equatorial regions. Grown primarily on plantations but also on smallholder farms, the tree produces its fruit in bunches over a lifespan of 20–25 years.[643] Eight to 12 bunches are cut from the tree annually[644] and undergo the initial stages of processing, including steaming the fruit, stripping the fruit from the bunches, and extracting and purifying the fruit pulp, at mills usually located on or very near the plantations or small farms (figure 8.1).[645] This process, which produces the crude palm oil, begins within 24 hours of harvesting under the best circumstances, as the fruit begins to degrade quickly once cut from the tree.[646] The crude palm oil is then refined immediately nearby or transported to a refinery for refining. Many ASEAN countries refine various crude vegetable oils, including palm oil, and agricultural products at the same refinery. Refined palm oil can be consumed as cooking oil or used in a variety of processed food products or in the chemicals or personal care products industries.

The equatorial climate is one of the most defining characteristics of oil palm tree growth, with the requirements of high temperatures, generous rainfall, and fertile soil naturally limiting the regions that can grow oil palm trees. Beyond 10 degrees latitude north or south of the equator, palm oil production is not commercially viable.[647] Nearly all of Indonesia's and Malaysia's land area devoted to oil palm tree production is located within this equatorial range, while Burma, Laos, and Cambodia lie wholly outside it; elsewhere in ASEAN, only the southernmost portions of Thailand and the Philippines lie within this range.[648] Efforts to create a type of oil palm tree that would be commercially productive in climates further from the equator have not yet proved sustainable.[649] Consequently, only Indonesia and Malaysia are able to grow large numbers of oil palm trees—and they are doing so on ever-larger numbers and sizes of plantations that are becoming increasingly efficient.[650]

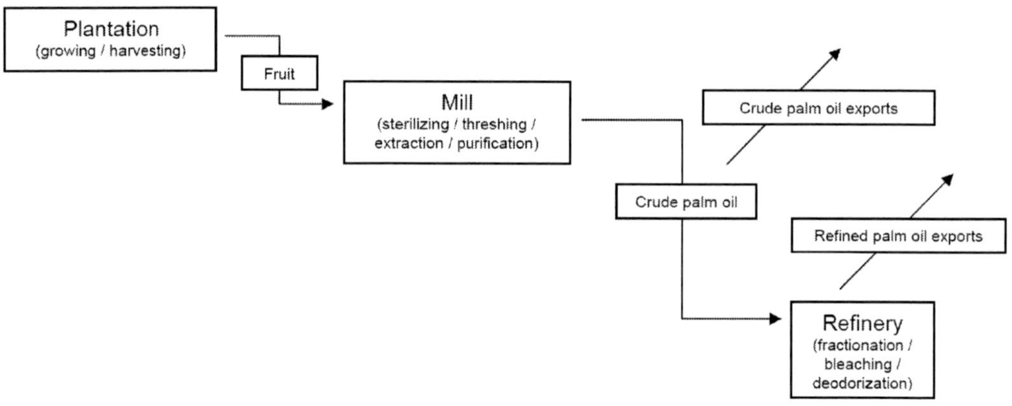

Source: Compiled by USITC staff.

Figure 8.1. Palm oil production chain.

Table 8.1. Palm Oil: Production, by ASEAN Country and Marketing Year (1,000 MT)

Country	2005/06	2006/07	2007/08	2008/09	2009/10 (est.)
Indonesia	15,560	16,600	18,000	20,500	21,500
Malaysia	15,485	15,290	17,567	17,259	18,500
Thailand	784	1,170	1,050	1,200	1,300
Philippines	61	60	65	70	70

Source. USDA, FAS, Production, Supply and Distribution database (accessed May 4, 2010).

Note. The marketing year for Indonesia and Malaysia is October to September; for the Philippines and Thailand, it is January to December.

Producers in both Indonesia and Malaysia strive to meet growing global demand for palm oil.651 Although the oil palm tree was originally brought to Southeast Asia from Africa,652 Indonesian palm oil production accounts for 46 percent of global production, and Malaysian production accounts for 41 percent (table 8.1). Thailand is a distant third, with 3 percent, while there are smaller production locations in other countries situated around the equator, such as Nigeria, Colombia, and Ghana.[653] Because of their size,[654] as well as their importance to consumers worldwide desiring affordable vegetable oil, the palm oil industries of Indonesia and Malaysia continue to be the most competitive suppliers globally and, consequently, attract the largest amounts of intra-ASEAN and foreign direct investment in palm oil.[655]

ASEAN countries produce approximately 90 percent of the global palm oil supply, but represent less than 9 percent of the world's population. Accordingly, most ASEAN palm oil exports (95 percent in 2008) go to non-ASEAN markets. The sheer size of Indonesian and Malaysian production means that their competition in the global market comes only from other vegetable oils. However, much of this competition comes from suppliers far from ASEAN, in places where other vegetable oils have a long-standing commercial presence or cultivation and marketing advantages.[656] Nevertheless, in ASEAN markets and markets throughout Asia, palm oil's low price and abundant supply, as well as its nearby production location, make it the vegetable oil of choice.[657]

Among the goods covered within the Roadmap, palm oil had the largest share by value in intra-ASEAN and ASEAN external agricultural trade. In 2008, trade in palm oil represented almost 80 percent of intra-ASEAN trade and more than 65 percent of ASEAN external trade in the products covered in the Roadmap.[658] It appears that Burma's role as facilitator of the Roadmap has not visibly changed the direction or pace of palm oil industry integration within ASEAN. Several factors likely contribute to this circumstance, including Burma's preoccupation with its domestic social and political issues; its lack of a domestic palm oil industry; and its low per capita use of palm oil.[659]

Regional Integration, Export Competitiveness, and Inbound Investment

Leading competitive factors

Refined palm oil and products derived from it can be produced in any ASEAN country, despite limitations on where oil palm trees can be cultivated and crude palm oil produced. The primary factors behind the industry's increased export competitiveness and regional value-chain integration were the increase in refining and processing capacity in ASEAN countries other than Indonesia and Malaysia (as well as other countries worldwide) to satisfy local demand; the price-competitiveness of refined palm oil against other vegetable oils in the ASEAN market; increased productivity; and opportunities for infrastructure investment.[660]

Regional Integration

ASEAN countries that do not produce enough crude palm oil to satisfy domestic market needs, such as Thailand and Vietnam, have increased their refining capacity, occasionally with foreign investment assistance from Malaysia in particular, creating new demand for crude palm oil from Indonesia and Malaysia in these countries.[661] ASEAN countries generally have also increased their food processing capacity (cookies, cooking oil) in recent years, creating demand for palm oil as an input.[662]

Table 8.2. Palm Oil: Exports by ASEAN Country, 2004–08 (million $)

	2004	2005	2006	2007	2008
		Crude and refined			
Malaysia	5,038	4,905	5,326	7,665	11,784
Indonesia	3,978	3,646	4,756	6,064	8,674
Singapore	110	266	213	273	371
Thailand	16	5	47	107	212
Vietnam	2	0	0	4	11
Other	1	4	3	10	8
Total ASEAN exports	9,145	8,826	10,345	14,123	21,060
			Crude		
Indonesia	1,899	1,778	2,548	3,446	4,641
Malaysia	890	1,104	1,394	1,619	2,083
Thailand	5	1	44	95	208
Singapore	41	128	122	183	196
Cambodia	1	1	2	2	7
Other	0	0	0	2	1
Total ASEAN exports	2,835	3,012	4,109	5,347	7,136
			Refined		
Malaysia	4,072	3,801	3,830	6,046	9,701
Indonesia	2,076	1,869	2,196	2,618	4,033
Singapore	57	137	90	90	175
Vietnam	2	0	0	4	11
Thailand	11	4	3	12	4
Other	1	2	0	7	1
Total ASEAN exports	6,219	5,813	6,119	8,777	13,925

Source. WITS, Integrated Data Warehouse (accessed January 5, 2010).

Notes. Brunei, Burma, Indonesia, and Laos do not report trade data. Figures for ASEAN exports are based on those for partner- country imports. Data for 2008 are understated, because numerous small countries and one identifiably important market (Bangladesh) have not reported 2008 import figures. Figures may not add to totals shown because of rounding.

Export Competitiveness

Trade

Indonesia and Malaysia are not only the largest ASEAN exporters of palm oil, they also dominate world trade, primarily because only Indonesia and Malaysia have sizable quantities available for export (tables 8.2 and 8.3). Indonesia and Malaysia each account for 44–45 percent of global palm oil exports; Benin and Papua New Guinea are next, with 3 percent in total. As with most of the other, smaller palm oil–producing countries, ASEAN members Thailand and the Philippines consume as much palm oil as they produce domestically, although some trade occurs.

Table 8.3. Palm Oil: Intra-ASEAN Trade, Total and Select Countries, 2004–08 (million $)

	2004	2005	2006	2007	2008
	Crude and refined				
Intra-ASEAN imports	601	315	407	386	1,012
Malaysia	356	157	282	291	614
Singapore	148	58	37	71	349
Thailand	39	9	1	1	34
	Crude				
Intra-ASEAN imports	358	156	233	206	518
Malaysia	355	152	230	205	475
Thailand	0	0	0	0	33
Singapore	2	3	2	2	10
	Refined				
Intra-ASEAN imports	237	159	174	179	494
Singapore	146	55	35	69	340
Malaysia	1	5	52	87	139
Philippines	50	85	80	21	10
Thailand	39	9	1	1	1

Source:. WITS, Integrated Data Warehouse (accessed January 5, 2010).

Notes. Brunei, Burma, Indonesia, and Laos do not report trade data. ASEAN exports based on partner-country imports. Data for 2008 are understated, because numerous small countries and one identifiably important market (Bangladesh) have not reported 2008 import figures. Figures may not add to totals shown because of rounding.

The largest markets for ASEAN palm oil lie outside ASEAN.[663] ASEAN countries devoted 97 percent of their palm oil exports to non-ASEAN markets, including four markets larger than ASEAN: China, the EU, India, and Pakistan. Note that the U.S. market for palm oil ($1.0 billion in 2008) (table 8.4) is roughly equivalent to intraASEAN import amounts ($1.0 billion in 2008) (table 8.3).

Notable differences exist between exports of ASEAN crude and refined palm oil. ASEAN crude palm oil is primarily exported to the EU and India, which have sufficient refining capacity available (table 8.3). ASEAN refined palm oil is primarily exported to China, which grows and imports soybeans and primarily devotes its refining capacity to that oilseed (table 8.4).

Within ASEAN, crude palm oil is shipped overwhelmingly to Malaysia (91.7 percent in 2008) as feedstock for its refineries and downstream processing, and refined palm oil is primarily destined for Singapore, to be processed into food products and for re-export.[664] More than half of intra-ASEAN palm oil trade is of crude palm oil, indicating the presence of refinery facilities in countries other than Indonesia and Malaysia supplying global markets with finished palm oil products.

The value of intra-ASEAN palm oil trade has increased by two-thirds during 2004–08, from $601 million to $1.0 billion, but the value of ASEAN palm oil exports to non-ASEAN markets increased far more, rising 133 percent to $20.4 billion. Tariffs are not a barrier to intra-ASEAN palm oil trade. Under ASEAN's Common Effective Preferential Tariff, import tariffs on palm oil trade within ASEAN are 5 percent or less, with the largest importers of crude and refined palm oil (Malaysia and Singapore) having duty rates of free (table 8.5). Tariffs on palm oil from sources outside ASEAN can be higher, although ASEAN countries import very little palm oil from outside ASEAN, reflecting Indonesia's and Malaysia's dominant positions as global suppliers. Nontariff measures (NTMs) do not appear to be trade impediments currently, as no ASEAN palm oil- exporting country has highlighted any restrictions of this type on its palm oil exports. However, smaller ASEAN palm oil-producing countries may be more likely to turn to NTMs to protect their less-efficient and higher-cost palm oil industries from lower-priced Indonesian and Malaysian imports now that intra-ASEAN palm oil tariffs have been reduced to free.[665]

Table 8.4. Palm Oil: ASEAN Exports to select non-ASEAN Partners, 2004–08 (million $)

	2004	2005	2006	2007	2008
	Crude and refined				
ASEAN exports to non-ASEAN partners	8,789	8,668	10,062	13,832	20,446
China	1,868	1,781	2,251	3,638	5,206
EU-27	1,447	1,474	1,756	2,427	3,866
India	1,707	1,281	1,163	1,434	2,425
Pakistan	640	752	762	1,146	1,674
USA	137	181	298	545	1,001
Total ASEAN exports	9,145	8,826	10,345	14,123	21,060
	Crude				
EU-27	790	833	1,092	1,487	2,463
India	739	829	1,068	1,336	1,909
Pakistan	56	59	159	329	592
China	3	41	259	269	521
Bangladesh	448	482	694	903	n/a
Total ASEAN exports	2,835	3,012	4,109	5,347	7,136
	Refined				
China	1,864	1,740	1,992	3,370	4,685
EU-27	657	641	664	940	1,403
Pakistan	584	694	603	816	1,082
USA	137	181	298	545	1,001
Japan	233	211	233	369	586
Russia	172	262	237	279	542
Total ASEAN exports	6,219	5,813	6,119	8,777	13,925

Source. WITS, Integrated Data Warehouse (accessed January 5, 2010).

Notes. Brunei, Burma, Indonesia, and Laos do not report trade data. Figures for ASEAN exports are based on those for partner- country imports. Data for 2008 are understated, because numerous small countries and one identifiably important market (Bangladesh) have not reported 2008 import figures. Figures may not add to totals shown because of rounding.

Indonesia and Malaysia impose export taxes on palm oil, affecting the supply and composition of their palm oil exports. The Indonesian government announced in December 2009 that its export tax on crude palm oil would be 3 percent ad valorem beginning January 2010. In the past, the Indonesian government has used its palm oil export tax to maintain domestic supply and lower prices, but it has not used these tax revenues to assist the palm oil industry.[666] The effects of these export taxes vary because the tax rate fluctuates with the international price of crude palm oil.[667]

Table 8.5. ASEAN Country Tariffs for Palm Oil, 2004, 2008, and 2010 (Ad Valorem Percent or Specific)

ASEAN members		2004		2008	2010[a]
	Imports from other ASEAN countries	Imports from non- ASEAN countries (most-favored nation rate {MFN})	Imports from other ASEAN countries	Imports from non-ASEAN countries (MFN)	Imports from other ASEAN countries
Brunei	0	0	0	0	0
Burma	1	1	1	1	1
Cambodia	7	7	5	7	5
Indonesia	0	0	0	0	0
Laos	5	10	0	10	0
Malaysia (crude)	0	0	0	0	0
Malaysia (refined)	5	5	0	5	0
Philippines (crude)	3	15	3	15	3
Philippines (refined)	5	15	0	15	0
Singapore	0	0	0	0	0
Thailand[(b)]	5	B1.32/liter or B2.50/liter	5	B1.32/liter or B2.50/liter	0
Vietnam (crude)	5	5	3	3	3
Vietnam (refined)	15	30 or 50	5	3 or 25	5

Source. Agreement on the Common Effective Preferential Tariff (CEPT) Scheme for the ASEAN Free Trade Area, Consolidated Package of Tariff Reductions for 2008, http://www.aseansec.org/19802.htm; Royal Malaysia Customs Department, HS-Explorer (accessed April 2010).

Note. Information reflects tariff rates for crude and refined palm oil imports unless specified otherwise.

[a] 201 0 MFN rates for ASEAN imports from non-ASEAN countries are not available to USITC staff for every ASEAN member country.

[b] Thailand's currency is the baht.

The Malaysian government imposes a differential export tax on palm oil to encourage the domestic production of downstream palm oil products. Malaysia uses the funds generated to support Malaysian agricultural research and development and palm oil distribution to the poor.[668] For example, neutralized, bleached, and deodorized palm olein (a refined palm oil product) is fully exempt from export tax while crude palm oil is subject to a 10– 30 percent export tax, depending on its market price. The Malaysian government waives the export tax when the palm oil is shipped to refineries having Malaysian investors, which can provide an advantage for Malaysian-owned refineries.[669]

Increased demand

Palm oil is one of the world's most competitive vegetable oils by price because of abundant supply and the low cost of production.[670] Palm oil maintains this market position despite increased demand from the largest overseas markets, such as China, the EU, and India, as well as the United States. Within ASEAN, this price-competitiveness has encouraged nonproducing countries to satisfy their vegetable oil demand through imports and producer countries to consume this natural resource.[671]

Commercial-scale palm oil production in Indonesia and Malaysia is influenced by large, ASEAN-based multinational corporations, such as Sime Darby and Wilmar, that own the regionally integrated production chain from plantation to crusher to refinery and, frequently, to factories in which the refined palm oil is further transformed (e.g., biodiesel, lubricants) or incorporated into other goods (e.g., cookies).[672] Each of these facilities may be located in different ASEAN countries, and a number of these corporations have holdings in multiple ASEAN countries, including agro-businesses in Singapore, which has no indigenous raw oil palm supplies.[673]

Palm oil is marketed throughout ASEAN under labels owned by the major producers. Sime Darby is a prime example of a fully integrated palm oil producer, growing the palm fruit on its own plantations and selling its refined palm cooking oil under its own name.[674] Many palm oil producers that do not sell their own refined oil directly to consumers under their own cross-border brand will sell refined oil, or crude oil to be refined, to domestic processors in each ASEAN member country that have established their own national brands.[675]

Indonesia's large and lower-income population generates a heavy demand for vegetable oils. As a result, that country already consumes a great portion of its domestic palm oil production as food (box 8.1).[676] The Indonesian government has, in fact, taken actions to promote domestic food consumption of palm oil for the economic benefit of its population. These actions have affected the amounts available for export in less- and more-processed forms, from crude palm oil to health care products.[677]

Increased productivity

The production of biodiesel is one area in which companies in ASEAN countries may be able to create great value by using palm oil for nonfood purposes. Indonesia and Malaysia already have well-established and growing biodiesel industries resulting from government plans to increase revenue streams from palm oil production, but even smaller palm oil-producing countries, such as the Philippines and Thailand, have biodiesel policies and production facilities that use those countries' much smaller palm oil supplies to much higher value effect than for food consumption.

Palm oil is widely used and traded throughout ASEAN, with four countries producing commercially significant amounts and all ASEAN countries consuming palm oil for food and nonfood purposes. Most ASEAN members, however, grow few or no oil palm trees, having too little land (Brunei, Singapore) or too unfriendly a climate (Burma, Cambodia, Laos, Vietnam) for commercial-scale palm oil production.

Box 8.1. Palm Oil: Profile of ASEAN Integration[A]

Natural limits restrict the ability of ASEAN nations that are not palm oil producers to become more integrated with existing ASEAN palm oil production industries. One reason is the need to locate mills close to plantations so that the mills can begin processing the oil palm fruit very shortly after harvest (usually within 24 hours) before the fruit and oil begin to spoil.[b]

Palm Oil: Domestic consumption, by ASEAN Country, Type, and Marketing Year (Thousand MT)

Country	2005/06	2006/07	2007/08	2008/09	2009/10 (est.)
Burma	304	330	365	375	390
Food	304	330	365	375	390
Industrial	0	0	0	0	0
Indonesia	4,255	4,523	4,651	4,467	4,914
Food	3,710	3,852	3,987	3,801	4,200
Industrial	465	586	579	581	629
Malaysia	2,926	3,109	2,986	2,995	3,277
Food	720	804	820	840	910
Industrial	1,986	2,080	1,936	1,920	2,080
Philippines	223	130	89	79	78
Food	200	110	68	64	63
Industrial	23	20	21	15	15
Singapore	50	50	76	105	105
Food	50	50	76	105	105
Industrial	0	0	0	0	0
Thailand	671	694	747	1,002	990
Food	388	350	397	500	500
Industrial	227	250	260	400	400
Vietnam	326	415	428	451	485
Food	326	415	428	451	485
Industrial	0	0	0	0	0

Source. USDA, FAS, Production, Supply and Distribution database (accessed May 4, 2010).
Note. The marketing year for Indonesia, Malaysia, and Singapore is October to September; for the Philippines, Thailand, and Vietnam, January to December; for Burma, September to August. Figures may not add to totals shown because of feed waste consumption.

[a] For a more detailed discussion of economic integration in free trade areas, such as that envisioned for the ASEAN Economic Community, see chapter 1. See also Sally, "Regional Economic Integration in Asia," 2010, 4– 5.
[b] USITC, hearing transcript, February 3, 2010, 43 (testimony of Rosidah Radzian, Embassy of Malaysia); government official, telephone interview by USITC staff, April 16, 2010.
[c] Brunei, Cambodia, and Laos do not have confirmed consumption amounts during 2004–08, although their proximity to price-competitive palm oil from Indonesia and Malaysia and the regional preference for vegetable oils as a part of the diet strongly suggest that these countries, at a minimum, consume palm oil as a food. For Brunei, its participation with Indonesia, Malaysia, and the Philippines in the EAGA, an initiative that encompasses production, processing, and trade in palm oil, lends support for this conclusion.
[d] Industry representatives, interview by USITC staff, Kuala Lumpur, Malaysia, March 12, 2010

Although growing oil palm trees and producing crude palm oil may not be an option for these ASEAN members, they consume it and are capable of refining crude palm oil and processing it into food (usually the majority of consumption for poorer ASEAN countries) or industrial products (see table).[c] Vietnam, in particular, has a growing refining capacity and is a draw for foreign refining investment, including from Malaysia, because of a growing domestic market.[d] Most of the palm oil refined in Vietnam is likely to be consumed domestically, as that country routinely experiences shortfalls in vegetable oil production compared with consumption. But Vietnam reportedly sold several million dollars worth of palm oil to China in 2008, indicating that, as with other, much smaller ASEAN palm oil producers, some trade occurs outside of the Indonesian and Malaysian markets.

Increases in demand for biodiesel within ASEAN countries, among other industrial and oleochemical products,[678] would result in more production facilities and similar increases in demand for palm oil supplies throughout ASEAN from ASEAN palm oil-producing countries.

Improved crop maintenance and species cultivation, better soil management techniques, conversion of existing land from other agricultural purposes, reclamation of degraded lands, and creation of new plantations from uncultivated fore stland have all contributed to higher levels and lower costs of palm oil production, increasing the amounts available for export.[679] The government of Malaysia, along with several multinational corporations, has led the effort to improve yields and shares the results of those attempts with Indonesian producers in which Malaysian producers have invested.[680] Indonesia and Malaysia have each increased the amount of cultivated land devoted to growing oil palm—Indonesia through the expansion of new planting areas and Malaysia through the conversion of existing agricultural land to palm oil production.[681]

Roundtable on Sustainable Palm Oil (RSPO)

In the absence of ASEAN policies noticeably affecting the palm oil industry and in tandem with the influence of multinational corporations,[682] the RSPO is the single most identifiable organizational influence on the ASEAN palm oil production and marketing chain and will likely grow steadily from its comparatively small volume and low company-participation numbers.[683] The RSPO includes representatives of banks, social and environmental nongovernmental organizations, and all levels of the palm oil value chain, from plantations to retailers.[684] The majority of the 400 individual participants in the RSPO in 2010 are Indonesian and Malaysian.[685] Created in 2004 to address concerns by palm oil consumers that palm oil production was harmful to the environment, the RSPO's existence is intended to encourage producers to seek RSPO certification attesting that their processes meet certain sustainable criteria. This certification would then make their palm oil more desirable in foreign markets willing to pay a price premium for such oil.

Despite their dominant position in the global palm oil industry, which limits customers' ability to source low-priced vegetable oils elsewhere, Indonesian and Malaysian producers have perceived RSPO certification to bestow economic advantages in very profitable markets over noncertified palm oil.[686] This recognition appears in the palm oil investment chain as well, as Malaysia transfers its environmentally sensitive technology and capital to plantations in Indonesia.[687] Nevertheless, only six Malaysian and three Indonesian producers had produced RSPO-certified palm oil through November 2009, and this oil represented only a small portion of each country's overall production.[688]

RSPO certification of certain ASEAN palm oil suppliers and supplies has had a demonstrable effect on sales to some major markets. Consumers in the EU and the United States are at the forefront of palm oil sustainability requirements; they have proven they are willing to pay a higher premium for such palm oil[689] and have taken action when these expectations are not met.[690] For example, Unilever, one of the world's largest corporate consumers of palm oil, stopped sourcing palm oil from two Indonesian suppliers within a span of months because of environmental shortcomings.[691] A number of multinational corporations have announced that they will use only RSPO-certified palm oil no later than 2015.[692]

Other leading markets, such as China, India, and Pakistan, are less influenced by RSPO efforts. Largely for price reasons, they have continued to provide ready markets for uncertified palm oil, hampering sustainability standardization efforts, and hindering the increase in values that certification can bring.[693] Each of these countries has a great need for low-cost vegetable oils as food for their large, poorer populations and has demonstrated a willingness to purchase palm oil that has not been RSPO certified.[694] This demand has lessened the pressure on palm oil producers to undertake the additional costs necessary for RSPO certification, and because these markets are obtaining enough of the low-cost oil they want, they have little incentive to pay higher prices for RSPO-certified oil.[695] This situation hampers the uniform application of RSPO standards and efforts by palm oil producers to increase the value of their export product.

Nonetheless, the RSPO certification structure has served to increase the export competitiveness of palm oil in markets (e.g., the EU and United States) where palm oil production is perceived as being more environmentally destructive (primarily by deforestation) than production of other vegetable oils, as well as to draw investment by companies desiring to ensure supplies of such sustainable oil. U.S. companies with a significant presence in palm oil-producing countries, such as Cargill in Indonesia, operate their own plantations adhering to RSPO standards, as do Malaysian companies transferring their environmentally sensitive production practices to investments in Indonesia.[696]

Inbound Investment

Indonesia is the destination for the largest number of palm oil investment activities among ASEAN countries (table 8.6). As the largest producer of palm oil in the world, and the country with the greatest amount of land under cultivation (7.2 million hectares in 2009), Indonesia is a natural destination for investment in establishing, enlarging, and managing palm oil plantations.[697] Investments in refining operations, on the other hand, occur in multiple ASEAN countries because construction of a refinery is not affected by the climate and growing conditions that limit where plantations are located.

The level of intra-ASEAN palm oil investment activities varies from millions of dollars to less than $100,000. The data in table 8.6 are representative of all types of palm oil industry investment activities in ASEAN, and the value trend indicates that intra-ASEAN investment activities are increasing somewhat, particularly as Malaysia continues to seek increases in the supply of crude palm oil for its refining facilities and postrefining commercial consumption.[698]

Table 8.6. Palm Oil: Intra-ASEAN Investment, Select Activities, 2004–09

Year	Investor country	Investor company	Target country	Target company	Nature of investment	Value (est. million $)
2009	Malaysia	IOI Corp.	Indonesia	(a)	Plantation services	48.8
2009	Singapore	Abaca Enterprise PTE	Indonesia	Teguh Swakarsa Sejahtera	Plantation operating services	4.3
2009	Malaysia	Kwantas Corp.	Indonesia	Kinabalu Invesdag Indonesia	Plantation operating services	2.4
2009	Malaysia	Dynasive Enterprise SDN	Indonesia	Prima Alumga	Planter	0.1
2008	Singapore	Pacific Agriculture Holdings PTE	Indonesia	PT Tunas Sejati Abadi	Plantation operating services	<0.1
2007	Malaysia	IOI Corp.	Indonesia	Bumitama Gunajaya Agro	Plantation services	82.9
2006	Malaysia	TH Plantations	Indonesia	(a)	Refinery expansion	7.0
2006	Malaysia	Ayer Molek Rubber Co.	Indonesia	PT Varita Majutama	Plantation operating services	0.6
2005	Malaysia	Golden Hope Plantations	Vietnam	(a)	Refinery expansion	13.5
2005	Malaysia	Maxillion PTE	Indonesia	PT Pukun Mandiri Lestari	Plant operations	0.1
2004	Malaysia	Delloyd Plantations SDN	Indonesia	Rebinmas Jaya	Plantation operating services	11.3
2004	Malaysia	Taipan Hectares SND	Indonesia	Rebinmas Jaya	Plantation operating services	6.6
2004	Malaysia	Ahmad Zaki Resources	Indonesia	PT Ichtiar Gusti Pudi	Cultivation services	1.8

Sources. Bureau van Djik, Zephyr database (accessed January 9, 2010); fDi Markets.

Note. This table does not include corporate mergers or takeovers involving ASEAN palm oil interests, such as the 2006 purchase by Singapore's Wilmar International of Malaysia's PPB Oil Palms for $1.1 billion.

[a] Not available.

Wealthier ASEAN countries (Malaysia and Singapore) have parlayed the initial and continuing financial benefits of processing palm oil into the production and marketing of higher-valued products. They have tended to focus their intra-ASEAN investment strategies on guaranteeing supply for their processing facilities, with purchases of plantations in the prime growing areas (Indonesia) facilitating industrial integration.[699] This is particularly important for Singapore, which has almost no domestic palm oil industry other than refining and processing facilities.[700] On the other hand, Malaysia has a more integrated palm oil industry infrastructure than Indonesia but less land for possible production expansions because Indonesia has a land mass five times that of Malaysia (and has a larger labor pool). The majority of the available land in Malaysia has already been either converted to palm oil production or protected as conservation areas.[701]

Among ASEAN countries, Indonesia and Malaysia have more liberal investment rules for foreign capital in the agricultural sector. They have encouraged investment from ASEAN (Malaysia, Singapore) and non-ASEAN (China, Japan, Korea) countries.[702] Indonesian law permits foreign ownership of companies up to 95 percent, which is beneficial for foreign investors unless they encounter difficulties finding Indonesian partners for the remaining 5 percent stake, which has occurred.[703] Malaysian law allows complete foreign ownership of companies with export-oriented production, but has restrictions on foreign ownership of land, which can hinder foreign investments and industry improvement.[704]

Most ASEAN countries, though, significantly restrict the investment of foreign capital in agriculture. Restrictions range from bans on foreign ownership of land (such as palm oil plantations) to requirements for majority local ownership of manufacturing facilities (such as palm oil refineries). The tabulation below reflects the general restrictions on land and facilities investment by ASEAN countries:

Brunei	Foreign investment stake may reach 100 percent, but the government may impose restrictions.
Burma	Foreign partner share in joint ventures must be at least 35 percent, but can reach 100 percent. The government reserves the absolute right to cultivate the land.
Cambodia	Foreign investment stake may reach 100 percent, except for land ownership.
Indonesia	Foreign partner stake must be no more than 95 percent.
Laos	Foreign partner share in joint ventures must be at least 30 percent, but can reach 100 percent.
Malaysia	Foreign investment in export enterprises may reach 100 percent. Acquisitions of land restricted.
Philippines	Foreign partner stake limited to 40 percent on public lands. Foreign investment in export enterprises may reach 100 percent. Nationality requirement may be suspended for certain ASEAN-related activities. Foreign ownership of land prohibited.
Singapore	Foreign partner stake must be no more than 49 percent.
Thailand	Foreign partner stake must be no more than 49 percent.
Vietnam	Must be in the form of a joint venture, with the foreign partner holding no more than a 49 percent stake. Foreign ownership of land prohibited.

Sources. Jenny Corbett and So Umezaki, eds., "Deepening East Asian Economic Integration," ERIA Research Project Report 2008, no. 1, chap. 5, table 6; USTR, "National Trade Estimate," 2010, 185, 289; industry representative, interview by Commission staff, Singapore, March 5, 2010.

Some ASEAN countries are considering economic policies that would effectively make their country less attractive to palm oil investment. For example, Vietnam is considering a price control measure that would affect the price of palm oil, which Vietnam imports in comparatively large quantities. Government-mandated pricing for palm oil would create disincentives for foreign companies seeking to enter Vietnam's palm oil processing and related food products industries.[705]

Trade Facilitation, Logistics Services, and E-commerce

E-commerce is an important commercial element in the palm oil trade, particularly in reference to the palm oil exchange in Malaysia. Bursa Malaysia Derivatives (BMD) is one component of Bursa Malaysia (previously known as the Kuala Lumpur Stock Exchange); it maintains most widely cited contract pricing information for palm oil and is a spot price source. Palm oil futures and options are also traded on the BMD.[706] In addition, the palm oil industry has expanded its exchange reach with palm oil futures recently beginning to be traded in Indonesia and the United States: this development is increasing the opportunities for more global traders to participate in the palm oil trade and facilitating palm oil consumption vis-à-vis other vegetable oils worldwide.[707]

The Commission found little evidence that trade facilitation issues and logistics services strongly impacted ASEAN regional integration, export competitiveness, and inbound investment in the palm oil industry.[708]

APPENDIX A. USTR REQUEST LETTER

EXECUTIVE OFFICE OF THE PRESIDENT
THE UNITED STATES TRADE REPRESENTATIVE
WASHINGTON, D.C. 20508

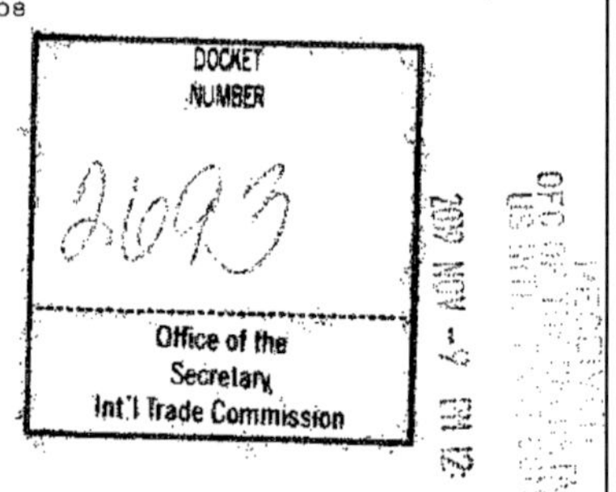

DOCKET NUMBER 2693

Office of the Secretary Int'l Trade Commission

November 6, 2009

The Honorable Shara L. Aranoff
Chairman
U.S. International Trade Commission
500 E Street, S.W.
Washington, D.C. 20436

Dear Chairman Aranoff:

In 2006, the United States and the member countries of the Association of Southeast Asian Nations (ASEAN) signed the U.S.-ASEAN Trade and Investment Framework Arrangement (TIFA). In addition to enhancing trade and investment relations between the United States and ASEAN countries, the TIFA is designed to support ASEAN's stated goal of full regional economic integration by 2015. Achieving this goal would make the region a more attractive trade and investment partner for U.S. business.

ASEAN is a key market for the United States. With robust economies and a combined population of about 550 million, the ASEAN market provides significant potential opportunities for U.S. companies. The ten-member countries of ASEAN together already comprise the fourth largest export market of the United States and its fifth largest two-way trading partner. Trade continues to grow steadily, and two-way goods trade totaled $177 billion in 2008. ASEAN integration would allow businesses to operate more efficiently in one large, regional market rather than small, fragmented ones.

As part of the ASEAN Economic Community Blueprint intended to chart and achieve economic integration by 2015, ASEAN has identified 12 priority integration sectors for accelerated economic integration. To assist U.S. trade policymakers in assessing progress in this integration and possible opportunities, we are asking the U.S. International Trade Commission (the "Commission") to provide certain information on the six sectors of most interest to U.S. exporters and investors. The six sectors, using ASEAN definitions, are: electronics, automotives, agro-based products, healthcare, textiles and apparel, and wood-based products.

Pursuant to authority delegated by the President to the United States Trade Representative under Section 332(g) of the Tariff Act of 1930, I therefore request that the Commission prepare a report that provides a brief overview of regional trends in economic integration, export competitiveness, and inbound investment for each of the following sectors: electronics, automotives, agro-based products, healthcare, textiles and apparel, and wood-based products. In addition, I request that the Commission identify ASEAN industries within these sectors that have experienced significant changes in regional economic integration, export competitiveness, and inbound investment in recent years and provide profiles of these industries in its report. To the extent possible, the report should include illustrative country information and should be based on the most recent five year period for which data are available.

Each industry profile should be concise and, to the extent possible, include the following information:

1. A description of the selected industry(ies) within ASEAN, including the industry's position relative to global competitors, as relates to export competitiveness and inbound investment flows;
2. Identification of the leading ASEAN exporting countries and their key markets;
3. Identification of the leading ASEAN country recipients of inbound investment and source countries of that investment;
4. Identification of pairs or groups of countries within ASEAN that have experienced significant integration related to industry production and/or marketing; and
5. An analysis of leading competitive factors that have contributed to changes in industry regional integration, export competitiveness, and inbound investment. Although all contributing factors should be identified and analyzed to the extent practicable, I am particularly interested in how regional improvements in trade facilitation, logistics services, and e-commerce have contributed to these changes.

The Commission is requested to deliver the report no later than August 2, 2010.

I anticipate that the Commission's report will be made available to the public in its entirety. Therefore, the report should not contain any confidential business or national security information.

The Commission's assistance in this matter is greatly appreciated.

Sincerely,

Ambassador Ronald Kirk

APPENDIX B. FEDERAL REGISTER NOTICE

International Trade Commission

[Investigation No. 332–511] ASEAN: Regional Trends in Economic Integration, Export Competitiveness, and Inbound Investment for Selected Industries

Agency: United States International Trade Commission.

Action: Institution of investigation and scheduling of hearing.

Summary: Following receipt on November 9, 2009, of a request from the United States Trade Representative (USTR) under section 332(g) of the Tariff Act of 1930 (19 U.S.C. 1332(g)), the U.S. International Trade Commission (Commission) instituted investigation No.

332–511, *ASEAN: Regional Trends in Economic Integration, Export Competitiveness, and Inbound Investment for Selected Industries.*

Dates: *December 30, 2009:* Deadline for filing requests to appear at the public hearing. *January 5, 2010:* Deadline for filing pre-hearing briefs and statements. *February 3, 2010:* Public hearing. *February 10, 2010:* Deadline for filing post-hearing briefs and statements. *March 10, 2010:* Deadline for filing all other written submissions. *August 2, 2010:*
Transmittal of Commission report to the USTR.

Addresses: All Commission offices, including the Commission's hearing rooms, are located in the United States International Trade Commission Building, 500 E Street, SW., Washington, DC. All written submissions should be addressed to the Secretary, United States International Trade Commission, 500 E Street, SW., Washington, DC 20436. The public record for this investigation may be viewed on the Commission's electronic docket (EDIS) at *http://www.usitc.gov/ secretary/edis.htm.*

For further Information Contact: Project leader John Fry (202–708–4157 or *john.fry@usitc.gov*) or deputy project leader Vincent Honnold (202–205–3314 or *vincent.honnold@usitc.gov*) for information specific to this investigation. For information on the legal aspects of this investigation, contact William Gearhart of the Commission's Office of the General Counsel (202–205–3091 or *william.gearhart@usitc.gov*). The media should contact Margaret O'Laughlin, Office of External Relations (202–205–1819 or *margaret.olaughlin@usitc.gov*). Hearing-impaired individuals may obtain information on this matter by contacting the Commission's TDD terminal at 202–205–1810. General information concerning the Commission may also be obtained by accessing its Internet server *http://www.usitc.gov*).

Persons with mobility impairments who will need special assistance in gaining access to the Commission should contact the Office of the Secretary at 202–205–2000. *Background:* As requested by the USTR, the Commission will conduct an investigation and prepare a report that provides a brief overview of regional trends in economic integration, export competitiveness, and inbound investment in six industry sectors in Southeast Asia. The sectors are electronics, automotives, agro-based products, healthcare, textiles and apparel, and wood-based products. In addition, the Commission will identify ASEAN industries within these sectors that have experienced significant changes in regional economic integration, export competitiveness, and inbound investment in recent years and provide profiles of these industries in its report. To the extent possible, the report will include illustrative country information and be based on the most recent five year period for which data are available. As requested, each industry profile, to the extent possible, will include the following information:

1. A description of the selected industry(ies) within ASEAN, including the industry's position relative to global competitors, as it relates to export competitiveness and inbound investment flows;
2. Identification of the leading ASEAN exporting countries and their key markets;
3. Identification of the leading ASEAN country recipients of inbound investment and source countries of that investment;

4. Identification of pairs or groups of countries within ASEAN that have experienced significant integration related to industry production and/or marketing; and
5. An analysis of leading competitive factors that have contributed to changes in industry regional integration, export competitiveness, and inbound investment, with a focus on how regional improvements in trade facilitation, logistics services, and ecommerce have contributed to these changes.

The USTR requested that the Commission deliver its report no later than August 2, 2010.

Public Hearing: A public hearing in connection with this investigation will be held at the U.S. International Trade Commission Building, 500 E Street, SW., Washington, DC, beginning at 9:30 a.m. on Wednesday, February 3, 2010. Requests to appear at the public hearing should be filed with the Secretary no later than 5:15 p.m., December 30, 2009, in accordance with the requirements in the ''Submissions'' section below. All pre-hearing briefs and statements should be filed no later than 5:15 p.m., January 5, 2010; and all post-hearing briefs and statements responding to matters raised at the hearing should be filed no later than 5:15 p.m., February 10, 2010. In the event that, as of the close of business on December 30, 2009, no witnesses are scheduled to appear at the hearing, the hearing will be canceled. Any person interested in attending the hearing as an observer or nonparticipant may call the Office of the Secretary (202–205–2000) after December 30, 2009, for information concerning whether the hearing will be held. *Written Submissions:* In lieu of or in addition to participating in the hearing, interested parties are invited to file written submissions concerning this investigation. All written submissions should be addressed to the Secretary, and all such submissions (other than pre- and post-hearing briefs and statements) should be received no later than 5:15 p.m., March 10, 2010. All written submissions must conform with the provisions of section 201.8 of the Commission's Rules of Practice and Procedure (19 CFR 201.8). Section 201.8 requires that a signed original (or a copy so designated) and fourteen (14) copies of each document be filed. In the event that confidential treatment of a document is requested, at least four (4) additional copies must be filed, in which the confidential information must be deleted (see the following paragraph for further information regarding confidential business information). The Commission's rules authorize filing submissions with the Secretary by facsimile or electronic means only to the extent permitted by section 201.8 of the rules (see Handbook for Electronic Filing Procedures, *http:// www.usitc.gov/secretary/ fed*B*reg*B*notices/rules/documents/han dbook*B*on*B*electronic*B*filing.pdf*). Persons with questions regarding electronic filing should contact the Office of the Secretary (202–205–2000). Any submissions that contain confidential business information must also conform with the requirements of section 201.6 of the Commission's Rules of Practice and Procedure (19 CFR 201.6). Section 201.6 of the rules requires that the cover of the document and the individual pages be clearly marked as to whether they are the ''confidential'' or ''nonconfidential'' version, and that the confidential business information be clearly identified by means of brackets. All written submissions, except for confidential business information, will be made available for inspection by interested parties. In its request letter, the USTR stated that it intends to make the Commission's report available to the public in its entirety, and asked that the Commission not include any confidential business information in the report it sends to the USTR. Any confidential business information received by the

Commission in this investigation and used in preparing this chapter will not be published in a manner that would reveal the operations of the firm supplying the information.

Issued: December 1, 2009. By order of the Commission.

William R. Bishop,

Acting Secretary to the Commission.

[FR Doc. E9–29025 Filed 12–4–09; 8:45 am]

BILLING CODE P

APPENDIX C. CALENDAR OF HEARING PARTICIPANTS

Calendar of Public Hearing

Those listed below appeared as witnesses at the United States International Trade Commission.s hearing:

Subject: ASEAN: Regional Trends in Economic Integration, Export Competitiveness, and Inbound Investment for Selected Industries

Inv. No.: 332-511

Date and Time: February 3, 2010, 9:30 a.m.

Sessions were held in connection with this investigation in the Main Hearing Room (room 101), 500 E Street, S.W., Washington, D.C.

Embassy Witnesses

Embassy of Malaysia Washington, D.C.

Rosidah Radzian, Science Attache, Embassy of Malaysia, *and* Regional Manager—Americas, Malaysian Palm Oil Board (MPOB)

Organization and Witness

American Palm Oil Council (APOC) Washington, D.C.

Mohd Salleh Kassim, Executive Director

U.S. Chamber of Commerce Washington, D.C.

Murray Hiebert, Senior Director, Asia

APPENDIX D. POSITIONS OF INTERESTED PARTIES

Summary of Positions of Interested Parties[709]

Thomas G. Aquino, philippine senior undersecretary of trade and industry[710]

Mr. Aquino stated that, in an attempt to fast-track the realization of the ASEAN Economic Community by 2015, ASEAN selected 12 economic sectors to undergo accelerated integration under the Priority Integration Program. He explained that ASEAN countries were designated to lead the sectoral integration efforts; the Philippines was designated the country coordinator for the electronics sector. Mr. Aquino noted that ASEAN developed sectoral protocols and roadmaps to help the 12 sectors rapidly integrate. He indicated that ASEAN has made progress in integration in a number of areas, including (1) the development and implementation of sectoral mutual recognition arrangements (MRAs) for certain goods and services, (2) bilateral agreements on visa exemption, (3) cooperative efforts to combat illegal trade in forest products, and (4) tariff liberalization.

Mr. Aquino stated that the Philippines has an important role to play in ASEAN's regional economic integration by virtue of the fact that it is the country coordinator for the electronics sector. He noted that the electronics sector is a vital component of the ASEAN economy, accounting for a large portion of the region's exports, and reported that Singapore, Malaysia, Thailand, and the Philippines were the region's top four of exporters of electronic goods in 2008; Indonesia and Vietnam were also exporters of electronic goods. Mr. Aquino noted a recent study which indicated that the most significant challenge to the integration of ASEAN's electronics sector was for each of the 10 member countries to attain the capacity to meet international standards of competitiveness. Another challenge to the integration of the region's electronics sector was the state of the global economy, which had turned downward in the recent past but now appeared to be on the road to recovery, with a consequent upturn in demand for electronics products.

Mr. Aquino stated that the Philippines government and the semiconductor and electronics industries in the Philippines have taken steps to integrate with ASEAN's electronics sector. These steps included (1) enlisting the cooperation of ASEAN electronics producers in regional integration efforts through the ASEAN Electronics Forum, (2) working with other ASEAN members to harmonize standards for safety and electromagnetic capability, (3) working with other ASEAN members on MRAs and harmonization of regulatory regimes, and (4) hosting an ASEAN Electronics Business Opportunities Exhibition and Conference. Mr. Aquino noted that these regional integration efforts and others demonstrated ASEAN's commitment to remaining an important player in the global electronics industry, promoted intra-ASEAN trade and investments, and improved the electronics sector's capabilities and quality.

Mr. Aquino noted that the Philippines government has cooperated in various regional integration efforts for other economic sectors, including wood-based products, healthcare, and automotives. Finally, he stated that the semiconductor and electronics industries in the Philippines has expressed an interest in hosting a meeting of chief executives of five leading U.S. electronics firms and five leading ASEAN electronics firms to identify issues and initiatives that could lead to increased regional integration of the ASEAN electronics sector.

Regular follow-up meetings of these chief executives could ensure a continuing strategic partnership of U.S. and ASEAN electronics firms.

American Palm Oil Council[711]

The American Palm Oil Council (APOC), a trade association funded by a portion of the assessments on Malaysian palm oil production, states that its mission is the promotion of palm oil in food and nonfood applications in the United States. In his hearing testimony, Mohd Salleh Kassim, APOC executive director, noted that palm oil contains no trans fats and that its use in U.S. processed food products as part of a vegetable oil blend has increased in recent years because of U.S. labeling and other requirements intended to diminish the presence of trans fats in the U.S. diet.

He presented various exhibits showing the increase in U.S. palm oil imports and in palm oil's share of global vegetable oils production and consumption markets. He listed palm oil importing and exporting countries and commented that the world could face a shortage in oils and fats.

Mr. Kassim described the sustainability activities that the Malaysian palm oil industry has undertaken, including subscribing to the standards and certification system under the Roundtable on Sustainable Palm Oil (RSPO), comprising eight principles, 39 criteria underlying the principles, and 120 auditable points. He noted that six Malaysian producers, as well as other ASEAN and non-ASEAN producers, have become RSPO certified, and provided estimates on the amount of palm oil expected to be RSPO certified in 2010.

Mr. Kassim provided information about the comparatively high productivity of oil palm compared to other oilseed crops. He also recounted the growing interest in oil palm cultivation on the part of developing countries in the tropics, many of which have poverty alleviation and food security concerns, because of the oil palm's long and productive lifespan, high oil yield, and cost-effectiveness.

Malaysian Palm Oil Board[712]

The Malaysian Palm Oil Board (MPOB) is the primary government agency for the promotion and development of national policies and priorities for the Malaysian palm oil industry and is funded by assessments on Malaysian palm oil production. In her hearing testimony, Rosidah Radzian, the science attaché at the Embassy of Malaysia in Washington, DC, and MPOB regional manager for the Americas, stated that she would speak about the importance of the trade relationship between Malaysia and the United States in palm oil and highlighted the Malaysian government's goal of raising the standard of living and upgrading the infrastructure in rural Malaysia through agricultural economic development. She noted that the Malaysian oil palm industry is one of the country's biggest foreign export earners, contributing $18 billion in 2008, compared to $4.5 billion in 2000. She stated that four decades of research and development has given Malaysia a competitive advantage in the palm oil trade and made Malaysia the number one palm oil exporter in the world. Although acknowledging that the Malaysian government has played a key role in the palm oil industry, she credited agencies such as the MPOB, the Malaysian Palm Oil Council, the Malaysian Palm Oil Association, and other stakeholders with the success of palm oil in Malaysia.

In explaining why Malaysia should be the primary U.S. trading partner for palm oil, Ms. Radzian noted, among other things, Malaysia's emphasis on food quality and safety standards and on environmental considerations without any form of subsidy from the government. She also stressed Malaysia's activities as chair of the Codex Alimentarius Commission's Committee on Oil and Fats beginning in 2008 and as a member of the International Maritime Organization regarding international shipping of bottled liquids and commodities.

Ms. Radzian explained that Malaysian palm oil farmers and their families are provided with free housing, medical care, and schooling, helping to reduce Malaysian poverty. She described the environmental efforts that the Malaysian palm oil industry has undertaken, including carbon footprint life cycle analysis and peatland management.

She stated that oil palm farming in Malaysia represents about 67 percent of the agricultural area in Malaysia and that conservation and forest protection policies prevent additional farm land from being cultivated. She cited a genome project between the MPOB and a U.S. company to facilitate the production of palm oil varieties with a higher yield.

National Milk Producers Federation (NMPF) and U.S. Dairy Export Council (USDEC)[713]

National Milk Producers Federation (NMPF) is the national farm commodity organization that represents U.S. dairy farmers and the dairy cooperative marketing organizations they own and operate throughout the United States. U.S. Dairy Export Council (USDEC) is a nonprofit, independent membership organization that represents the export interests of U.S. milk producers, dairy cooperatives, proprietary processors, and export traders. USDEC's mission is to increase the volume and value of U.S. dairy exports.

NMPF and USDEC stated in their joint submission that a bilateral free trade agreement between the United States and ASEAN would provide significant new export opportunities in important Asian markets. They noted that U.S. dairy industry is eager to increase sales in the region, particularly considering the fact that an FTA was concluded in 2008 between major dairy exporters Australia and New Zealand and ASEAN member countries. NMPF and USDEC explained that under the ASEAN-Australia-New Zealand

FTA (AANZFTA), Australian and New Zealand dairy exporters are receiving preferential access to the ASEAN market through lower tariffs. They expressed a belief that as the AANZFTA becomes fully implemented and tariffs decline, the competitiveness of U.S. dairy companies in ASEAN markets will erode. In particular, they asserted the United States will face unfavorable dairy tariff differentials against Australia and New Zealand in the Philippines (the sixth largest export destination for U.S. dairy products in 2009) and Indonesia (the seventh largest destination).

NMPF and USDEC went on to say that these circumstances will create a difficult competitive situation for U.S. exports to ASEAN, a region of 580 million people, a growing middle class, and a combined GDP of $1.5 trillion. They pointed out that by 2020, the AANZFTA will provide duty-free treatment for dairy products shipped from Australia and New Zealand to the Philippines and Vietnam; in addition, all ASEAN member countries are to grant Australia and New Zealand deep cuts in tariffs on dairy products over the next decade. NMPF and USDEC added that other dairy competitors are also moving into the ASEAN market; the European Union, for example, has recently launched free trade negotiations with ASEAN.

NMPF and USDEC said they believe that the only way to remedy the competitive disadvantage facing U.S. dairy exporters is through a U.S.-ASEAN FTA, particularly if such an agreement eliminates tariff discrimination against U.S. dairy products. In addition, an FTA with ASEAN would help grow a high-value market area, and additional U.S. exports to the region would bolster milk prices for America's dairy farms and support additional jobs in the dairy processing and transportation sectors.

US-ASEAN Business Council[714]

The US-ASEAN Business Council is a trade association of over 100 American companies which has worked on behalf of American firms doing business in the ASEAN region for over 25 years. The council stated that it has supported ASEAN's regional economic integration efforts for many years. The council noted that in certain aspects ASEAN is more important economically for the United States than China or India, and that greater economic integration for ASEAN will lead to increased American investment in the area and greater export opportunities for American businesses.

The council stated that currently ASEAN is a long way from achieving the level of regional economic integration that has occurred in the European Union; progress in regional economic integration has been slow. The council noted that despite reductions in tariffs among ASEAN countries, it is sometimes easier and cheaper to move goods between Singapore and Los Angeles than between Malaysia and Vietnam. Rules and regulations to facilitate greater foreign investment in the region have also been slow to materialize, according to the council. The council stated that the single biggest obstacle to ASEAN's regional economic integration is that its institutions are not developed enough to carry out the cross-border linking projects and structural policy reforms needed to create an economic community.

The council indicated that ASEAN was originally formed in 1967 in response to political and security concerns. As time passed, ASEAN member countries began to realize that regional economic integration would help them to attract foreign investment and develop economically. The council noted, however, that this increased economic cooperation was not accompanied by the structural and institutional reforms necessary to bring about real regional economic integration. It was reported that one factor that had a positive impact on ASEAN's regional economic integration efforts was the Asian financial crisis in 1997, which prompted stronger efforts by ASEAN countries to coordinate economic policies. The council stated that the recent global recession has led to greater recognition by ASEAN countries that regional economic integration is critical to succeeding in the global economy, to competing successfully with India and China, and creating ASEAN multinational companies.

The council noted that, although ASEAN's regional economic integration presents opportunities for U.S. companies, it also presents challenges for them. These challenges include lost opportunities for U.S. firms to export to ASEAN nations as a result of preferential trade agreements between ASEAN and other countries such as China, as well as greater competition from ASEAN firms as they grow and become regional and global businesses.

U.S. Chamber of Commerce[715]

The U.S. Chamber of Commerce (Chamber) is the world's largest business federation, representing more than three million businesses and organizations in every major classification of U.S. business—manufacturing, retailing, services, construction, wholesaling, and finance. In addition, the Chamber has a substantial international reach, with 112 American Chambers of Commerce abroad and an increasing number of members engaged in importing and exporting goods and services. The Chamber favors strengthened international competitiveness and opposes artificial U.S. and foreign barriers to international business.

With 582 million consumers, ASEAN is the third largest market in Asia after China and India. The Chamber's submission states that not only would increased regional economic integration benefit U.S. companies, but it would also boost growth, increase trade, and create jobs in Southeast Asia. According to the Chamber, integration would stimulate new consumer demand that U.S. firms would seek to meet, as well as provide new investment opportunities for U.S. companies and increase ties between ASEAN member countries and the United States.

The Chamber stated that through the ASEAN Free Trade Agreement (AFTA), the 10 member countries are moving toward an integrated regional market, principally by eliminating tariffs on industrial and agricultural products. It asserted that fully implementing AFTA is critical to the entry and long-term growth of business in the region and providing U.S. businesses with the scale of production that will offer them competitive cost efficiencies without having to establish a manufacturing presence in each market.

The Chamber's submission states that its member companies tell them that regional integration is still at an early stage, and more work is required to remove barriers to trade in goods and the free flow of capital and labor. Specifically, ASEAN needs to address cumbersome rules of origin, customs procedures need to be simplified, and nontariff barriers need to be eliminated. The Chamber asserted that Southeast Asia risks being left behind unless it speeds up economic integration, and ASEAN's self-imposed deadline of five more years does not leave much time to solve the necessary technical and political challenges.

The Chamber expressed its opinion that investment in the ASEAN region is held back by differences between countries in regulations and business environments, as well as political and business culture. The 2009 ASEAN Comprehensive Investment Agreement (ACIA) consolidated earlier ASEAN investment agreements, but the AICA only covers a limited number of sectors and does not include logistics services. According to the submission, U.S. companies canvassed by the Chamber urge ASEAN to undertake efforts to enhance the transparency of investment legislation, and address restrictions from overlapping and contradictory legislation, regulations, and licensing criteria, which effectively negate the benefits of the liberalization measures.

The U.S. business community, as represented by the Chamber, supports efforts to implement the ASEAN Single Window (ASW) because it believes the ASW will reduce the cost and complexities of doing business in the region. The Chamber added that for ASW to be effective, some ASEAN countries will need to make improvements in their legal and administrative infrastructure, such as implementing clear and transparent import and export procedures, increasing cooperation among government bodies, and providing facilities for e-commerce. The ASW has made only slow progress in recent years at both the national and ASEAN levels: at the regional level, it is unclear to U.S. companies what ASW is supposed to do and whether it is merely a collection of individual country national single windows (NSWs) or intended to move toward an integrated ASW that will give traders from outside ASEAN a portal to access ASEAN NSWs. The Chamber would like more regular updates from ASEAN member countries on the development of NSWs and regular discussions between U.S. companies in the region and the ASW Steering Committee.

The Chamber's submission reported that its member companies believe that the focus of ASEAN should shift to the progressive elimination of nontariff barriers (NTBs) now that ASEAN tariff elimination has largely taken place. According to the Chamber, the elimination of NTBs by ASEAN should be considered as important as eliminating tariffs. One of the top concerns of U.S. companies doing business in ASEAN is the lack of transparency in government regulation and rules; in particular, unique technical standards and differing technical regulations and processes increase investment costs, both at the national and regional levels.

The Chamber noted that ASEAN Coordinating Committee on Standards and Quality (ACCSQ) is working to address different standards between ASEAN member countries, but the Chamber is requesting more dialogue between government entities and private sector interests to identify key areas of focus for standards and identify mechanisms for improving the regional standards regime. Two other Chamber recommendations to ASEAN members include developing national Websites for publishing regulations and standards and partnering with the American National Standards Institute (ANSI) for technology transfers.

Finally, the Chamber's submission and hearing testimony provided specific comments from their members in the electronics, logistics services, automotives, agro-based products, and healthcare sectors. In the area of electronics, member companies are seeking a harmonized ASEAN tariff nomenclature; a comprehensive, harmonized information technology-based platform in areas such as customs and licensing procedures that would increase efficiency and transparency in dealing with ASEAN governments; and an ASEAN single window that would help electronics firms increase supply chain efficiencies.[716] In the automobile sector, members of the Chamber are concerned about domestic content laws, such as those that exist in Malaysia, which protect locally made products.[717]

Appendix E. Country Profiles

Brunei Darussalam

Economic Profile

Economic Indicators

	2004	2008
Population (million)	0.4	0.4
Nominal GDP (billion US$)	7.9	14.3
Real GDP growth (%)	0.5	−1.9
Goods exports (million US$)	5,067	10,449
Goods imports (million US$)	1,428	2,550
Trade balance (million US$)	3,639	7,899
GDP per capita (US$ at PPP)	n/a	n/a
Inward FDI stock (million US$)	9,139	10,361

Source: Economist Intelligence Unit, *Country Report: Brunei Darussalam*, June 2009 and March 2010; U.S. Central Intelligence Agency, *The World Factbook: Brunei Darussalam,* March 2010; United Nations Conference on Trade and Development, FDISTAT Interactive Database. PPP = purchasing power parity.

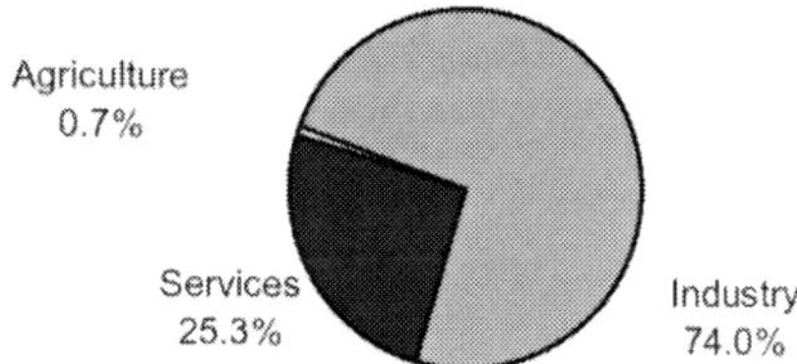

Principal imports: machinery and transport equipment, manufactured goods, food, and chemicals.
Principal exports: crude oil, natural gas, and garments.
Source. U.S. Central Intelligence Agency, The World Factbook: Brunei Darussalam, March 2010.

Sources of GDP (2008)

Main Trade Partners, Percent of Total, 2008

Export markets		Suppliers	
Japan	43.2	ASEAN countries	47.0
ASEAN countries	22.7	United States	13.7
South Korea	14.8	Japan	8.5
Australia	10.4	China	6.9

Source. Economist Intelligence Unit, *Country Report: Brunei Darussalam,* March 2010.

Brunei Darussalam's trade agreements include:

- ASEAN-China, preferential trade agreement covering goods (eff. July 2003);
- ASEAN-China, economic integration agreement covering services (eff. July 2007);
- ASEAN-Japan, free trade agreement covering goods (eff. Dec. 2008);

- ASEAN Free Trade Area (AFTA), free trade agreement covering goods (eff. Jan. 1992);
- Brunei Darussalam-Japan, free trade agreement and economic integration agreement covering goods and services (eff. July 2008); and
- Trans-Pacific Strategic Economic Partnership, free trade agreement and economic integration agreement covering goods and services (eff. May 2006).

Burma

Economic Profile

Economic Indicators

	2004	2008
Population (million)	48.0	49.6
Nominal GDP (billion US$)	10.0	22.6
Real GDP growth (%)	13.6	0.9
Goods exports (million US$)	2,927	6,348
Goods imports (million US$)	1,999	3,427
Trade balance (million US$)	928	2,921
GDP per capita (US$ at PPP)	491	435
Inward FDI stock (million US$)	4,791	5,546

Source. Economist Intelligence Unit, *Country Report: Burma,* June 2009; United Nations Conference on Trade and Development, FDISTAT Interactive Database. PPP = purchasing power parity.

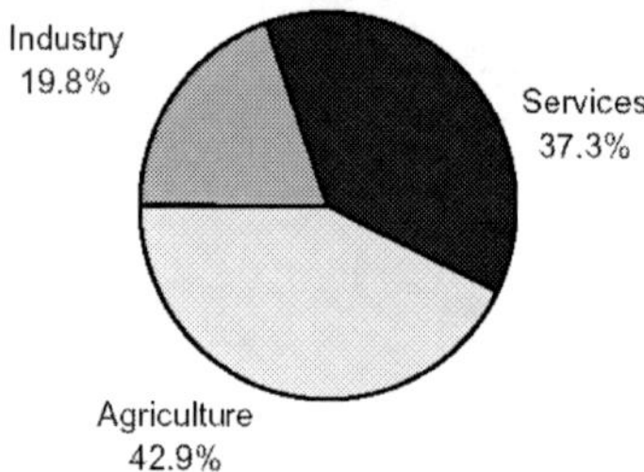

Principal imports: fabric, petroleum products, fertilizer, plastics, machinery, transport equipment, cement, construction materials, crude oil, food products, and edible oil.
Principal exports: natural gas, wood products, pulses, beans, fish, rice, clothing, and jade and gems.
Source: U.S. Central Intelligence Agency, The World Factbook: Burma, March 2010.

Sources of GDP (2009)

Main Trade Partners, Percent of Total, 2007

Export markets		Suppliers	
Thailand	44.7	China	35.9
India	14.1	Thailand	20.7
China	6.9	Singapore	16.8
Japan	5.6	Malaysia	4.4
Malaysia	2.6	India	3.5

Source: Economist Intelligence Unit, *Country Report: Burma,* June 2009.

Burma's trade agreements include:

- ASEAN-China, preferential trade a greement covering goods (eff. July 2003);
- ASEAN-China, economic integration a greement covering services (eff. July 2007);
- ASEAN-Japan, free trade a greement covering goods (eff. Dec. 2008); and
- ASEAN Free Trade Area (AFTA), free trade agreement covering goods (eff. Jan. 1992).

Source: World Trade Organization, Regional Trade Agreements, Burma, http://rtais.wto. org (accessed March 17, 2010). Eff. = effective.

Cambodia

Economic Profile

Economic Indicators

	2004	2008
Population (million)	13.7	14.6
Nominal GDP (billion US$)	5.3	10.6
Real GDP growth (%)	10.3	5.0
Goods exports (million US$)	2,589	4,312
Goods imports (million US$)	3,270	6,370
Trade balance (million US$)	–681	–1,335
GDP per capita (US$ at PPP)	1,257	1,921
Inward FDI stock (million US$)	2,090	4,637

Source. Economist Intelligence Unit, *Country Report: Cambodia,* June 2009; United Nations Conference on Trade and Development, FDISTAT Interactive Database. PPP = purchasing power parity.

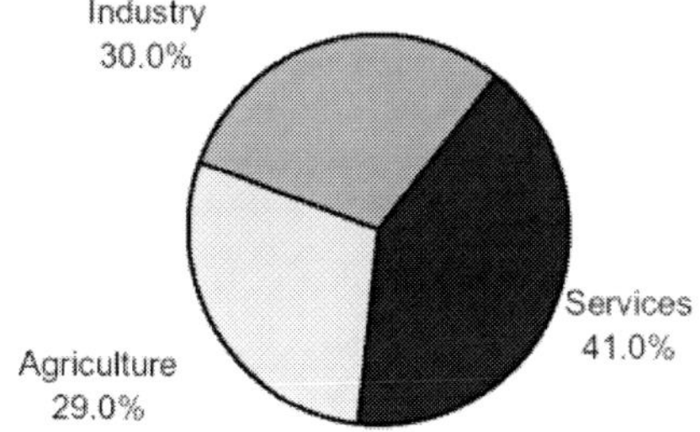

Principal imports: petroleum products, cigarettes, gold, construction materials, machinery, motor vehicles, and pharmaceutical products.

Principal exports: clothing, timber, rubber, rice, fish, tobacco, and footwear. Source: U.S. Central Intelligence Agency, The World Factbook: Cambodia, March 2010.

Source: U.S. Central Intelligence Agency, *The World Factbook: Cambodia*, March 2010.

Sources of GDP (2007)

Main Trade Partners, Percent of Total, 2008

Export markets		Suppliers	
United States	53.7	Thailand	34.9
Israel	26.7	China	20.5
Germany	7.6	Vietnam	19.5
Canada	5.8	Hong Kong	11.2

Source: Economist Intelligence Unit, Country Report: Cambodia, June 2009

Cambodia's trade agreements include:

- ASEAN-China, preferential trade agreement covering goods (eff. July 2003);
- ASEAN-China, economic integration agreement covering services (eff. July 2007);
- ASEAN-Japan, free trade agreement covering goods (eff. Dec. 2008); and
- ASEAN Free Trade Area (AFTA), free trade agreement covering goods (eff. Jan. 1992)

Source: World Trade Organization, Regional Trade Agreements, Cambodia, http://rtais.wto.org (accessed March 17, 2010). Eff. = effective.

Indonesia

Economic Profile

Economic Indicators

	2004	2008
Population (million)	226.0	237.5
Nominal GDP (billion US$)	256.8	510.8
	2004	**2008**
Real GDP growth (%)	5.0	6.1
Goods exports (million US$)	70,766	139,290
Goods imports (million US$)	50,615	115,981
Trade balance (million US$)	20,151	23,309
GDP per capita (US$ at PPP)	2,850	3,824
Inward FDI stock (million US$)	15,858	67,044

Source: Economist Intelligence Unit, Country Report: Indonesia, June 2009; United Nations Conference on Trade and Development, FDISTAT Interactive Database. PPP = purchasing power parity.

Main Trade Partners, Percent of Total, 2008

Export markets		Suppliers	
Japan	21.6	Singapore	28.8
Singapore	11.7	China	13.7
United States	11.1	Japan	10.8
China	10.1	Malaysia	6.2

Source. Economist Intelligence Unit, *Country Report: Indonesia,* June 2009.

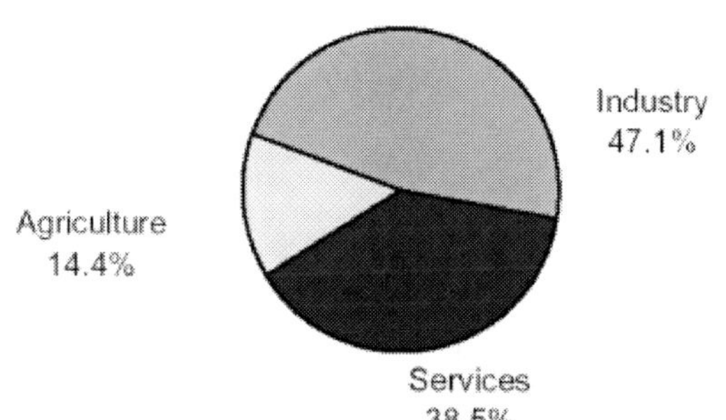

Principal imports: machinery and equipment, chemicals, fuels and foodstuffs.
Principal exports: oil and gas, electrical appliances, plywood, textiles, and rubber.
Source. U.S. Central Intelligence Agency, The World Factbook: Indonesia, March 2010.

Sources of GDP (2009)

Indonesia's trade agreements include:

- ASEAN-China, preferential trade agreement covering goods (eff. July 2003);
- ASEAN-China, economic integration agreement covering services (eff. July 2007);
- ASEAN-Japan, free trade agreement covering goods (eff. Dec. 2008);
- ASEAN Free Trade Area (AFTA), free trade agreement covering goods (eff. Jan. 1992); and
- Japan-Indonesia, free trade agreement and economic integration agreement covering goods and services (eff. July 2008).

Source. World Trade Organization, Regional Trade Agreements, Indonesia, http://rtais.wto.org (accessed March 17, 2010). Eff. = effective.

Laos

Economic Profile

Economic Indicators

Population (million)	5.8	6.2
Nominal GDP (billion US$)	2.5	5.2
Real GDP growth (%)	6.9	7.5
Goods exports (million US$)	341	1,161
Goods imports (million US$)	614	1,387
Trade balance (million US$)	–273	–226
GDP per capita (US$ at PPP)	n/a	n/a
Inward FDI stock (million US$)	641	1,408

Source. Economist Intelligence Unit, Country Report: Laos, June 2009; United Nations Conference on Trade and Development, FDISTAT Interactive Database. PPP = purchasing power parity.

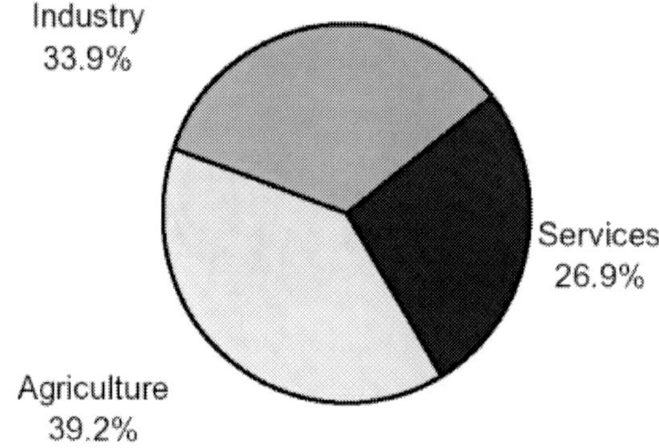

Principal imports: machinery and equipment, vehicles, fuel, and consumer goods.
Principal exports: wood products, coffee, electricity, tin, copper, and gold.
Source. U.S. Central Intelligence Agency, The World Factbook: Laos, March 2010.

Sources of GDP (2009)

Main Trade Partners, Percent of Total, 2008

Export markets		**Suppliers**	
Thailand	49.0	Thailand	68.6
Vietnam	18.6	China	11.3
China	12.1	Vietnam	4.7
South Korea	6.3	South Korea	2.5
United Kingdom	3.3	Japan	2.5

Source. Economist Intelligence Unit, *Country Report: Laos*, June 2009.

Laos' trade agreements include:

- ASEAN-China, preferential trade agreement covering goods (eff. July 2003);
- ASEAN-China, economic integration agreement covering services (eff. July 2007);
- ASEAN-Japan, free trade agreement covering goods (eff. Dec. 2008);
- ASEAN Free Trade Area (AFTA), free trade agreement covering goods (eff. Jan. 1992); and
- Laos-Thailand, preferential trade agreement covering goods (eff. June 1991).

Source: Laos is not a member of the World Trade Organization but information about its trade agreements can be determined from the World Trade Organization, Regional Trade Agreements, Thailand, http://rtais.wto.org (accessed March 17, 2010). Eff. = effective.

Malaysia

Economic Profile

Economic Indicators

	2004	2008
Population (million)	25.6	27.7
Nominal GDP (billion US$)	124.7	222.2
Real GDP growth (%)	6.8	4.6
Goods exports (million US$)	126,817	198,919
Goods imports (million US$)	99,244	154,708
Trade balance (million US$)	27,573	44,211
GDP per capita (US$ at PPP)	10,854	13,852
Inward FDI stock (million US$)	43,047	73,262

Source. Economist Intelligence Unit, *Country Report: Malaysia,* June 2009; United Nations Conference on Trade and Development, FDISTAT Interactive Database. PPP = purchasing power parity.

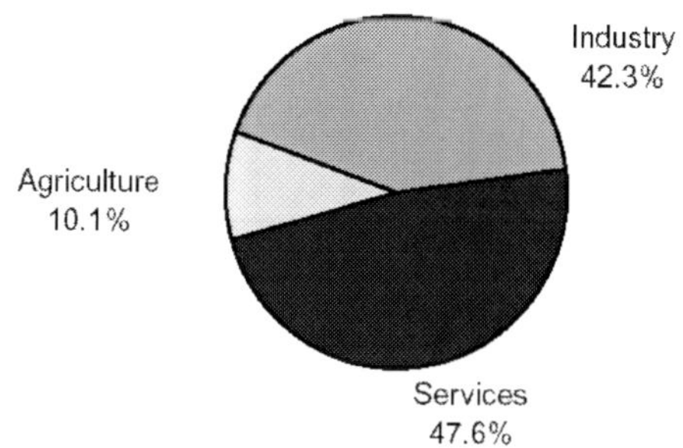

Principal imports: electronics, machinery, petroleum products, plastics, vehicles, iron and steel products, and chemicals.

Principal exports: electronic equipment, petroleum and liquefied natural gas, wood and wood products, palm oil, rubber, textiles, and chemicals.

Source. U.S. Central Intelligence Agency, The World Factbook: Malaysia, March 2010.

Sources of GDP (2009)

Main Trade Partners, Percent of Total, 2008

Export markets		Suppliers	
Singapore	14.7	China	12.8
United States	12.5	Japan	12.5
Japan	10.8	Singapore	11.0
China	9.5	United States	10.8
Thailand	4.8	Thailand	5.6

Source: Economist Intelligence Unit, *Country Report: Malaysia*, June 2009.

Malaysia's trade agreements include:

- ASEAN-China, preferential trade agreement covering goods (eff. July 2003);
- ASEAN-China, economic integration agreement covering services (eff. July 2007);
- ASEAN-Japan, free trade agreement covering goods (eff. Dec. 2008);
- ASEAN Free Trade Area (AFTA), free trade agreement covering goods (eff. Jan. 1992);
- Japan-Malaysia, free trade agreement and economic integration agreement covering goods and services (eff. July 2006); and
- Pakistan-Malaysia, free trade agreement and economic integration agreement covering goods and services (eff. Jan. 2008).

Source. World Trade Organization, Regional Trade Agreements, Malaysia, http://rtais.wto.org (accessed March 17, 2010). Eff. = effective.

Philippines

Economic Profile

Economic Indicators

	2004	2008
Population (million)	86.2	92.7
Nominal GDP (billion US$)	86.9	168.6
Real GDP growth (%)	6.4	4.6
Goods exports (million US$)	38,794	48,202
Goods imports (million US$)	44,478	60,784
Trade balance (million US$)	−5,684	−12,582
GDP per capita (US$ at PPP)	2,678	3,457
Inward FDI stock (million US$)	12,737	21,470

Source. Economist Intelligence Unit, Country Report: Philippines, June 2009; United Nations Conference on Trade and Development, FDISTAT Interactive Database. PPP = purchasing power parity.

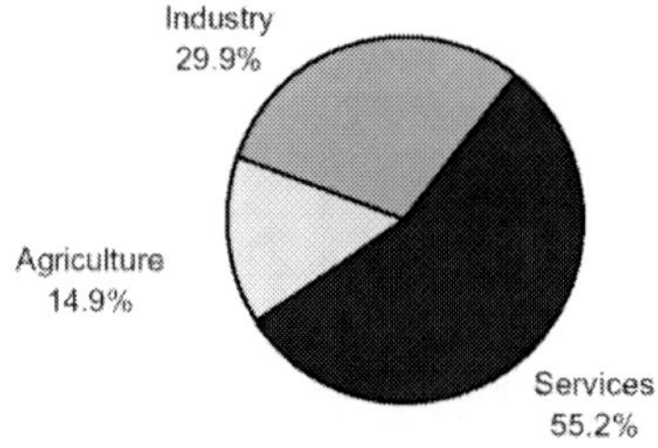

Principal imports: electronic products, mineral fuels, machinery and transport equipment, iron and steel, textile fabrics, grains, chemicals, and plastic.

Principal exports: semiconductors and electronic products, transport equipment, garments, copper products, petroleum products, coconut oils, and fruits.

Source: U.S. Central Intelligence Agency, The World Factbook: Philippines, March 2010.

Sources of GDP (2009)

Main Trade Partners, Percent of Total, 2008

Export markets		Suppliers	
China	34.0	Japan	18.3
United States	16.9	United States	15.2
Japan	16.4	China	14.8
Hong Kong	12.0	Singapore	13.8

Source. Economist Intelligence Unit, *Country Report: Philippines*, June 2009.

Philippines' trade agreements include:

- ASEAN-China, preferential trade agreement covering goods (eff. July 2003);
- ASEAN-China, economic integration agreement covering services (eff. July 2007);
- ASEAN-Japan, free trade agreement covering goods (eff. Dec. 2008);
- ASEAN Free Trade Area (AFTA), free trade agreement covering goods (eff. Jan. 1992); and
- Japan-Philippines, free trade agreement and economic integration agreement covering goods and services (eff. Dec. 2008).

Source. World Trade Organization, Regional Trade Agreements, Philippines, http://rtais.wto.org (accessed March 17, 2010). Eff. = effective.

Singapore

Economic Profile

Economic Indicators

	2004	2008
Population (million)	4.2	4.8
Nominal GDP (billion US$)	109.7	181.9
Real GDP growth (%)	9.3	1.1
Goods exports (million US$)	199,393	342,708
Goods imports (million US$)	168,330	309,564
Trade balance (million US$)	31,063	33,144
GDP per capita (US$ at PPP)	33,037	40,319
Inward FDI stock (million US$)	169,433	326,142

Source: Economist Intelligence Unit, Country Report: Singapore, June 2009; United Nations Conference on Trade and Development, FDISTAT Interactive Database. PPP = purchasing power parity.

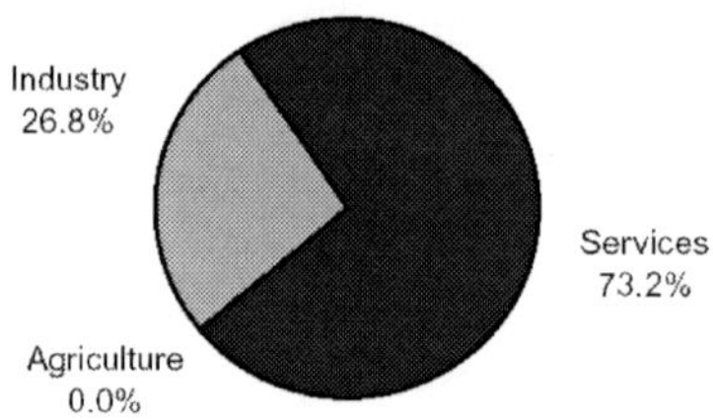

Principal imports: machinery and equipment, mineral fuels, chemicals, foodstuffs, and consumer goods.

Principal exports: machinery and equipment (including electronics), consumer goods, pharmaceuticals and other chemicals, and mineral fuels.

Source: U.S. Central Intelligence Agency, The World Factbook: Singapore, March 2010.

Sources of GDP (2009)

Main Trade Partners, Percent of Total, 2009

Export markets		Suppliers	
Hong Kong	11.6	United States	14.7
Malaysia	11.5	Malaysia	11.6
United States	11.2	China	10.5
Indonesia	9.7	Japan	7.6
China	9.7	Indonesia	5.8

Source. U.S. Central Intelligence Agency, *The World Factbook: Singapore,* March 2010.

Singapore's trade agreements include:

- ASEAN-China, preferential trade agreement covering goods (eff. July 2003);
- ASEAN-China, economic integration agreement covering services (eff. July 2007);
- ASEAN-Japan, free trade agreement covering goods (eff. Dec. 2008);
- ASEAN Free Trade Area (AFTA), free trade agreement covering goods (eff. Jan. 1992);
- China-Singapore, free trade agreement and economic integration agreement covering goods and services (eff. Jan. 2009);
- EFTA-Singapore, free trade agreement and economic integration agreement covering goods and services (eff. Jan. 2003);
- India-Singapore, free trade agreement and economic integration agreement covering goods and services (eff. Aug. 2005);
- Japan-Singapore, free trade agreement and economic integration agreement covering goods and services (eff. Nov. 2002);
- Jordan-Singapore, free trade agreement and economic integration agreement covering goods and services (eff. Aug. 2005);
- Korea-Singapore, free trade agreement and economic integration agreement covering goods and services (eff. Mar. 2006);
- New Zealand-Singapore, free trade agreement and economic integration agreement covering goods and services (eff. Jan. 2001);

- Panama-Singapore, free trade agreement and economic integration agreement covering goods and services (eff. July 2006);
- Peru-Singapore, free trade agreement and economic integration agreement covering goods and services (eff. Aug. 2009);
- Singapore-Australia, free trade agreement and economic integration agreement covering goods and services (eff. July 2003);
- Trans-Pacific Strategic Economic Partnership, free trade agreement and economic integration agreement covering goods and services (eff. May 2006); and
- US-Singapore, free trade agreement and economic integration agreement covering goods and services (eff. Jan. 2004).

Source. World Trade Organization, Regional Trade Agreements, Singapore, http://rtais.wto.org (accessed March 17, 2010). Eff. = effective.

Thailand

Economic Profile

Economic Indicators

	2004	2008
Population (million)	65.1	66.3
Nominal GDP (billion US$)	161.3	273.3
Real GDP growth (%)	6.3	2.6
Goods exports (million US$)	94,979	174,822
Goods imports (million US$)	84,194	157,274
Trade balance (million US$)	10,785	17,548
GDP per capita (US$ at PPP)	6,353	8,235
Inward FDI stock (million US$)	53,187	104,850

Source. Economist Intelligence Unit, Country Report: Thailand, June 2009; United Nations Conference on Trade and Development, FDISTAT Interactive Database. PPP = purchasing power parity.

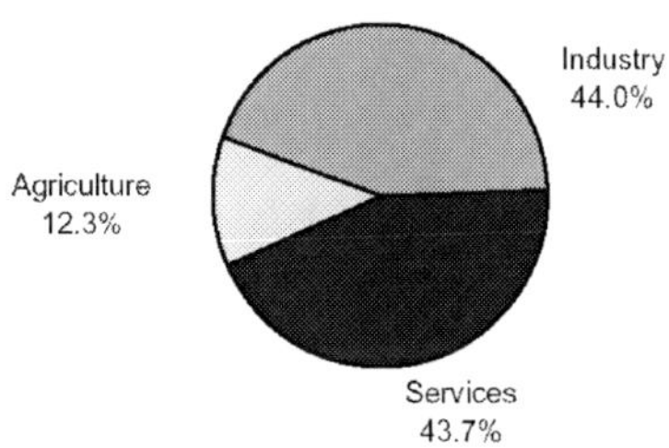

Principal imports: capital goods, intermediate goods and raw materials, consumer goods, and fuels.

Principal exports: textiles and footwear, fishery products, rice, rubber, jewelry, automobiles, computers, and electrical appliances.

Source. U.S. Central Intelligence Agency, *The World Factbook: Thailand*, March 2010.

Sources of GDP (2009)

Main Trade Partners, Percent of Total, 2008

Export markets		Suppliers	
United States	11.2	Japan	18.8
Japan	11.2	China	11.2
China	9.1	United States	6.4
Singapore	5.6	United Arab Emirates	6.0
Hong Kong	5.6	Malaysia	5.5

Source. Economist Intelligence Unit, *Country Report: Thailand,* June 2009.

Thailand's trade agreements include:

- ASEAN-China, preferential trade agreement covering goods (eff. July 2003);
- ASEAN-China, economic integration agreement covering services (eff. July 2007);
- ASEAN-Japan, free trade agreement covering goods (eff. Dec. 2008);
- ASEAN Free Trade Area (AFTA), free trade agreement covering goods (eff. Jan. 1992);
- Japan-Thailand, free trade agreement and economic integration agreement covering goods and services (eff. Nov. 2007);
- Laos-Thailand, preferential trade agreement covering goods (eff. June 1991);
- Thailand-Australia, free trade agreement and economic integration agreement covering goods and services (eff. Jan. 2005); and
- Thailand-New Zealand, free trade agreement and economic integration agreement covering goods and services (eff. July 2005).

Vietnam

Economic Profile

Economic Indicators

	2004	2008
Population (million)	82.7	86.1
Nominal GDP (billion US$)	45.4	91.3
Real GDP growth (%)	7.8	6.2
Goods exports (million US$)	26,485	61,600
Goods imports (million US$)	28,772	77,606
Trade balance (million US$)	−2,287	−16,006
GDP per capita (US$ at PPP)	1,932	2,793
Inward FDI stock (million US$)	29,115	48,325

Source. Economist Intelligence Unit, *Country Report: Vietnam,* June 2009; United Nations Conference on Trade and Development, FDISTAT Interactive Database. PPP = purchasing power parity.

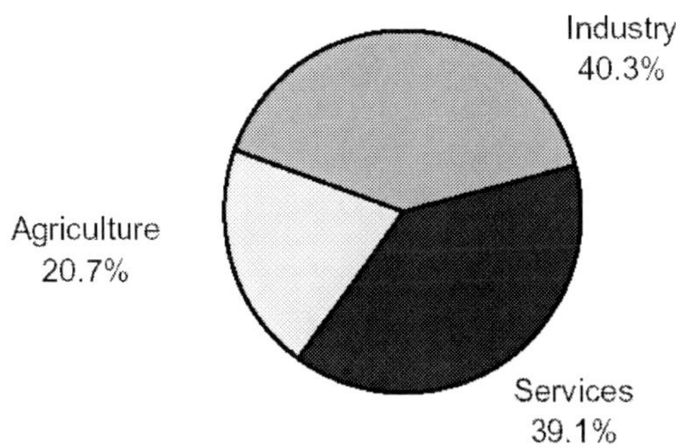

Principal imports: machinery and equipment, petroleum products, fertilizer, steel products, raw cotton, grain, cement, and motorcycles.
Principal exports: crude oil, marine products, rice, coffee, rubber, tea, garments, and shoes.
Source: U.S. Central Intelligence Agency, *The World Factbook*: Vietnam, March 2010.

Sources of GDP (2009)

Main Trade Partners, Percent of Total, 2007

Export markets		Suppliers	
United States	21.5	China	20.4
Japan	10.8	Singapore	11.8
Australia	7.0	Japan	9.6
China	5.9	South Korea	7.7

Source. Economist Intelligence Unit, *Country Report: Vietnam*, June 2009.

Vietnam's trade agreements include:

- ASEAN-China, preferential trade agreement covering goods (eff. July 2003);
- ASEAN-China, economic integration agreement covering services (eff. July 2007);
- ASEAN-Japan, free trade agreement covering goods (eff. Dec. 2008);
- ASEAN Free Trade Area (AFTA), free trade agreement covering goods (eff. Jan. 1992); and
- Japan-Vietnam, free trade agreement and economic integration agreement covering goods and services (eff. Oct. 2009).

Source: World Trade Organization, Regional Trade Agreements, Vietnam, http://rtais. wto.org (accessed March 17, 2010). Eff. = effective.

Appendix F. Harmonized System Numbers for the Priority Sectors and Selected Profiled Industries

Table F.1. Agro-based Products: Priority Sector and Profile HS Selection

0702.00	1205.10	1515.29
0708.10	1205.90[a]	2002.10
0708.20	1206.00[a]	2002.90
0708.90	1207.10	2005.40
0710.21	1207.20	2005.51
0710.22	1207.30	2005.59
0713.10	1207.40	2005.90
0713.20	1207.50	2008.11
0713.31	1207.60	2008.19
0713.32	1207.99	2008.20
0713.33	1208.10	2009.41
0713.39	1208.90	2009.49
0713.40	1507.10	2009.50
0713.50	1507.90	2302.10
0713.90	1511.10	2303.10
0801.11	1511.90	2304.00
0801.19	1512.11	2306.10
0804.50	1512.19	2306.20
1005.10	1512.21	2306.30
1005.90	1512.29	2306.41
1102.20	1513.11	2306.49
1102.30	1513.19	2306.50
1201.00	1513.21	2306.60
1203.00	1513.29	2306.70
1204.00	1515.21	2306.90

Source. Roadmap for Integration of Agro-Based Products Sector, Attachment 1.
[a] Profile selection: Palm oil.

Table F.2. Automotive: Priority Sector and Profile HS Selection

4011.10	8482.10	8701.20	8708.92[a]
4011.20	8482.20	8702.10	8708.93[a]
4011.40	8482.30	8702.90	8708.94[a]
4011.69	8482.40	8703.10	8708.99[a]
4011.99	8482.50	8703.21	8709.11
4012.11	8482.80	8703.22	8709.19
4012.12	8482.91	8703.23	8709.90
4012.19	8482.99	8703.24	8710.00
4012.20	8483.10	8703.31	8711.10
4012.90	8483.20	8703.32	8711.20
4013.10	8483.30	8703.33	8711.30
4013.90	8483.40	8703.90	8711.40

7315.11	8483.50	8704.10	8711.50
7315.19	8483.60	8704.21	8711.90
7315.89	8483.90	8704.22	8713.10
7315.90	8484.10	8704.23	8713.90
7320.10	8484.20	8704.31	8714.11
7320.20	8484.90	8704.32	8714.19
7320.90	8485.90	8704.90	8714.20
8407.31	8507.10	8705.10	8714.91
8407.32	8507.20	8705.20	8714.92
8407.33	8507.30	8705.30	8714.93
8407.34	8507.40	8705.40	8714.94
8408.20	8507.80	8705.90	8714.95
8409.91	8507.90	8706.00	8714.96
8409.99	8511.10 [a]	8707.10	8714.99
8413.30	8511.20 [a]	8707.90	8716.10
8415.20	8511.30 [a]	8708.10 [a]	8716.20
8415.81	8511.40 [a]	8708.21 [a]	8716.31
8415.82	8511.50 [a]	8708.29 [a]	8716.39
8415.83	8511.80 [a]	8708.31 [a]	8716.40
8415.90	8511.90 [a]	8708.39 [a]	
8421.23	8512.10 [a]	8708.40 [a]	
8421.31	8512.20 [a]	8708.50 [a]	
8421.99	8512.30 [a]	8708.60 [a]	
8427.10	8512.40 [a]	8708.70 [a]	
8427.20	8512.90 [a]	8708.80 [a]	
8427.90	8544.30 [a]	8708.91 [a]	

Source. Roadmap for Integration of Automotive Sector, Attachment 1.
[a] Profile selection: Motor vehicle parts.

Table F.3. Electronics: Priority Sector and Profile HS Selection

3909.10	8462.29	8475.21	8514.40	8519.92
3909.20	8464.10	8475.29	8514.90	8519.93
3909.30	8464.20	8475.90	8515.11	8519.99
3909.40	8464.90	8477.10	8515.19	8520.10
7108.13	8465.91	8477.40	8515.29	8520.20
7111.00	8465.92	8477.59	8515.31	8520.32
7321.11	8465.95	8477.80	8515.39	8520.33
7321.12	8465.99	8477.90	8515.80	8520.39
7321.13	8466.10	8479.50	8515.90	8520.90
7321.81	8466.20	8479.82	8516.10	8521.10
7321.82	8466.30	8479.89	8516.21	8521.90
7321.83	8466.91	8479.90	8516.29	8522.10
7321.90	8466.94	8480.71	8516.31	8522.90
8414.30	8469.11	8501.10	8516.32	8523.11
8414.51	8469.12	8504.10	8516.33	8523.12
8415.10	8469.20	8504.21	8516.40	8523.13
8415.20	8469.30	8504.22	8516.50	8523.20

Table F.3. (Continued)

8415.81	8470.10	8504.23	8516.60	8523.30
8415.82	8470.21	8504.31	8516.71	8523.90
8415.83	8470.29	8504.32	8516.72	8524.10
8415.90	8470.30	8504.33	8516.79	8524.31
8418.10	8470.40	8504.34	8516.80	8524.32
8418.21	8470.50	8504.40	8516.90	8524.39
8418.22	8470.90	8504.50	8517.11	8524.40
8418.29	8471.10	8504.90	8517.19	8524.51
8418.30	8471.30	8506.10	8517.21	8524.52
8418.40	8471.41	8506.30	8517.22	8524.53
8418.50	8471.49	8506.40	8517.30	8524.60
8418.61	8471.50 [a]	8506.50	8517.50	8524.91
8418.69	8471.60 [a]	8506.60	8517.80	8524.99
8418.91	8471.70 [a]	8506.80	8517.90	8525.10
8418.99	8471.80 [a]	8506.90	8518.10	8525.20
8419.39	8471.90	8509.10	8518.21	8525.30
8419.81	8472.10	8509.20	8518.22	8525.40
8419.89	8472.20	8509.30	8518.29	8526.10
8419.90	8472.30	8509.40	8518.30	8526.91
8420.99	8472.90	8509.80	8518.40	8527.12
8422.11	8473.10	8509.90	8518.50	8527.13
8424.89	8473.21	8510.10	8518.90	8527.19
8450.11	8473.29	8510.20	8519.10	8527.21
8450.12	8473.30 [a]	8510.30	8519.21	8527.29
8450.19	8473.40	8510.90	8519.29	8527.31
8450.20	8473.50	8514.10	8519.31	8527.32
8450.90	8474.31	8514.20	8519.39	8527.39
8462.21	8474.32	8514.30	8519.40	8527.90
8528.12	8540.71	9001.90	9011.10	9030.89
8528.13	8540.72	9002.11	9011.20	9030.90
8528.21	8540.79	9002.19	9011.80	9031.10
8528.22	8540.81	9002.20	9011.90	9031.20
8528.30	8540.89	9002.90	9012.10	9031.30
8529.10	8540.91	9004.90	9012.90	9031.41
8529.90	8540.99	9006.10	9013.10	9031.49
8531.20	8541.10	9006.20	9013.20	9031.80
8531.80	8541.21	9006.30	9013.80	9031.90
8531.90	8541.29	9006.40	9013.90	9032.10
8532.10	8541.30	9006.51	9016.00	9032.20
8532.21	8541.40	9006.52	9017.10	9032.81
8532.22	8541.50	9006.53	9017.20	9032.89
8532.23	8541.60	9006.59	9017.30	9032.90
8532.24	8541.90	9006.61	9017.80	9033.00
8532.25	8542.10	9006.62	9017.90	9101.11
8532.29	8542.21	9006.69	9018.11	9101.12
8532.30	8542.29	9006.91	9018.12	9101.19

8532.90	8542.60	9006.99	9018.13	9101.21
8533.10	8542.70	9007.11	9018.14	9101.29
8533.21	8542.90	9007.19	9018.19	9101.91
8533.29	8543.11	9007.20	9018.20	9101.99
8533.31	8543.19	9007.91	9022.12	9102.11
8533.39	8543.20	9007.92	9022.13	9102.12
8533.40	8543.30	9008.10	9022.14	9102.19
8533.90	8543.40	9008.20	9022.19	9102.21
8534.00	8543.81	9008.30	9022.21	9102.29
8536.10	8543.89	9008.40	9022.29	9102.91
8536.20	8543.90	9008.90	9022.30	9102.99
8536.30	8544.11	9009.11	9022.90	9108.11
8536.50	8544.19	9009.12	9023.00	9108.12
8536.61	8544.20	9009.21	9027.10	9108.19
8536.69	8544.30	9009.22	9027.20	9108.20
8536.90	8544.41	9009.30	9027.30	9108.90
8537.10	8544.49	9009.91	9027.40	9504.10
8537.20	8544.51	9009.92	9027.50	
8538.10	8544.59	9009.93	9027.80	
8538.90	8544.60	9009.99	9027.90	
8539.31	8544.70	9010.10	9030.10	
8540.11	8548.90	9010.41	9030.20	
8540.12	9001.10	9010.42	9030.31	
8540.20	9001.20	9010.49	9030.39	
8540.40	9001.30	9010.50	9030.40	
8540.50	9001.40	9010.60	9030.82	
8540.60	9001.50	9010.90	9030.83	

Source. Roadmap for Integration of Electronics Sector, Attachment 1.
[a] Profile selection: Computer components.

Table F.4. Healthcare: Priority Sector and Profile HS Selection

1211.10	3004.10	3401.19
1211.20	3004.20	3401.20
1211.30	3004.31	3401.30
1211.90	3004.32	4014.10
2936.10	3004.39	4818.40
2936.21	3004.40	4818.50
2936.22	3004.50	5208.21
2936.23	3004.90	6307.90
2936.24	3005.10	8419.20
2936.25	3005.90	8713.10
2936.26	3006.10	8714.20
2936.27	3006.20	9001.30
2936.28	3006.30	9018.11
2936.29	3006.40	9018.12
2936.90	3006.50	9018.13
2937.12	3006.60	9018.14

Table F.4. (Continued)

2937.19	3006.70	9018.19
2937.21	3006.80	9018.20
2937.22	3303.00	9018.31
2937.23	3304.10	9018.32
2937.29	3304.20	9018.39
2937.31	3304.30	9019.10
2937.39	3304.91	9019.20
2937.40	3304.99	9021.10
2937.50	3305.10	9021.29
3001.10	3305.20	9021.31
3001.20	3305.30	9021.39
3001.90	3305.90	9021.40
3002.10	3306.10	9021.50
3002.20	3306.20	9021.90
3002.30	3306.90	9022.12
3002.90	3307.10	9022.13
3003.10	3307.20	9022.14
3003.20	3307.30	9022.19
3003.31	3307.41	9022.21
3003.39	3307.49	9022.29
3003.40	3307.90	9402.90
3003.90	3401.11	

Source. Roadmap for Integration of Healthcare Sector, Attachment 1.

Table F.5. Textiles and Apparel: Priority Sector and Profile HS Selection

5001.00	5113.00	5206.33	5209.59 [a]	5301.30	5403.20	5408.33	5510.90
5002.00	5201.00	5206.34	5210.11	5302.10	5403.31	5408.34	5511.10
5003.10	5202.10	5206.35	5210.12	5302.90	5403.32	5501.10	5511.20
5003.90	5202.91	5206.41	5210.19	5303.10	5403.33	5501.20	5511.30
5004.00	5202.99	5206.42	5210.21	5303.90	5403.39	5501.30	5512.11
5005.00	5203.00	5206.43	5210.22	5304.10	5403.41	5501.90	5512.19
5006.00	5204.11	5206.44	5210.29	5304.90	5403.42	5502.00	5512.21
5007.10	5204.19	5206.45	5210.31	5305.11	5403.49	5503.10	5512.29
5007.20	5204.20	5207.10	5210.32	5305.19	5404.10	5503.20	5512.91
5007.90	5205.11	5207.90	5210.39	5305.21	5404.90	5503.30	5512.99
5101.11	5205.12	5208.11 [a]	5210.41	5305.29	5405.00	5503.40	5513.11
5101.19	5205.13	5208.12 [a]	5210.42	5305.90	5406.10	5503.90	5513.12
5101.21	5205.14	5208.13 [a]	5210.49	5306.10	5406.20	5504.10	5513.13
5101.29	5205.15	5208.19 [a]	5210.51	5306.20	5407.10	5504.90	5513.19
5101.30	5205.21	5208.21 [a]	5210.52	5307.10	5407.20	5505.10	5513.21
5102.11	5205.22	5208.22 [a]	5210.59	5307.20	5407.30	5505.20	5513.22
5102.19	5205.23	5208.23 [a]	5211.11	5308.10	5407.41	5506.10	5513.23
5102.20	5205.24	5208.29 [a]	5211.12	5308.20	5407.42	5506.20	5513.29
5103.10	5205.26	5208.31 [a]	5211.19	5308.90	5407.43	5506.30	5513.31
5103.20	5205.27	5208.32 [a]	5211.21	5309.11	5407.44	5506.90	5513.32
5103.30	5205.28	5208.33 [a]	5211.22	5309.19	5407.51	5507.00	5513.33
5104.00	5205.31	5208.39 [a]	5211.29	5309.21	5407.52	5508.10	5513.39
5105.10	5205.32	5208.41 [a]	5211.31	5309.29	5407.53	5508.20	5513.41

5105.21	5205.33	5208.42 [a]	5211.32	5310.10	5407.54	5509.11	5513.42
5105.29	5205.34	5208.43 [a]	5211.39	5310.90	5407.61	5509.12	5513.43
5105.40	5205.35	5208.49 [a]	5211.41	5311.00	5407.69	5509.21	5513.49
5106.10	5205.41	5208.51 [a]	5211.42	5401.10	5407.71	5509.22	5514.11
5106.20	5205.42	5208.52 [a]	5211.43	5401.20	5407.72	5509.31	5514.12
5107.10	5205.43	5208.53 [a]	5211.49	5402.10	5407.73	5509.32	5514.13
5107.20	5205.44	5208.59 [a]	5211.51	5402.20	5407.74	5509.41	5514.19
5108.10	5205.46	5209.11 [a]	5211.52	5402.31	5407.81	5509.42	5514.21
5108.20	5205.47	5209.12 [a]	5211.59	5402.32	5407.82	5509.51	5514.22
5109.10	5205.48	5209.19 [a]	5212.11	5402.33	5407.83	5509.52	5514.23
5109.90	5206.11	5209.21 [a]	5212.12	5402.39	5407.84	5509.53	5514.29
5110.00	5206.12	5209.22 [a]	5212.13	5402.41	5407.91	5509.59	5514.31
5111.11	5206.13	5209.29 [a]	5212.14	5402.42	5407.92	5509.61	5514.32
5111.19	5206.14	5209.31 [a]	5212.15	5402.43	5407.93	5509.62	5514.33
5111.20	5206.15	5209.32 [a]	5212.21	5402.49	5407.94	5509.69	5514.39
5111.30	5206.21	5209.39 [a]	5212.22	5402.51	5408.10	5509.91	5514.41
5111.90	5206.22	5209.41 [a]	5212.23	5402.52	5408.21	5509.92	5514.42
5112.11	5206.23	5209.42 [a]	5212.24	5402.59	5408.22	5509.99	5514.43
5112.19	5206.24	5209.43 [a]	5212.25	5402.61	5408.23	5510.11	5514.49
5112.20	5206.25	5209.49 [a]	5301.10	5402.62	5408.24	5510.12	5515.11
5112.30	5206.31	5209.51 [a]	5301.21	5402.69	5408.31	5510.20	5515.12
5112.90	5206.32	5209.52 [a]	5301.29	5403.10	5408.32	5510.30	5515.13
6201.99	6204.41	6208.92	6215.90	6304.93			
6202.11	6204.42 [b]	6208.99	6216.00	6304.99			
6202.12 [b]	6204.43	6209.10	6217.10	6305.10			
6202.13	6204.44	6209.20 [b]	6217.90	6305.20			
6202.19	6204.49	6209.30	6301.10	6305.32			
6202.91	6204.51	6209.90	6301.20	6305.33			
6202.92 [b]	6204.52 [b]	6210.10	6301.30	6305.39			
6202.93	6204.53	6210.20	6301.40	6305.90			
6202.99	6204.59	6210.30	6301.90	6306.11			
6203.11	6204.61	6210.40	6302.10	6306.12			
6203.12	6204.62 [b]	6210.50	6302.21	6306.19			
6203.19	6204.63	6211.11	6302.22	6306.21			
6203.21	6204.69	6211.12	6302.29	6306.22			
6203.22 [b]	6205.10	6211.20	6302.31	6306.29			
6203.23	6205.20 [b]	6211.31	6302.32	6306.31			
6203.29	6205.30	6211.32	6302.39	6306.39			
6203.31	6205.90	6211.33	6302.40	6306.41			
6203.32 [b]	6206.10	6211.39	6302.51	6306.49			
6203.33	6206.20	6211.41	6302.52	6306.91			
6203.39	6206.30 [b]	6211.42	6302.53	6306.99			
6203.41	6206.40	6211.43	6302.59	6307.10			
6203.42 [b]	6206.90	6211.49	6302.60	6307.20			
6203.43	6207.11 [b]	6212.10	6302.91	6307.90			
6203.49	6207.19	6212.20	6302.92	6308.00			
6204.11	6207.21 [b]	6212.30	6302.93	6309.00			
6204.12 [b]	6207.22	6212.90	6302.99	6310.10			
6204.13	6207.29	6213.10	6303.11	6310.90			
6204.19	6207.91	6213.20 [b]	6303.12				
6204.21	6207.92	6213.90	6303.19				
6204.22 [b]	6207.99	6214.10	6303.91				

Table F.5. (Continued)

6204.23	6208.11	6214.20	6303.92				
6204.29	6208.19[b]	6214.30	6303.99				
6204.31	6208.21	6214.40	6304.11				
6204.32 [b]	6208.22	6214.90	6304.19				
6204.33	6208.29	6215.10	6304.91				
6204.39	6208.91[b]	6215.20	6304.92				

Source. Roadmap for Integration of Textiles and Apparel Sector, Attachment 1.

[a] Profile selection: Cotton woven fabrics.

[b] Profile selection: Cotton woven apparel.

Table F.6. Wood-based Products: Priority Sector and Profile HS Selection

4401.10	4409.20 [a]	4412.29 [a]	9401.50
4401.21	4410.21	4412.92 [a]	9401.61
4401.22	4410.29	4412.93 [a]	9401.69
4401.30	4410.31	4412.99 [a]	9403.30
4402.00	4410.32	4413.00	9403.40
4405.00	4410.33	4414.00	9403.50
4406.10	4410.39	4415.10	9403.60
4406.90	4410.90	4415.20	
4407.10 [a]	4411.11 [a]	4416.00	
4407.24	4411.19	4417.00	
4407.25	4411.21 [a]	4418.10	
4407.26	4411.29 [a]	4418.20	
4407.29	4411.31	4418.30 [a]	
4407.91	4411.39	4418.40	
4407.92	4411.91	4418.50	
4407.99	4411.99	4418.90 [a]	
4408.10	4412.13 [a]	4419.00	
4408.31	4412.14 [a]	4420.10	
4408.39	4412.19	4420.90	
4408.90	4412.22 [a]	4421.10	
4409.10 [a]	4412.23 [a]	4421.90	

Source. Roadmap for Integration of Wood-Based Products Sector, Attachment 1.

[a] Profile selection: Hardwood plywood and flooring.

References

[1] Acoca, Brigitte. *Empowering E-Consumers: Strengthening Consumer Protection in the Internet Economy: Background Report.* Organisation for Economic Co-operation and Development (OECD), November 2009.

[2] Association of Southeast Asian Nations (ASEAN). "*ASEAN Economic Community (AEC) Blueprint.*" November 20, 2007. http://www.aseansec.org/21083.pdf.

[3] Asian Development Bank (ADB). *Designing and Implementing Trade Facilitation in Asia and the Pacific.* Mandaluyong City, Philippines: Asian Development Bank, 2009. http://www.adb.org/Documents/Books/Trade-Facilitation-Reference-Book/default.asp.

[4] Bureau van Dijk. Zephyr database. https://zephyr.bvdep.com/version-2010211/cgi/template.dll?product=24 (accessed various dates).

[5] Capannelli, Giovanni; Jong-Wa, Lee; Peter, Petri. Developing Indicators for Regional Economic Integration and Cooperation. *Asian Development Bank Working Paper on Regional Economic Integration*, No. 33, September 2009. http://papers.ssrn.com/sol3/papers.cfm?abstract id=1 546235.

[6] European Commission (EC), Enterprise and Industry Directorate-General. "*European Competitiveness Report*, 2009." EC SEC(2009)1657 final, March 2010. http://ec.europa.eu/enterprise/newsroom/cf/itemlongdetail.cfm?item id=3908.

[7] Financial Times. FDIMarkets database. http://www.fdimarkets.com (accessed various dates).

[8] International Monetary Fund. *Balance of Payments and International Investment Position Statistics*. http://www.imf.org/external/np/sta/bop/bop.htm (accessed April 22, 2010).

[9] International Monetary Fund. *International Investment Position (IIP) Statistics*. http://www.imf.org/external/np/sta/iip/iip.htm (accessed April 22, 2010).

[10] Kim, Soyoung; Jong-Wha, Lee; Cyn-Young, Park. "*The Ties that Bind Asia, Europe, and the United States*." ADB Economics Working Paper 192, February 2010. http://www.adb.org/Documents/Working-Papers/2010/Economics-WP192.pdf.

[11] OECD. *Benchmark Definition of Foreign Direct Investment* (Paris: OECD, 2008).

[12] Perdikis, Nicholas. "*Overview of Trade Agreements: Regional Trade Agreements*." In Handbook on International Trade Policy, edited by A; William, Kerr, D. James, Gaisford, 82-91. Northhampton, MA: Edward Elgar Publishing, Inc., 2007.

[13] Pierson, Mary, Y. "East Asia: Regional Economic Integration and Implications for the United States." *Law and Policy in International Business*, 1994, 25, 1161-85.

[14] United Nations University. "*Different Forms of Integration*." UNU Open Course Ware. http://ocw.unu.edu/programme-for-comparative-regional-integration-studies/introducing-regionalintegration/different-forms-of-integration/ (accessed February 16, 2009).

[15] U.S. International Trade Commission (USITC). East Asia: *Regional Economic Integration and Implications for the United States*. USITC Publication 2621. Washington, DC: USITC, May 1993.

[16] ———. *Logistic Services: An Overview of the Global Market and Potential Effects of Removing Trade Impediments*. USITC Publication 3770. Washington, DC: USITC, May 2005.

[17] ———. *Wood Flooring and Hardwood Plywood: Competitive Conditions Affecting the U.S.* Industries. USITC Publication 4032. Washington, DC: USITC, August 2008.

[18] ———. *Sub-Saharan Africa: Effects of Infrastructure Conditions on Export Competitiveness*, Third Annual Report. USITC Publication 4071. Washington, DC: USITC, April 2009.

[19] World Economic Forum (WEF). *Global Competitiveness Report* 2009–2010. Geneva: World Economic Forum, 2009. http://www.weforum.org/en/initiatives/gcp/index.htm.

[20] Acoca, Brigitte. *Empowering E-Consumers: Strengthening Consumer Protection in the Internet Economy*. Paris: Organisation for Economic Co-operation and Development (OECD), December 2009.

[21] Aldaba, Rafaelita, M; Josef T, Yap; Peter, A. Petri. "Investment and Capital Flows: Implications of the AEC." In *Realizing the AEC: A Comprehensive Assessment*, edited by G. Michael, Plummer and Chia Siow Yue, prepared for the ASEAN Secretariat, 2009.

[22] Amin, Ruzita Mohd; Zarina Hamid; Norma MD Saad. "*Economic Integration among ASEAN Countries: Evidence from Gravity Model.*" East Asian Development Network (EADN) Working Paper 40, February 2009.

[23] Aquino, Thomas, G. Philippine Senior Undersecretary of Trade and Industry. Written testimony submitted to the U.S. International Trade Commission in connection with inv. no. 332-511, ASEAN: *Regional Trends in Economic Integration, Export Competitiveness, and Inbound Investment for Selected Industries*, March 3, 2010.

[24] Asian Development Bank (ADB). *Designing and Implementing Trade Facilitation in Asia and the Pacific.* Mandaluyong City, Philippines: Asian Development Bank, 2009.

[25] ———. Emerging Asian Regionalism: *A Partnership for Shared Prosperity*, 2008.

[26] Association of Southeast Asian Nations (ASEAN). "*Agreement to Establish and Implement the ASEAN Single Window*," December 9, 2005. http://www.aseansec.org/18005.htm.

[27] ———. "*ASEAN Accelerates Integration of Priority Sectors.*" Media release, November 29, 2004. http://www.aseansec.org/16620.htm.

[28] ———. "*ASEAN Economic Community Blueprint.*" November 20, 2007. http://www.aseansec.org/21083.pdf.

[29] ———. "*ASEAN Economic Community Scorecard*: *Charting Progress towards Regional Economic Integration*," March 2010.

[30] ———. ASEAN Foreign Direct Investment Statistics Database. http://www.asean.org/18144.htm (accessed March 25, 2010).

[31] ———. "*ASEAN Integration in Services*," August 2009. http://www.aseansec.org/20440.htm.

[32] ———. "*ASEAN Investment Area: An Update.*" http://www.aseansec.org/6480.htm (accessed April 22, 2010).

[33] ———. *ASEAN Investment Report 2007*, n.d. http://www.aseansec.org/21406.pdf (accessed April 22, 2010).

[34] ———. *ASEAN Investment Report 2008,* n.d. http://www.aseansec.org/22111.pdf (accessed April 23, 2010).

[35] *ASEAN Investment Report 2009,* n.d. http://www.aseansec.org/documents/AIR2009.pdf (accessed April 23, 2010).

[36] ———. "*ASEAN Leaders Sign ASEAN Charter.*" News release, November 20, 2007. http://www.aseansec.org/21085.htm.

[37] ———. "*ASEAN Protocol on Enhanced Dispute Settlement Mechanism*," 2004. http://www.aseansec.org/16754.htm.

[38] ———. "*ASEAN Trade Facilitation Work Programme* 2007–2015, Revised 08 July 2009," July 8, 2009.

[39] ———. "*Cebu Declaration on the Acceleration of the Establishment of an ASEAN Community by 2015*," January 13, 2007. http://www.aseansec.org/19260.htm.

[40] ———. *Declaration of ASEAN Concord, Indonesia.* February 24, 1976. http://www.aseansec.org/1649.htm.

[41] ———. *Declaration of ASEAN Concord II (Bali Concord II), Indonesia,* October 7, 2003. http://www.aseansec.org/15159.htm.

[42] ———. "e-ASEAN Framework Agreement," November 24, 2000. http://www.aseansec.org/6267.htm. "*Economic Achievement.*" http://www.aseansec.org/11832.htm (accessed March 10, 2010). "Fact Sheet: An ASEAN Economic Community by 2015," August 20, 2008.

[43] ———. "*Fact Sheet: ASEAN Dispute Settlement System,*" February 24, 2009.

[44] ———. "*Fact Sheet: ASEAN Framework Agreement on Services,*" February 26, 2009.

[45] ———. "*Fact Sheet: Initiative for ASEAN Integration and Narrowing the Development Gap,*" January 12, 2010.

[46] ———. "*Highlights of the ASEAN Comprehensive Investment Agreement (ACIA).*" August 26, 2008. http://www.aseansec.org/20632.htm (accessed March 10, 2010).

[47] ———. "*Informal ASEAN Economic Ministers Meeting.*" Media statement, January 19-20, 2004. http://www.aseansec.org/15661.htm.

[48] ———. *Investment Guarantee Agreement*, 1987. http://www.aseansec.org/6464.htm.

[49] ———. "*Memorandum of Understanding on the Implementation of the ASEAN Single Window Pilot Project,*" February 5, 2010. http://www.customs.go.th/UploadFile/News/N0310009.pdf

[50] ———. *Roadmap for an ASEAN Community 2009–2015*, 2009. http://www.aseansec.org/publications/RoadmapASEANCommunity.pdf

[51] ———. "*Roadmap for the Integration of Logistics Services,*" August 24, 2007. http://www.aseansec.org/20883.pdf.

[52] ASEAN Comprehensive Investment Agreement, 2009. http://www.aseansec.org/documents/FINALSIGNED-ACIA.pdf (accessed April 23, 2010).

[53] ASEAN Economic Ministers. "*Joint Media Statement of the 41st ASEAN Economic Ministers' (AEM) Meeting*, Bangkok," August 13–14, 2009. http://www.aseansec.org/JMS-41st-AEM.pdf.

[54] ———. "*Joint Media Statement of the Fortieth ASEAN Economic Ministers' (AEM) Meeting.*" News Release, August 25–26, 2008. http://www.aseansec.org/22738.htm.

[55] ———. "*Joint Media Statement of the Thirty-Seventh ASEAN Economic Ministers' (AEM) Meeting.*" News Release, September 28, 2005. http://www.aseansec.org/17778.htm.

[56] ASEAN Single Window (ASW) Project. "*ASW Legal Memorandum of Understanding Agreed.*" ASW Project Web site, September 27, 2009. http://advanceiqc.com/advance/asw/legal/asw-legalmemorandum-of-understanding-agreed.

[57] ASEAN Trade In Goods Agreement, February 26, 2009. http://www.aseansec.org/22223.pdf.

[58] Atje, Raymond. "ASEAN Economic Community: In Search of a Coherent External Policy." In *Deepening Economic Integration: The ASEAN Economic Community and Beyond*, edited by Hadi Soesastro, 156–72. Economic Research Institute for ASEAN and East Asia (ERIA) Research Project Report 2007, no. 1–2, 2007. http://www.eria.org/research/images/pdf/PDF%20No.1-2/No.1-2-part2- 10.pdf.

[59] Benjelloun, Rachid. "*Technical Instructions: Consultancy Services on the Technical Architecture for the ASEAN Single Window Pilot Project.*" ASW Project Web site, March 31, 2010. http://advanceiqc.com/uploads/rfp-asw-pilot-project-component-1.pdf.

[60] Calvo-Pardo, Hector; Caroline Freund; Emanuel Ornelas. "*The ASEAN Free Trade Agreement: Impact on Trade Flows and External Trade Barriers*." The World Bank. Policy Research Working Paper 4960, June 2009.

[61] Capannelli, Giovanni; Jong-Wha, Lee; Peter, Petri. "Developing Indicators for Regional Economic Integration and Cooperation." Asian Development Bank. *Working Paper Series on Regional Economic Integration*, 33, September 2009.

[62] Chia, Siow Yue. "*Regional Trade Policy Cooperation and Architecture in East Asia*." Asian Development Bank Institute. ADBI Working Paper Series 191, February 2010.

[63] Conference of Asia Pacific Express Carriers (CAPEC). "*Express Delivery Services (EDS): ASEAN Regulatory Matrix and International Best Practices*," May 2009.

[64] Corbett, Jenny; So Umezaki. "Overview: Deepening East Asian Economic Integration." In *Deepening East Asian Economic Integration*, edited by Jenny Corbett; So Umezaki, 1-57. Economic Research Institute for ASEAN and East Asia (ERIA) Research Project Report 2008, no. 1, March 2009. http://www.eria.org/pdf/research/y2008/no1/DEI-00-Front Cover.pdf.

[65] Cordenillo, Raul L. "*The Future of the ASEAN Free Trade Area and the Free Trade Areas between ASEAN and Its Dialogue Partners*," n.d. http://www.aseansec.org/18425.htm (accessed March 10, 2010).

[66] De Souza; Robert, Mark Goh; Sumeet Gupta; Luo Lei. *An Investigation into the Measures Affecting the Integration of ASEAN's Priority Sectors (Phase 2): The Case of Logistics.* Jakarta: ASEAN- Australia Development Cooperation Program, 2007. http://www.aseansec.org/aadcp/repsf/publications.html.

[67] Díaz-Pinés, Agustín. *Indicators of Broadband Coverage*. Paris: OECD, December 10, 2009. http://www.oecd.org/dataoecd/41/39/44381795.pdf.

[68] Duek, Ana. "*ASEAN Economic Integration: Evolving Patterns of Trade in AFTA*." Asia Competitiveness Institute (ACI) Working Paper Series, December 2008.

[69] Economic and Social Commission for Asia and the Pacific (ESCAP). "*ASEAN and Trade Integration*." Trade and Investment Division. Staff Working Paper 01/09, April 8, 2009.

[70] ESCAP. "*Trade Statistics in Policymaking*," n.d. http://www.unescap.org/tid/publication/tipub2559.pdf (accessed March 10, 2010).

[71] European Commission. "*Overview of FTA and Other Trade Negotiations*," updated April 21, 2010. http://trade.ec.europa.eu/doclib/docs/2006/december/tradoc 118238.pdf.

[72] Galexia. *Harmonisation of E-Commerce Legal Infrastructure in ASEAN*, April 2008. http://www.galexia.com/public/research/articles/research articles-art53 .html.

[73] Greenwald, Alyssa. "The ASEAN-China Free Trade Area (AFTA): A Legal Response to China's Economic Rise?" *Duke Journal of Comparative and International Law*, 16, no. 1, 2006, 193-217.

[74] Guerrero, Rosabel, B. "Regional Integration: The ASEAN Vision in 2020." *IFC Bulletin*, no. 32 (January 2010), 52-58.

[75] Hiebert, Murray. U.S. Chamber of Commerce. Written testimony submitted to the U.S. International Trade Commission in connection with inv. no. 332-511, *ASEAN: Regional Trends in Economic Integration, Export Competitiveness, and Inbound Investment for Selected Industries*, March 12, 2010.

[76] ———. U.S. Chamber of Commerce. Written testimony submitted to the U.S. International Trade Commission in connection with inv. no. 332-511, ASEAN:

Regional Trends in Economic Integration, Export Competitiveness, and Inbound Investment for Selected Industries, March 10, 2010.

[77] Hollweg, Claire; Marn-Heong, Wong. "*Measuring Regulatory Restrictions in Logistics Services*." Economic Research Institute for ASEAN and East Asia (ERIA) Discussion Paper Series ERIADP-2009-14, May 2009. http://www.eria.org/pdf/ERIA-DP-2009-14.pdf.

[78] International Telecommunications Union (ITU). ICT Eye database. http://www.itu.int/ITUD/ICTEYE/Indicators/Indicators.aspx (accessed April 21, 2010).

[79] Katsiak, Ann. "*ASW Project Improves ASEAN Capacity to Harmonize Data Requirements for Clearing Goods across Borders*." ASW Project Web site, January 2, 2010. http://advanceiqc.com/advance/asw/technical/asw-project-improves-asean-capacity-toharmonize-data-requirements-for-clearing-goods-across-borders.

[80] Le, Quang Anh. "*Briefing on ASEAN Single Window to US-ASEAN Business Council*." Presentation for a roundtable discussion at the US-ASEAN Business Council, Singapore, March 9, 2010.

[81] Malaysian Institute of Economic Research, ADBI, and East-West Center. "Regional Market for Goods, Services, and Skilled Labor." In *Realizing the AEC: A Comprehensive Assessment*, edited by G; Michael, Plummer, Chia Siow Yue. Prepared for the ASEAN Secretariat. Singapore: Institute of Southeast Asian Studies, 2009.

[82] Malaysian Ministry of International Trade and Industry. "*16th ASEAN Economic Ministers' Retreat*." Media release, February 28, 2010. http://www.miti.gov. my/cms/content.jsp?id=com.tms.cms.article.Article_14d32d58-c0a81573- 2f1 d2f1 d-4e630e07.

[83] Mealy, Mark. US-ASEAN Business Council. Written testimony submitted to the U.S. International Trade Commission in connection with inv. no. 332-511, *ASEAN: Regional Trends in Economic Integration, Export Competitiveness, and Inbound Investment for Selected Industries*, February 4, 2010.

[84] Narjoko, Dionisius; Pratiwi Kartika. "Narrowing the Development Gap in ASEAN through the AEC." In *Realizing the AEC: A Comprehensive Assessment*, edited by G; Michael, Plummer, Chia Siow Yue. Prepared for the ASEAN Secretariat. Singapore: Institute of Southeast Asian Studies, 2009.

[85] Organisation for Economic Co-operation and Development (OECD). *OECD Information Technology Outlook, 2008*. Paris: OECD, 2008.

[86] Petri, Peter A. "Competitiveness and Leverage: Benefits from an ASEAN Economic Community." In *Realizing the AEC: A Comprehensive Assessment*, edited by G; Michael, Plummer, Chia Siow Yue. Prepared for the ASEAN Secretariat. Singapore: Institute of Southeast Asian Studies, 2009.

[87] Pitsuwan, Dr. Surin, Secretary-General of ASEAN. "*Progress in ASEAN Economic Integration since the Adoption of the ASEAN Charter*." Speech at the Business Dialogue with the Federation of Japanese Chambers of Commerce and Industry in ASEAN. June 29, 2009.

[88] Plummer, Michael, G; Chia Siow, Yue. "Assessing the Economics of the ASEAN Economic Community: Introduction." In *Realizing the AEC: A Comprehensive Assessment*, edited by G; Michael, Plummer, Chia Siow Yue. Prepared for the ASEAN Secretariat. Singapore: Institute of Southeast Asian Studies, 2009.

[89] Prime Minister's Office of Malaysia. "*Liberalisation of the Services Sector*." News release, April 22, 2009. http://www.mida.gov.my/en v2/uploads/images/invest/invest-pdf/liberaliseservicePS .pdf.

[90] Severino, Rodolfo, C. "The ASEAN Developmental Divide and the Initiative for ASEAN Integration." *ASEAN Economic Bulletin*, 24, no. 1, 2007, 35-44.

[91] Soesastro, Hadi. "*Accelerating ASEAN Economic Integration: Moving beyond AFTA*." CSIS Working Paper Series WPE-091, March 2005.

[92] ———. "Implementing the ASEAN Economic Community (AEC) Blueprint." In *Deepening Economic Integration: The ASEAN Economic Community and Beyond*, edited by Hadi, Soesastro, 47–59. ERIA Research Project Report 2007, no. 1-2, 2007. http://www.eria.org/research/images/pdf/PDF%20No.1-2/No.1-2-part2-3.pdf.

[93] Sussangkarn, Chalongphob. "*Revitalizing ASEAN Competitiveness*." *East-West Dialogue* 2, (September 2008):10–11.http://forum.eastwestcenter.org/eastwestdialogue/ewd2/commentary-chalongphob/.

[94] Thangavelu, Shandre, M. "Non-Tariff Barriers, Integration and Export Growth in ASEAN." *Preliminary draft prepared for GEP Conference at University of Nottingham*, Malaysia, January 2010.

[95] U.S. International Trade Commission (USITC). Logistic Services: An Overview of the Global Market and Potential Effects of Removing Trade Impediments. USITC Publication 3770. Washington, DC: USITC, 2005.

[96] Vietnam E-Commerce and IT Agency (VECITA). *Vietnam E-Commerce Report 2008*. Hanoi: VECITA, 2009.

[97] World Bank. Doing Business database. http://www.doingbusiness.org (accessed April 20, 2010).

[98] ———. Logistics Performance Index database. http://info.worldbank.org/etools/trade survey/mode1b.asp (accessed April 20, 2010).

[99] World Trade Organization (WTO). "*Report of the Working Party on the Accession of Cambodia: Addendum; Part II—Schedule of Specific Commitments in Services; List of Article II MFN Exemptions*." WT/ACC/KHM/2 1/Add.2, August 19, 2003.

[100] ———. "Working Party on the Accession of Viet Nam: Schedule CLX—Viet Nam; Part II—Schedule of Specific Commitments in Services; List of Article II MFN Exemptions, Addendum." WT/ACC/VNM/48/Add.2, October 27, 2006.

[101] Asian Productivity Organization. *APO Productivity Databook 2009*. Tokyo: Keio University Press, 2009.

[102] Association of Southeast Asian Nations (ASEAN). "*Roadmap for Integration of Electronics Sector*." November 29, 2004. http://www.aseansec.org/16656.htm.

[103] Bradsher, Keith. "*As Labor Costs Soar in China, Manufacturers Look to Vietnam*." New York Times, June 18, 2008. http://www.nytimes.com/2008/06/18/business/worldbusiness/18invest.html?_r=3&pagewanted=1 (accessed April 26, 2010).

[104] Bureau van Dijk. Zephyr database. Available for a fee at http://zephyr2.bvdep.com/version201055/Home.serv?product=zephyrneo (accessed May 3, 2010).

[105] Corbett, Jenny; So Umezaki, eds. "Table 6: FDI Policy." *In Deepening East Asian Economic Integration*. ERIA Research Project Report 2008, No. 1, March 2009. http://www.eria.org/research/y2008-no1.html (accessed March 16, 2010).

[106] Doner, Rick; Bryan Ritchie. "*Economic Crisis and Technological Trajectories: Hard Disk Drive Production in Southeast Asia.*" MIT Japan Program, Working Paper 01.06, 2001.

[107] Ernst, Dieter. "*Global Production Networks in East Asia's Electronics Industry and Upgrading Prospects in Malaysia.*" In Global Production Networking and Technological Change in East Asia, edited by M; Shahid Yusuf, Anjum Altaf; Kaoru Nabeshima. Washington, D.C.: World Bank and Oxford University Press, 2004.

[108] Foster, William; Zhang Cheng; Jason Dedrick; Kenneth, L. Kraemer. *Technology and Organizational Factors in the Notebook Industry Supply Chain.* Tempe, AZ: Institute for Supply Management and the W.P. Carey School of Business, 2006.

[109] Fukase, Emiko, Will, Martin. "*A Quantitative Evaluation of Vietnam's Accession to the ASEAN Free Trade Area.*" World Bank Policy Research Working Paper no. 2220, November 1999. http://ideas.repec.org/p/wbk/wbrwps/2220.html.

[110] IBIS World. "*Computer Body Manufacturing in China.*" IBIS World and ACMR China Industry Report 4041, May 21, 2009. Available for a fee at http://www.ibisworld.com/ (accessed June 22, 2009).

[111] ———. "*Global Computer Hardware Manufacturing.*" IBIS World Industry Report C2523-GL, May 5, 2009. Available for a fee at http://www.ibisworld.com/ (accessed June 22, 2009).

[112] Intel Web site. "RosettaNet." https://supplier.intel.com/static/B2Bi/RosettaNet.htm (accessed April 26, 2010).

[113] Irwin, Douglas, A. *Free Trade under Fire.* 3rd ed. Princeton: Princeton University Press, 2009.

[114] ISuppli. "*Seagate Maintains Top Ranking in Hard Disk Drive Market in Q4,*" March 22, 2010. http://www.isuppli.com/News/Pages/Seagate-Maintains-Top-Ranking-in-Hard-Disk-DriveMarketin-Q4.aspx.

[115] Martin, Susan Taylor. "*For Jabil Circuit, Vietnam Is the New China.*" St. Petersburg Times, June 8, 2008. http://www.tampabay.com/news/for-jabil-circuit-vietnam-is-the-new-china/609847 (accessed April 28, 2010).

[116] Mirza, Hafiz; Axele, Giroud. "*Regional Integration and Benefits from Foreign Direct Investment in ASEAN Economies: The Case of Viet Nam.*" Asian Development Review 21, no. 1, 2004.

[117] Mitarai, Hisami. "*Issues in the ASEAN Electric and Electronics Industry and Implications for Vietnam.*" Working paper, Nomura Research Institute, February 2005. http://www.grips.ac.jp/vietnam/KOarchives/doc/EB02_IIPF/7EB02_Chapter6.pdf (accessed April 15, 2010).

[118] Montevirgen, Clyde. "*Industry Surveys: Semiconductors.*" Standard & Poor's Industry Surveys. November 2009.

[119] Multilateral Investment Guarantee Agency. *Benchmarking FDI Competitiveness in Asia.* Washington, D.C.: World Bank Group, 2003.

[120] Parsons, David. "*An Investigation into the Measures Affecting the Integration of ASEAN's Priority Sectors (Phase 2): The Case of Electronics.*" With the assistance of Mawardi Maghfuri, Bintoro Ariyanto, and Rina Oktaviani. Final Report, REPSF Project No. 06/001b, June 2007.

[121] Pecht, Michael; Sanjay, Tiku. "*Bogus!*" IEEE Spectrum, May 2006. http://spectrum.ieee.org/computing/hardware/bogus (accessed May 4, 2010).

[122] Rasiah, Rajah. "Expansion and Slowdown in Southeast Asian Electronics Manufacturing." *Journal of the Asia Pacific Economy*, 14, no. 2, 2009.

[123] RosettaNet Web site. http://www.rosettanet.org/ (accessed May 7, 2010).

[124] Santiago, Ernie, B. "*Development of ASEAN Framework for Trade Negotiations: Electronics Industry*." Paper for the United States Agency for International Development, March 2007. http://pdf.usaid.gov/pdf docs/PNADJ685 .pdf.

[125] Stoler, Andrew, L; Phan Van, Sam. "*Vietnam: Intel and the Electronics Sector*." In Trade and Poverty Reduction in the Asia-Pacific Region, edited by L; Andrew, Stoler, Jim Redden; Lee Ann, Jackson. Cambridge: Cambridge University Press and the World Trade Organization, 2009.

[126] United States Trade Representative. "*Indonesia*." 2009 National Trade Estimate Report on Foreign Trade Barriers. Washington, DC: USTR, 2009. http://www.ustr.gov/about-us/press-office/reports-andpublications/2009/2009-national-trade-estimate-report-foreign-trad.

[127] U.S. Bureau of Labor Statistics (BLS). "*Computer and Electronic Product Manufacturing*." Career Guide to Industries, 2010–11 Edition. http://www.bls.gov/oco/cg/cgs010.htm (accessed various dates).

[128] World Bank. Worldwide Governance Indicators database. http://info.worldbank.org/governance/wgi/mc_countries.asp (accessed April 22, 2010).

[129] ———. *WITS*, Integrated Data Warehouse. http://wits.worldbank.org/witsweb (accessed January 5, 2010).

[130] World Trade Organization (WTO). *Information Technology Agreement: Schedules of Concessions*. http://www.wto.org/english/tratop e/inftec e/itscheds e.htm (accessed April 27, 2010).

[131] Association of Southeast Asian Nations (ASEAN) Competitiveness Enhancement (ACE) Project. "*Textiles and Apparel Supply Chain, Corridor Diagnostics, SWOT Analysis, and Remediation Action Plan for Light-Weight Woven Cotton Fabric (HS 5208) Exported from Thailand to Vietnam*." Prepared for United States Agency for International Development (USAID) Asia, May 2009.

[132] Ayling, Joe. "*Analysis: Emerging Exporters Eye China Market Share*." Just Style, April 16, 2010. http://www.just-style.com/article.aspx?id=107441&lk=s (accessed April 27, 2010).

[133] ———. "Speaking with Style: Janet Fox, JC Penney." *Just Style*, April 27, 2010. http://www.just-style.com/article.aspx?id=107496 (accessed April 27, 2010).

[134] BDLINK Cambodia and ACE Project Team. "*Textile and Apparel Supply Chain Corridors Light-Weight Woven Cotton Fabric (HS 5208) Thailand to Cambodia, SWOT Analysis, and Remediation Action Plan*." Prepared for United States Agency for International Development (USAID) Asia, July 2009.

[135] Blake, Chris. "Blaming China: Indonesian Garment Makers Say Free Trade Pact Leaves Them on Brink of Collapse." *Los Angeles Times*, April 26, 2010. http://www.latimes.com/business/nationworld/wire/sns-ap-as-indonesia-china-tradefears,0,3780726.story.

[136] Fee, Jonathan; Shannon, BJ. "*Free Trade Agreements: Japan's Free Trade Network*." Cotton Council, Inc. http://www.cottonusasupplychain.com/sourcing/fta-japan.htm (accessed March 16, 2010).

[137] Galvez, James Konstantin. "*RP Eyes Duty-Free Trade Deal with US to Boost Textile Industry*." Manila Times, April 20, 2010. http://www.manilatimes.net/index.php/news/nation/15616-rp-eyes-dutyfree-trade-deal-with-us-to-boost-textile-industry.

[138] Greenwood, Tom. "*WFP Laos-Logistics Overview*." World Food Programme. http://www.wfplogistics.org/content/wfp-laos-logistics-overview (accessed May 7, 2010).

[139] Likhikiatikul, Jittdarat. "*Global Fashion Brands Sign 23 MOU with ASEAN Suppliers*." ScottAsia Communications, June 15, 2010. http://scottasia.net/site/20 1 0/06/global-fashion-brands-sign-23- mou-with-asean-suppliers/.

[140] Minor, Peter. "*Upgrading ASEAN Textiles and Apparel Supply Chains*." Nathan Associates, Inc. Prepared for United States Agency for International Development (USAID), June 16, 2008.

[141] Textile Outlook International. "*World Textile and Apparel Trade and Production Trends: China, South- East Asia and South Asia*," September–October 2008, 17-36.

[142] U.S. Department of State. "*Background Note on Vietnam*," October 2009. http://www.state.gov/r/pa/ei/bgn/4130.htm (accessed April 29, 2010).

[143] Vietnam Textile Organization (VITAS). "*Garments Top Nation's Export Lists*," January 12, 2010.http://www.vietnamtextile.org/en/ChiTietTinTuc.aspx?MaTinTuc=493&Matheloai=5 (accessed April 27, 2010).

[144] ———. "*Vietnam Receives Just a Small Slice of the Garment Industry Pie*," January 26, 2010. http://www.vietnamtextile.org/en/ChiTietTinTuc.aspx?MaTinTuc=523&Matheloai=5 (accessed April 27, 2010).

[145] American Hardwood Export Council (AHEC). "*Southeast Asia Market Report*," November 2009.

[146] Association of Southeast Asian Nations (ASEAN). "*Agreement on the Common Effective Preferential Tariff (CEPT) Scheme for the ASEAN Free Trade Area*, Consolidated Package of Tariff Reductions for 2008," February, 25, 2009. http://www.asean.org/19802.htm.

[147] ———. "ASEAN Economic Community Blueprint." November 20, 2007. www.aseansec.org/21083.pdf.

[148] ———. "*ASEAN Economic Community Scorecard*." Jakarta: ASEAN Secretariat, March 2010.

[149] ———. ASEAN Investment Report 2009: *Sustained FDI Flows Dependent on Global Economic Recovery*, n.d. http://www.aseansec.org/18657.htm.

[150] ———. "CEPT-AFTA Rules of Origin for Wood-Based Products." http://www.miti.gov.my/cms/documentstorage/com.tms.cms.document.Document_7fa39e36-c0a81573-2d952d95-641ed0fc/ST%20Rules%20for%20Wood.doc (accessed March 20, 2010).

[151] ———. "Forest and Timber Certification." http://www.aseanforest- chm.org/issue_pages/sfm_implementation/forest_and_timber_certification.html (accessed March 8, 2010).

[152] ———. "Highlights of the ASEAN Comprehensive Investment Agreement (ACIA)." PowerPoint presentation, August 26, 2008. www.aseansec.org/21885.ppt.

[153] ———. "ASEAN-China Free Trade Area." Annex 2, "*Modality for Tariff Reduction/Elimination for Tariff Lines Placed in the Sensitive Track*." http://www.aseansec.org/19105.htm (accessed April 6, 2010).

[154] ———. *"Non-Tariff Barriers."* http://www.aseansec.org/10114.htm. Non-Tariff Measures Database. http://www.aseansec.org/16355.htm (accessed various dates).

[155] ———. "Roadmap for Integration of Wood-Based Products Sector." *Appendix 1 to the ASEAN Sectoral Integration Protocol for Wood-Based Products*, 2007. http://www.aseansec.org/16741.htm.

[156] ———. *Statistics of Foreign Direct Investment in ASEAN.* 7th Edition, 2005. http://www.aseansec.org/18177.htm.

[157] ———. Temporary Exclusion List and Sensitive List for Manufacturing, Agriculture, Fishery, Forestry, Mining and Quarrying Sectors, February 2008. http://www.aseansec.org/18657.htm.

[158] ———. "Terrestrial Ecosystems." Chapter 5, *Fourth ASEAN State of the Environment Report*. Jakarta: ASEAN Secretariat, 2009, 52-65. http://www.aseansec.org/18177.htm.

[159] Asia Forest Partnership (AFP). *"AFP in Brief,"* July 28, 2008. http://www.asiaforests.org/index.php.

[160] Austria, Myrna, S. The Pattern of Intra-A SEAN Trade in Priority Goods Sectors. Regional Economic Policy Support Facility (REPSF). *ASEAN-Australia Development Cooperation Program*, August 2004.

[161] Azarcon, Chulia J. "Comparative Tariff Policies of the ASEAN Member Countries." *Journal of Philippine Development* 43, Vol. 24, No. 1 (First Semester 1997).

[162] *Bamboo Flooring.* n.d. http://extension.ucdavis.edu/unit/green_building_and_sustainability/pdf/resources/bamboo_floori ng.pdf.

[163] Bibi, Muhd Safwan, Abdullah; Zaeidi Haji, Berudin. *"Forest Land Management in Brunei Darussalam."* Paper presented at symposium, "100 Years of Modern Land Administration," Brunei, November 11, 2009. http://www.tanah.gov.bn/Symposium/Forestry%20Department%20%5BFinal%5D.doc.

[164] Boungnakeo, K. *"Forest Law Enforcement and Governance in Lao PDR."* Forest Trends. http://www.forest-trends.org/~foresttr/documents/files/doc 832.pdf (accessed March 10, 2010).

[165] Buckley, Michael. "A Flooring Market Update." *Panels & Furniture Asia*, May/June 2009, 39-44.

[166] Congressional Research Service (CRS). *United States Relations with the Association of Southeast Asian Nations (ASEAN).* CRS Report for Congress, R40933. Washington, DC: CRS, November 16, 2009.

[167] Corbett, Jenny; So Umezaki, eds. *Deepening East Asian Economic Integration.* Economic Research Institute for ASEAN and East Asia (ERIA). Research Project Report 2008, No. 1. Jakarta: ERIA, 2008. http://www.eria.org/research/y2008-no1.html (accessed March 24, 1010).

[168] Diete, T; Tumaneng, Ian, S; Ferguson, Donald, MacLaren. "Log Export and Trade Policies in the Philippines: Bane or Blessing to Sustainable Forest Management?" *Forest Policy and Economics*, 2005, 7, 187-98.

[169] B. Durst, Patrick; et al., eds. *Forests Out of Bounds: Impacts and Effectiveness of Logging Bans in Natural Forests in Asia-Pacific.* Food and Agriculture Organization of the United Nations (FAO). Asia- Pacific Forestry Commission. Bangkok, Thailand: FAO, 2001.

[170] Djani, Dian Triansyah. "*ASEAN-EU Economic Relations: Maximizing Mutual Benefits*," July 19, 2007. http://www.kbriwina.at/downloads/Presentation Amb Triansyah.pdf.

[171] European Commission (EC). "*Forest Law Enforcement, Governance and Trade (FLEGT)*." http://ec.europa.eu/environment/forests/flegt.htm (accessed January 8, 2010).

[172] Food and Agriculture Organization of the United Nations (FAO). FAOSTAT Database. http://faostat.fao.org/site/626/default.aspx#ancor (accessed January 7, 2010).

[173] ———. National Forestry Programmes Update No. 34: ASIA and the Pacific, December 2000. ftp://ftp.fao.org/docrep/fao/003/x6900e/x6900e00.pdf.

[174] ———. *Global Forest Resources Assessment 2005: Progress towards Sustainable Forest Management*. Rome: FAO, 2006. http://www.fao.org/forestry/fra/fra2005/en/.

[175] Forest Trends. *Timber Markets and Trade between Laos and Vietnam: A Commodity Chain Analysis of Vietnamese-Driven Timber Flows*. January 2010.

[176] Global Trade Alert. "*Vietnam Tax Exemption for Exporters of Wood Products*." Center for Economic Policy Research, March 22, 2010. http://globaltradealert.org/measure/vietnam-tax-exemptionexporters-wood-planks.

[177] Global Trade Information Service, Inc. (GTIS). *Global Trade Atlas Data*base (accessed various dates).

[178] Ha, Ngyyet. "*Chips Down for Fabricated Woods*." VietnamNet, June 28, 2004 (accessed April 15, 2010). http://english.vietnamnet.vn/news/2004/06/170380/.

[179] Hunt, Glenn. Presentation to AMRC/AusAID Greater Mekong Conference, September 26-27, 2007.http://www.mekong.es.usyd.edu.au/events/past/conference07/images/papers%20&%20presentati ons/Glenn%20Hunt%20presentation.pdf.

[180] Illegal-Logging.Info. "EU FLEGT: Due Diligence." http://www.illegal-logging.info/approach.php?a id=120 (accessed January 16, 2010).

[181] International Tropical Timber Organization (ITTO). *Tropical Timber Market Report*, 14, no. 22 (November 16-30, 2009). http://www.itto.int/.

[182] ———. *New Directions for Tropical Plywood: Proceedings of an ITTO/FAO International Conference on Tropical Plywood, September* 26–28, 2005. Beijing, China. *ITTO Technical Series*, 2006, 26.

[183] *Jakarta Post*. "*Thousands Rally Against Free Trade Treaty in Surabaya, Semarang*," January 21, 2010. http://www.thejakartapost.com/news/2010/01/21/thousands-rally-against-free-trade-treatysurabaya-semarang.html (accessed March 6, 2010).

[184] Lang, Chris. "Deforestation in Vietnam, Laos and Cambodia." In *Deforestation, Environment, and Sustainable Development: A Comparative Analysis*, edited by D.K. Vajpeyi. Westport, CT and London: Praeger, 2001, 111–137. http://chrislang.org/2001/01/03/deforestation-in-vietnam-laosand-cambodia/.

[185] Lim, Hank, Giokhay; Kester TAY, Yi-Xun. "Regional Integration and Inclusive Development: Lessons from ASEAN Experience." Asia-Pacific Research and Training Network on Trade. Working Paper Series 59, December 2008. http://www.unescap.org/tid/artnet/pub/wp5908.pdf.

[186] Lopez-Casero, Federico. "EU-ASEAN Workshop on Trade in Wood Products in South East Asia, Kuala Lumpur, May 24–25 2006: Summary." *Institute for Global Environmental Strategies (IGES)*, June 20, 2006.

[187] Ma, Qiang; Jeremy, S. Broadhead, eds. *An Overview of Forest Products Statistics in South and Southeast Asia*. EC-FAO Partnership Programme, December 2002.

[188] Macek, Paul; Kalaya, Chareonying. "*Thailand Log Ban*." *TED Case Studies* 2, no. 1 (January 1993). http://www1.american.edu/TED/THAILOG.HTM.

[189] Malaysia, Government of. *Royal Malaysia Customs Department*, HS-Explorer (accessed April 10, 2010).

[190] Malaysia Timber Council. "*FAQs on Malaysia's Forestry & Timber Trade*." Kuala Lumpur, Malaysia, n.d.

[191] Malaysia Timber Industry Board (MTIB). Posted cess rates for wood products. http://www.mtib.gov.my/index.php?option=com content&view=article&id=93:cessrate &catid=40: service (accessed March 27, 2010).

[192] ———. "*Credit Opportunities for Forest Plantations*," n.d. http://www.mtib.gov.my/ index.php?option=com_content&view=article&id=94:forest-lantation&catid=40:service (accessed February 17, 2010).

[193] Nathan Associates. "*Toward a Roadmap for Integration of the ASEAN Logistics Sector*." Washington, DC: USAID, March 2007.

[194] Oliver, Rupert, Ben, Donkor. "*Monitoring the Competitiveness of Tropical Timber*." Yokohama, Japan: International Tropical Timber Organization (ITTO), forthcoming.

[195] Ofreneo, Rene, E. "Regional Integration in the Wood-Based Industry: Quo Vadis?" *Paper presented at ASEAN Business Summit, Cebu City, Philippines*, December 2006.

[196] Ozinga, Saskia; Iola, Leal. "Forest Watch Special Report: Update Report on FLEGT Voluntary Partnership Agreements." EU Forest Watch, February 2010. http://www.illegallogging.info/uploads/VPAupdate.pdf.

[197] Philippines Forest Management Bureau. "*2007 Forestry Statistics*." http://forestry.denr.gov.ph/stat2007.htm (accessed February 15, 2010).

[198] Phimmavong, S; Ozarska, B; Midgley, S; Keenan, R. "Forest and Plantation Development in Laos: History, Development and Impact for Rural Communities." *International Forestry Review*, 2009, 11(4).

[199] Purbawiyatna, Alan; Markku, Simula. Developing Forest Certification: Towards Increasing the Comparability and Acceptance of Forest Certification Systems Worldwide. International Tropical Timber Organization (ITTO). *ITTO Technical Series*, 2008, 29,

[200] Resosudarmo, Budy, P; Arief Anshory, Yusuf. "Is the Log Export Ban Effective? Revisiting the Issue through the Case of Indonesia." Australian National University. *Economics and Environment Network Working Paper EEN0602*, Canberra, Australia, June, 2006, 13.

[201] Sabah, Government of. "*Forest Industry in Sabah*." http://www.forest.sabah.gov.my/ conservation/ForestIndustry.pdf (accessed May 3, 2010).

[202] Saigon Times. "*Sumitomo Forestry to Build Plywood Factory in Long An*," January 13, 2010. http://english.thesaigontimes.vn/Home/business/corporate-news/8341/.

[203] Sally, Razeen. "Regional Economic Integration in Asia: The Track Record and the Prospects." *European Centre for International Political Economy*. Occasional Paper No. 2/2010. Brussels: ECIPE, 2010.

[204] Seneca Creek Associates. "*Illegal Logging and Global Wood Markets: The Competitive Impacts on the U.S. Wood Products Industry.*" Prepared for American Forest & Paper Association. Poolesville, MD, October 2004.

[205] Simula, Markku. "*The Pros and Cons of Procurement: Developments and Progress in Timber-Producing Policies as Tools for Promoting the Sustainable Management of Tropical Forests.*" International Tropical Timber Organization (ITTO). Technical Series 34, Yokohama, Japan, April 2010.

[206] Tachibana, Satoshi. "*Forest-Related Industries and Timber Exports of Malaysia: Policy and Structure.*" Research Paper, University of Tokyo, c1999. http://enviroscope.iges.or.jp/modules/envirolib/upload/1503/attach/3ws-12-tachibana.pdf.

[207] ———. "Impacts of Log Export Restrictions in Southeast Asia on the Japanese Plywood Market: An Econometric Analysis." *Journal of Forestry Research*, 5, no. 2 (2000), 51-57.

[208] U.S. Department of Agriculture Foreign Agricultural Service (FAS). *Burma: Solid Wood Products; Annual Report, 2005.* GAIN Report no. BM5005, March 4, 2005.

[209] ———. *Malaysia: Solid Wood Products; Annual Report, 2006.* GAIN Report no. MY6022, September 13, 2006.

[210] ———. *Thailand: Solid Wood Products; Annual Report*, 2006. GAIN Report no. MY6022, July 13, 2006.

[211] ———. *Vietnam: Solid Wood Products, Annual Report*, 2006. GAIN Report no. VM6075, December 8, 2006.

[212] U.S. Department of State. "*2010 Investment Climate Statements.*" March 2010. http://www.state.gov/e/eeb/rls/othr/ics/2010/index.htm (accessed April 15, 20 10, by country).

[213] U.S. International Trade Commission (USITC). *Wood Flooring and Hardwood Plywood: Competitive Conditions Affecting the U.S. Industries.* USITC Publication 4032. Washington, DC: USITC, 2008.

[214] U.S. Trade Representative (USTR). *Memorandum of Understanding between the Government of the United States of America and the Republic of Indonesia on Combating Illegal Logging and Associated Trade*, November 2006. http://www.ustr.gov/sites/default/files/US%20Indonesia%20MOU%20to%20Combat%20Illegal%20 Logging%20Signed%20Text _0.pdf (accessed January 8, 20 1 0).

[215] "USTR Announces New Asia-Pacific Initiative on Illegal Logging and Trade." Press release, September 2, 2009. http://www.ustr.gov/about-us/press-office/press-releases/2009/september/ustr-announces-new-asia-pacific-initiative-illeg.

[216] Wibowo, Prabianto. "ASEAN Roadmap for Integration of Wood-Based Products: Does It Provide a Suitable Framework to Guide the Development of Sustainable and Competitive Timber Trade?" *Paper presented at EU-ASEAN Workshop on Trade in Wood Products in South East Asia*, Kuala Lumpur, Malaysia, May 24-25, 2006. http://www.delmys.ec.europa.eu/en/special features/Prabianto.pdf.

[217] World Bank. Forest Law Enforcement and Governance (FLEG)—East Asia & Pacific. http://go.worldbank.org/Q3SXQDTSC0 (accessed January 8, 201 0).

[218] ———. *Indonesia Economic Quarterly: Building Momentum,* March 2010.

[219] World Trade Organization (WTO). *Tariff Analysis Online.* http://www.wto.org/english/tratope/tariffse/tao help e.htm (accessed April 2010).

[220] Airline Industry Information. "*Tiger Airways Increases Capacity on Singapore-Kuala Lumpur Route,*" September 29, 2008.

[221] Arunanondchai, Jutamus; Carsten, Fink. "*Trade in Health Services in the ASEAN Region.*" World Bank Policy Research Working Paper 4147, March 2007.

[222] Association of Private Hospitals Malaysia. "*Health Tourism in Malaysia*." Media release, October 2008; cited on Tourism Malaysia, October 2008. http://www.tourism.gov.my/corporate/mediacentre.asp?page=feature_malaysia&subpage=archive &news id=16.

[223] Association of Southeast Asian Nations (ASEAN). "*Appendix 1: Roadmap for Integration of Healthcare Sector*," n.d. http://www.aseansec.org/19429.pdf.

[224] ———. *ASEAN Yearbook of Statistics 2008*. Jakarta: ASEAN Secretariat, July 2009.

[225] ———. "*Eleventh ASEAN Transport Ministers Meeting, Vientiane, 17 November 2005*." Press release, November, 2005, 17.

[226] Badan Standardisasi Nasional (National Standardization Agency of Indonesia). "Indonesia Will Soon Ratify ASEAN Cosmetic Directive," September 9, 2009. http://www.bsn.go.id/news_detail.php?news_id=1278&language=en&language=en&language=id &language=en.

[227] Brekke; Sørgard. "*Public Versus Private Health Care in a National Health Service*," March 2006, 2.

[228] Bureau van Dijk. Zephyr Database (accessed April 27, 2010).

[229] Businessdictionary.com. "Memorandum of Understanding (MOU)," n.d. http://www.businessdictionary.com/definition/memorandum-of-understanding-MOU.html (accessed June 27, 2010).

[230] Chee, Heng, Leng. "*Medical Tourism in Malaysia: International Movement of Healthcare Consumers and the Commodification of Healthcare*." Asia Research Institute. Working Paper Series No. 83, January 2007.

[231] Danish Trade Council. "*The Healthcare Sector.*" *Royal Danish Embassy, Kuala Lumpur*, Malaysia, March 31, 2009.

[232] Davis, Lucy; Fredrik, Erixon. "*The Health of Nations: Conceptualizing Approaches to Trade in Health Care*." European Centre for International Political Economy. Policy Brief 04/2008.

[233] Deloitte Center for Health Solutions. *Medical Tourism: Update and Implications*, 2009.

[234] Deloitte Singapore. *Private Equity ASEAN*, 2008.

[235] Eucomed. "*Medical Technology Brief: Key Facts and Figures on the European Medical Technology Industry*," May 2007.

[236] Giusti, Daniele; Bart, Criel; Xavier de, Béthune. "Viewpoint: Public versus Private Healthcare Delivery: Beyond the Slogans." *Health Policy and Planning*, 12, no. 3, 1997, 193-98.

[237] Global Health Council. "*Private-Public Sector*," n.d. http://www.globalhealth.org/health_systems/private_public/ (accessed June 28, 2010).

[238] Goh, Jeffrey. "*Teleradiology Outsourcing: Rumblings from the Ground*." *SMA News* 38, no. 5 (May 2006), 22-23, 27.

[239] Gross, Ames; John, Minot. "ASEAN Medical Device Regulatory Integration: Countries Working to Modernize Healthcare Industry May Ease Medical Device Entry Into Southeast Asian Market." *Medical Product Outsourcing*, January 1, 2009.

[240] ———. "The Roar of Vietnam's Device Market." *Pacific Bridge Medical*, May 2009.

[241] Hanson, Kara; Peter, Berman. "Private Health Care Provision in Developing Countries: A Preliminary Analysis of Levels and Composition." Harvard School of Public Health, *Department of Population and International Health*, Data for Decision Making Project, n.d.

[242] Hospitals.sg. "*Singapore's Medical Tourism Figures Revealed by Health Minister*," January 28, 2009.

[243] International Enterprise Singapore. *Annual Report 2008/2009*, n.d. http://www.ie singapore.gov.sg/wps/portal/!ut/p/c4/04_SB8K8xLLM9MSSzPy8xBz9CP0os3gDC yNf30C _QA9Hg1APFz9ncycjAwjQL8h2VAQAC7xNtQ! !/ (accessed June 11, 2010).

[244] *International Medical Travel Journal.* "Indochina: Vietnam and Cambodia Make Hospital Developments," November 20, 2008.

[245] ———. "*Malaysia: Tracking Malaysian Medical Tourism Statistics*," September 8, 2009.

[246] ———. "*Malaysia Promoting Health Tourism from Brunei and Singapore*," March 25, 2010.

[247] ———. "*Thailand: Health and Medical Tourism Update,*" April 15, 2010.

[248] International Telecommunications Union and Inter-Americana Telecommunication Commission Organization of American States. *Tele-Education in the Americas.* OEA/Ser.LXVII.6.3. December 2001.

[249] Irrawaddy. "*Burma Licenses Private Hospitals, Clinics*," October 8, 2009.

[250] Jain, Sanjeev. "Malaysia Poised for a Leap of Growth—II." *BioSpectrum: Asia Edition*, May 11, 2010. http://www.biospectrumasia.com/content/110510MYS12639.asp.

[251] Lau, Jennifer. "*Medical Device Market: Mega Trends in Asia.*" Frost and Sullivan, Singapore, n.d.

[252] Lee-Brago, Pia. "ASEAN Agrees to Further Promote Integration of Traditional Medicine." *The Philippine Star*, January 7, 2010.

[253] Malaysian-German Chamber of Commerce and Industry. "*Market Watch 2009*, *The Healthcare Sector*," 2009.

[254] Malaysian Industrial Development Authority. "*Incentives for the Medical Devices Industry*," April 20, 2009. http://202. 190.126.187/en v2/index.php?page=medical-devices-industry.

[255] Malaysian Industrial Development Authority. "*Industries in Malaysia: Medical Devices Industry*," April 6, 2009. http://www.mida.gov.my/en v2/index.php?page=medical-devices.

[256] Ministry of Health Singapore. "*Healthcare Facilities: Overview*," n.d. http://www.moh.gov.sg/mohcorp/hcfacilities.aspx?id=106 (accessed June 28, 2010).

[257] ———. "Healthcare Institution Statistics," n.d. http://www.moh.gov.sg/mohcorp/ statistics.aspx?id=242 (accessed June 28, 2010).

[258] ———. "Memorandum of Understanding to Foster Greater Cooperation in Healthcare Between Singapore and Brunei." Press release, February 9, 2007.

[259] Noor, Narissa. "Thailand Looks to Set Up Private Hospital Here." *Borneo Bulletin*, March 30, 2010; cited in BruDirect.com, March 30, 2010.

[260] Pacific Bridge Medical. "Growing Foreign Investment in Vietnam's Pharmaceutical Industry." *Asia Medical eNewsletter*, February 4, 2008.

[261] Phouthonesy, Ekaphone. "Govt Grants First Private Hospital Licenses." *Vientiane Times*, September 22, 2009; cited in Lao Voices, September 22, 2009. http://laovoices. com/2009/09/22/govt-grantsfirst-private-hospital-licences/ (accessed April 14, 2010).

[262] Royal Malaysia Customs Department. HS-Explorer. http://tariff.customs.gov.my (accessed April 2010).

[263] Sabri, Reehan. "Working in Brunei Darussalam." *Pakistan Journal of Medical Science* 23, no. 5, 2007, 814-17.

[264] Seerngam, Selvaraja. "*ASEAN Pharmaceutical Harmonization*." Ministry of Health Malaysia. Presentation given at V Conference of the Pan American Network for Drug Regulatory Harmonization 2008, Argentina, 2008, November 17-19.

[265] Singapore Department of Statistics. "*Singapore's Investment Abroad 2008*," April 2010.

[266] Singapore Economic Development Board. "Industry Background: *Medical Technology*," May 3, 2010. http://www.sedb.com/edb/sg/en_uk/index/industry_sectors/medical_technology/industry_backgro und.html.

[267] Singapore Government. Central Provident Fund Board. "Healthcare Financing Framework in Singapore: *My CPF Providing for Your Healthcare Needs*," 2009, November 19, http://mycpf.cpf.gov.sg/CPF/my-cpf/Healthcare/PvdHC2.htm.

[268] Singapore Medicine. "*Growing Biomedical Manufacturing Industry*," n.d. http://www.singaporemedicine.com/leadingmedhub/biomed mfg.asp (accessed June 28, 2010).

[269] Smith, Richard, D. "Foreign Direct Investment and Trade in Health Services: A Review of the Literature." *Social Science & Medicine*, 2004, 59, 2313-23.

[270] Socio-economic and Environmental Research Institute (SERI). "Medical Tourism: A New Growth Frontier for Penang in 2009 and Beyond." *Penang Economic Monthly*, 11, issue 4 (April 2009), 10-15.

[271] Straits Times. "*Use of Medisave Overseas*," February 10, 2010.

[272] Timmons, Karen. "The Value of Accreditation." *Medical Tourism Magazine*, 2007, October 17.

[273] United Nations. Economic and Social Commission for Asia and the Pacific. "*E-Health in Asia and the Pacific*: *Challenges and Opportunities*," 2009.

[274] ———. "Medical Travel in Asia and the Pacific: *Challenges and Opportunities*," 2009.

[275] ———. Statistics Division. "CPC Ver. 2: Explanatory Notes," December 31, 2008. http://unstats.un.org/unsd/cr/registry/cpc-2.asp.

[276] Vietnews. "*Overseas Vietnamese Builds Hospital in Laos*," 2010, March 7.

[277] Whittaker, A. "Pleasure and Pain: Medical Travel in Asia." *Global Public Health*, 3, no. 3 (July 2008), 271-90.

[278] World Bank. *Data Indicators,* n.d. http://data.worldbank.org/indicator/SH.XPD.OOPC.ZS (accessed June 24, 2010).

[279] ———. World Development Indicators database, accessed May 3, 2010.

[280] World Eye Reports. "*Healthcare Tourism on the Rise*," 2009, November 14-15. http://www.worldeyereports.com/reports/2009/cd_malaysia_2009/2009_malaysia01.aspx.

[281] World Health Organization (WHO). Regional Office for the Western Pacific. "Cambodia." *Country Health Information Profiles*, 2009.

[282] ———. "Lao People's Democratic Republic." *Country Health Information Profiles*, 2009, 162-74.

[283] World Trade Organization. "Definition of Services Trade and Modes of Supply." *GATS Training Module: Chapter*, *1*, n.d. http://www.wto.org/english/tratop e/serv e/cbt course e/c 1s3p1 e.htm (accessed June 11, 2010).

[284] ———. Tariff Analysis Online. http://tariffanalysis.wto.org (accessed April 2010).

[285] Xinhua News Agency. "*Myanmar, Malaysian Airlines Increase Flight Capacity*," 2005, October 6.

[286] Aldaba, Rafaelita, M. "Assessing the Competitiveness of the Philippine Auto Parts Industry." Discussion Paper Series No. 2004-14. *Philippine Institute for Development Studies*, November 2007.

[287] Aldaba, Rafaelita, M; Josef, Yap, T. *Investment and Capital Flows: Implications of the ASEAN Economic Community*. Discussion Paper Series No. 2009-01. Philippine Institute for Development Studies, January 2009.

[288] American Chamber of Commerce in Thailand. "*Customs Meeting with Deputy Minister of Finance H.E. Mr. Pradit Phataraprasit*," 2009, April 24. www.amchamthailand.com/asp/view doc.asp?DocCID=2352.

[289] Association of Southeast Asian Nations (ASEAN). *ASEAN Industrial Cooperation Scheme*. http://www.aseansec.org/6402.htm (accessed May 6, 2010).

[290] ———. Questions and Answers on the CEPT. http://www.aseansec.org/10153.htm (accessed May 6, 2010).

[291] *ASIAtalk*. "Thai Auto-Electronics Spark Investment," August 2008.

[292] ———. "*Thailand: Auto Parts,*" September 2008.

[293] Austria, Myrna, S. *The Pattern of Intra-A SEAN Trade in the Priority Goods Sectors*. Final Report. Regional Economic Policy Support Facility, ASEAN-Australia Development Cooperation Program. REPSF Project No. 03/006e, August 2004.

[294] Auto Industry [UK] Web site. http://www.autoindustry.co.uk/features/qcd (accessed June 1, 2010). AutomotiveWorld.com. "*Malaysia: Govt Wants More Parts Companies to Be Tier 1*," April 22, 2010.

[295] *Bangkok Post*. "FTI Sees Winners and Losers," September 1, 2010. http://www.bangkokpost.com/business/economics/30737/fti-sees-winners-and-losers.

[296] Borhan, Mohamed Hairul. "Indonesia to Boost Automotive Industry." *International Enterprise Singapore*, April 16, 2010. http://www.iesingapore.gov.sg.

[297] Central Intelligence Agency (CIA). *The World Factbook: Indonesia*, April 21, 2010.

[298] Chrysler, Mark. "*New ASEAN-China Trade Pact Leaves Out Auto Industry, for Now*." Ward'sAuto.com, February 11, 2010. http://wardsauto.com/ar/asean_china_trade_100211.

[299] Doi, Ayako. "Faced with Shrinking U.S. Market, Japanese Suppliers—& Makers—Invest More in Thailand." *The Japan Automotive Digest*, September 29, 2008.

[300] ECORYS Research and Consulting; Institute for International and Development Economics; Centre for European Studies, Chulalongkorn University; Mekong Economics; PT Inacon Luhur Pertiwi; and Rajah & Tann for the European Commission. "*Trade Sustainability Impact Assessment*," April 3, 2009. http://ec.europa.eu/trade/analysis/sustainability-impact-assessments/assessments.

[301] Economist Intelligence Unit (EIU). *Indonesia: Automotive Report*, September 21, 2009.

[302] Embassy of the United States, Jakarta. Public Affairs Section. *Indonesia's Automotive Industry Confronts Global Economic Downturn*, February 12, 2009. http://jakarta.usembassy.gov/press rel/February09/industry auto.html.

[303] FDIMarkets database. http://www.fdimarkets.com (accessed January 10, 2010).

[304] Findlay, Christopher. "An Investigation into the Measures Affecting the Integration of ASEAN's Priority Sectors (Phase 2): Overview: Summary, Final Report." Regional Economic Policy Support Facility, *ASEAN-Australia Development Cooperation Program*. REPSF Project No. 06/001a, April 2007.

[305] Ford Motor Company. "*Ford, Key Suppliers Roll Out Innovative Business Model*," n.d.

[306] International Organization for Standardization (ISO). "*New Edition of ISO/TS 16949 Quality Specification for Automotive Industry Supply Chain*," July 2, 2009. http://www.iso.org/iso/pressrelease.htm?refid=Ref1234.

[307] International Organization of Motor Vehicle Manufacturers (OICA). "*World Motor Vehicle Production by Country and Type 2008–2009*." http://oica.net/wp-content/uploads/all-vehicles-2008-2009.pdf.

[308] Invest in China database. www.fdi.gov.cn (accessed January 20, 2010).

[309] The Japan Automotive Digest. "*Malaysia Opens Auto Industry to Foreign Investment*," 2009, November 2.

[310] Japan Automobile Manufacturers Association (JAMA). "*ASEAN and the Auto Supporting Industry*: Part 2 of 2." News from JAMA Asia (March 2010).

[311] Just-auto.com. "ASEAN Automotive Market Review." *Management briefing*, April 2007.

[312] Lau, Terence, J. "Distinguishing Fiction from Reality: The ASEAN Free Trade Area and Implications for the Global Auto Industry." *University of Dayton Law Review*, 13, no. 3, 2006, 453-76. http://law.udayton.edu/LawReview/documents/31-3/Lau.pdf.

[313] Ministry of Finance. "*Japan's FDI Flows to ASEAN and China by Country and Industry*, 2008," table.

[314] Ministry of International Trade and Industry (MITI). *Review of National Automotive Policy*, October 28, 2009. http://www.scribd.com/doc/21781137/Review-of-National-Automotive-Policy.

[315] Nag, Biswajit; Saikat, Banerjee; Rittwik, Chatterjee. "Changing Features of the Automobile Industry in Asia: Comparison of Production, Trade and Market Structure in Selected Countries." *Asia-Pacific Research Network on Trade*, Working Paper Series, No. 2007, 37, July.

[316] The Nation (Thailand). "*Major Firms see ASEAN as Second Investment Base*," October 27, 2009.

[317] Pratiwi, Dina Sih. "Automotive Industry." *Pefindo Credit Rating Indonesia*, n.d.

[318] Pugliese, Tony. "Analysis: ASEAN Q1 Vehicle Sales Jump 41% [includes data]." Just-auto.com, April 23, 2010.

[319] Thailand in the 2000's. "*Industry on the Move*," n.d.

[320] United States Agency for International Development/ADVANCE (USAID/ADVANCE). *Evaluation of Proposed Target Sector: ASEAN Supply Chain Competitiveness Project*, June 2008.

[321] United States Agency for International Development/Indonesia Competitiveness Forum (USAID/SENADA). "*Automotive Component Value Chain Overview*," August 2007.

[322] United States Chamber of Commerce. Written submission to the U.S. International Trade Commission in connection with inv. No. 332-511, *ASEAN: Regional Trends in Economic Integration, Export Competitiveness, and Inbound Investment for Selected Industries*, March 12, 2010.

[323] United States International Trade Commission (USITC). Hearing transcript in connection with inv. no. 332-511, ASEAN: Regional Trends in Economic Integration, *Export Competitiveness, and Inbound Investment for Selected Industries*, 2010, February 3.

[324] United States Trade Representative (USTR). *National Trade Estimate Report*, 2008. http://www.ustr.gov/sites/default/files/uploads/reports/2008/NTE/asset upload file356 14651.pd f.

[325] World Trade Organization (WTO). Report by the Secretariat. "*Indonesia Trade Policy Review*," 2003, May 28. http://docsonline.wto.org/GEN_viewerwindow.asp?http:// docsonline.wto.org:80/DDFDocuments /t/WT/TPR/S 117-3 .doc.

[326] ———."Indonesia Trade Policy Review," May 23, 2007. http://docsonline.wto.org/ GEN_viewerwindow.asp?http://docsonline.wto.org:80/DDFDocuments /t/WT/TPR/S 1 84-04.doc.

[327] ———. "*Malaysia Trade Policy Review (Revision)*," March 9, 2006. http://docsonline.wto.org/GENviewerwindow.asp?http://docsonline.wto.org:80/DDFDo cuments /t/WT/TPR/S 1 56R1-4.doc.

[328] ———. "*Malaysia Trade Policy Review*," December 14, 2009. http://docsonline.wto.org/GEN viewerwindow. asp?http:// docsonline. wto.org: 80/DDFDocuments/t/WT/TPR/S225-04.doc.

[329] ———. "Philippines Trade Policy Review," June 7, 2005. http://docsonline.wto.org/ GEN_viewerwindow.asp?http://docsonline.wto.org:80/DDFDocuments/t/WT/TPR/S 149R1 -3 .doc.

[330] ———. "Thailand Trade Policy Review," October 15, 2003. http://docsonline.wto.org/ GEN_viewerwindow.asp?http://docsonline.wto.org:80/DDFDocuments/t/WT/TPR/S 123-3 .doc.

[331] ———. "Thailand Trade Policy Review (Revision)," February 6, 2008. http://docs online.wto.org/GEN_viewerwindow.asp?http://docsonline.wto.org:80/DDFDocuments /t/WT/TPR/S 19 1R1-04.doc.

[332] Zephyr database. http://www.zephyrdeadata.com (accessed January 10, 2010).

[333] Agreement on the Common Effective Preferential Tariff (CEPT) Scheme for the ASEAN Free Trade Area, *Consolidated Package of Tariff Reductions for* 2008, http://www.aseansec.org/19802.htm.

[334] Agriprods.com. "*Anglo Eastern Plantations PLC*," n.d. (accessed April 1, 2010).

[335] American Palm Oil Council (APOC). "Palm Oil: Gift to Nature, Gift to Life," n.d. http://www.americanpalmoil.com/publications/Brief%20Palm%20Oil%20Story.pdf.

[336] Basiron, Yusof. "Palm Oil Production through Sustainable Plantations." *European Journal of Lipid Science and Technology*, 2007, 109. http://www.americanpalmoil. com/pdf/enviromental/Palm%20Oil%20Production%20Through%2 0Sustainable%20Plantations.pdf.

[337] BBC News. "*Orangutan Survival and the Shopping Trolley*." February 22, 2010. http://news.bbc.co.uk/panorama/hi/front page/newsid 8523000/8523999.stm.

[338] Biodiversity and Agricultural Commodities Program (BACP). "*Market Transformation Strategy for Palm Oil*." 2008, May 1. http://www.ifc.org/ifcext/sustainability.nsf/ AttachmentsByTitle/ref_Biodiversity_BACP_MTSPalmOil/$FILE/BACP MTS Palm SC Approved FINAL+new+format.pdf.

[339] Bromokusumo, Aji, K. Indonesia: Oilseeds and Products; Annual. GAIN Report no. ID9013. U.S. Department of Agriculture, *Foreign Agricultural Service*, 2009, May 5. http://gain.fas.usda.gov/Recent%20GAIN%20Publications/Commodity%20Report OILSEEDS% 20AND%20PRODUCTS%20ANNUAL Jakarta Indonesia 5-5-2009.pdf.

[340] Bromokusumo, Aji, K; Jonn Slette. Indonesia: Oilseeds and Products; Annual. GAIN Report, no. ID 1008. U.S. Department of Agriculture, *Foreign Agricultural Service*, 2010, March 18. http://gain.fas.usda.gov/Recent%20GAIN%20Publications/Oilseeds%20and%20Products%20An nual Jakarta Indonesia 3-18-2010.pdf.

[341] Brunei Darussalam Indonesia Malaysia Philippines East ASEAN Growth Area (BIMP-EAGA). "Report of the 9th Small & Medium Enterprises Development (SMED) Cluster Meeting," 2009. http://www.bimpbc.org/files/9th%20SMED%20Report.pdf.

[342] Bureau van, Djik. Zephyr database (accessed January 9, 2010).

[343] Chan, Yvonne. "*World Bank Arm Halts Financing to Palm Oil Industry*." BusinessGreen.com, September 10, 2009. http://www.businessgreen.com/business-green/news/2249248/world-bankarm-halts-financing.

[344] Corbett, Jenny; So, Umezaki. eds. "*Deepening East Asian Economic Integration.*" *ERIA Research Project Report 2008*, no. 1. http://www.eria.org/research/y2008-no1.html.

[345] Creagh, Sunanda. "*Cargill Considers Dropping Sinar Mas over Land Clearing Claims*." Reuters, March 25, 2010. http://www.reuters.com/article/idUSJAK50197620100325. fDi Markets database. http://www.fdimarkets.com/ (accessed January 9, 2010).

[346] Hickman, Martin. "Online Protest Drives Nestlé to Environmentally Friendly Palm Oil." The Independent, May 19, 2010. http://www.independent.co.uk/environment/green-living/online-protest-drivesnestl-to-environmentally-friendly-palm-oil- 1976443 .html.

[347] International Centre for Trade and Sustainable Development. "Malaysia Sees Possible WTO Case against EU Palm Oil Limits." Bridges Weekly Trade News Digest 14, no. 8 (May 19, 2010). http://ictsd.org/i/news/bridgesweekly/76269/.

[348] IOI Group. "*Agronomy Support*," 2009. http://www.ioigroup.com/business/busi_agritech.cfm.

[349] Law, Bill. "*Can Palm Oil Help Indonesia's Poor?*" BBC News, March 1, 2010. http://news.bbc.co.uk/2/hi/asia-pacific/8534031.stm.

[350] Chandran, MR. "*Commodity Sector Overview*." The Planter 86, no. 1008 (March 2010).

[351] Malaysian Palm Oil Council. "*Oil Palm: Tree of Life*," 2006. http://www.american palmoil.com/publications/TreeOfLife.pdf.

[352] Malaysian Palm Oil Promotion Council. "Business with Palm Oil," n.d. http://www.americanpalmoil.com/pdf/enviromental/business%20with%20palm%20oil.pdf (accessed February 11, 2010).

[353] Mongabay.com. "*BBC Documentary Leads Unilever to Blacklist Indonesian Palm Oil Company*," February 24, 2010. http://news.mongabay.com/2010/0224-palmoil.html.

[354] Oil Crop in Myanmar Project. "*Oil Crops in Myanmar*," March 17, 2010. http://ocdp-mas.org/OilCropsInMyanmar.php.

[355] Association of Southeast Asian Nations (ASEAN). *Roadmap for Integration of Agro-based Products*, November 29, 2004. http://www.aseansec.org/16656.htm (accessed various dates).

[356] Roundtable on Sustainable Palm Oil (RSPO). "*RSPO Certified Mills & CSPO Production*," 2009. http://rspo.org/?q=node/520.

[357] Royal Malaysia Customs Department. HS-Explorer. http://tariff.customs.gov.my (accessed April 2010). RSPO. "Who Is RSPO?" 2009. http://rspo.org/?q=page/9.

[358] Sally, Razeen. "*Regional Economic Integration in Asia: The Track Record and Prospects*." ECIPE Occasional Paper no. 2 (2010). http://www.ecipe.org/regional-economic-integration-in-asia-thetrack-record-and-prospects/PDF.
[359] Sime Darby Plantation. "Overview," 2008. http://www.simedarbyplantation.com/Overview.aspx.
[360] ———. "South East Asia," 2008. http://www.simedarby plantation.com/South East Asia.aspx. Southern Group. "Manufacturing: *Palm Oil Mill*," n.d. http://www.southern.com.my/palm-oil-mill.htm.
[361] Than, Tin Maung, Maung. "*Mapping the Contours of Human Security Challenges in Myanmar*." In Myanmar: State, Society and Ethnicity, edited by N. Ganesan and Kyaw Yin Hlaing (Singapore: Institute of Southeast Asian Studies), 2007.
[362] The Star Online. "*El Nino May Lift CPO Futures Prices*," March 10, 2010. http://biz.thestar.com.my/news/story.asp?file=/2010/3/10/business/5826696&sec=business.
[363] Thoenes, P. "Biofuels and Commodity Markets–Palm Oil Focus," Food and Agriculture Organization of the United Nations, 2006. http://www.rlc.fao.org/es/prioridades/bioenergia/pdf/commodity.pdf.
[364] United States Department of Agriculture (USDA). *Foreign Agriculture Service (FAS). Production*, Supply and Distribution database (accessed May 4, 2010).
[365] United States Trade Representative (USTR). "*2010 National Trade Estimate Report on Foreign Trade Barriers*," 2010.
[366] U.S. International Trade Commission (USITC). *Hearing transcript in connection with inv.* no. 332-511, ASEAN: Regional Trends in Economic Integration, *Export Competitiveness, and Inbound Investment for Selected Industries*, 2010, February 3.
[367] Valentine, Scott Victor. "*Reframing Global Warming: Toward a Strategic National Planning Framework*," n.d., 21.
[368] WITS. Integrated Data Warehouse (accessed various dates).

End Notes

[1] A copy of the letter from the USTR requesting this factfinding investigation is provided in app. A.
[2] United Nations University, "Different Forms of Integration."
[3] ASEAN, "ASEAN Economic Community Blueprint," November 20, 2007, section I.
[4] Ibid, section II.
[5] In general, trade creation is larger than trade diversion when the size of the initial tariffs reduced among members are large, demand patterns tend to be similar among members (allowing for greater specialization), and the tariffs on nonmembers are low. Perdikis, "Overview of Trade Agreements: Regional Trade Agreements," 2007, 88–89.
[6] Kim, Lee, and Park, "The Ties that Bind Asia, Europe, and the United States," February 2010, 1. Note that economic growth among regional economies is considered an important indicator of regional integration due to the association between trade openness and growth.
[7] USITC, *East Asia,* 1993. The countries included in this investigation were Singapore, China, South Korea, the Philippines, Taiwan, Hong Kong, Indonesia, Malaysia, and Thailand. The discussion of economic integration in this investigation was summarized in Pierson, "East Asia: Regional Economic Integration and Implications for the United States," 1994, 1168–85.
[8] Capannelli, Lee, and Petri, *Developing Indicators for Regional Economic Integration and Cooperation*, September 2009, 2.
[9] For example, see Capannelli, Lee, and Petri, "Developing Indicators for Regional Economic Integration and Cooperation," September 2009, 5; Plummer and Chia, "Assessing the Economics of the ASEAN Economic Community: Introduction," 2009, 8; and ADB, *Emerging Asian Regionalism*, 2008, 40–41.

[10] For more information about intraregional trade intensities, see ESCAP, "Trade Statistics in Policymaking," n.d., 52–53; Capannelli, Lee, and Petri, "Developing Indicators for Regional Economic Integration and Cooperation," September 2009, 6.

[11] These factors, among others, form the basis for the World Economic Forum (WEF)'s Global Competitiveness Index. The remaining factors not listed in the text above are institutions, infrastructure, and labor market efficiency. WEF, *Global Competitiveness Report 2009–2010*, 335.

[12] USITC, *Sub-Saharan Africa*, 2009, 1–2. The European Commission (EC), using a definition like that of the USITC, defined industry competitiveness as the position of an industry relative to its foreign rivals in terms of market shares, volume of trade, and relative cost, particularly in regard to labor productivity and the relative cost of labor. See EC, Enterprise and Industry Directorate-General, "European Competitiveness Report," 2009, 27.

[13] ASEAN, "ASEAN Economic Community Blueprint," November 20, 2007, section II.A3.

[14] For greenfield FDI projects, the Commission uses the FDIMarkets database provided by the Financial Times company. For M&A transactions, the Commission uses the Zephyr database provided by Bureau van Dijk.

[15] Official FDI statistics are compiled from individual country balance of payments (BOP) data and reported by each country to international organizations that report global FDI statistics, including the IMF, the OECD, and UNCTAD. IMF guidance for compiling BOP statistics can be found at http://www.imf.org/external/np/sta/op/bop.htm (accessed April 22, 2010), and guidance on compiling international investment position statistics can be found at http://www.imf.org/external/np/sta/iip/iip.htm (accessed April 22, 2010).

[16] As described by the OECD, "Users should be cautious when using the *FDI data* distinguishing *M&A type transactions,* described in this *Benchmark Definition,* which are different to what is generally referred to as M&A statistics produced by commercial and other sources. The former corresponds only to equity transactions that qualify as FDI (covering 10 percent–100 percent of ownership of voting power), while the latter represents the total capital of companies." OECD, *Benchmark Definition of Foreign Direct Investment,* 2008, 32 and Annex 9.

[17] A copy of the *Federal Register* notice is provided in app. B of this chapter.

[18] A list of hearing participants is provided in app. C of this chapter. A summary of the positions of interested parties can be found in app. D.

[19] For more background on ASEAN, see ASEAN Secretariat Web site, "About ASEAN," http://www.aseansec.org/bout_ASEAN.html (accessed January 5, 2010).

[20] East Timor and Papua New Guinea have shown an interest in joining ASEAN.

[21] Atje, "ASEAN Economic Community: In Search of a Coherent External Policy," 2007; Soesastro, "Implementing the ASEAN Economic Community (AEC) Blueprint," 2007.

[22] ASEAN Secretariat Web site, About ASEAN, http://www.aseansec.org/about_ASEAN.html (accessed January 5, 2010). According to one observer, the history and diversity of the ASEAN member states have resulted at times in a lack of political will and strong reluctance to cede national sovereignty to accelerate regional integration. Chia, "Trade and Investment Policies and Regional Economic Integration in East Asia," April 2010, 8.

[23] See, for example, ADB, *Emerging Asian Regionalism*, 2008, 22; ASEAN Secretariat, "Fact Sheet: An ASEAN Economic Community by 2015," August 20, 2008; and Petri, "Competitiveness and Leverage: Benefits from an ASEAN Economic Community," 2009, 216–217.

[24] *Declaration of ASEAN Concord*, 1976; ASEAN Secretariat, "Economic Achievement," n.d.

[25] The ASEAN-6 countries are Brunei, Indonesia, Malaysia, the Philippines, Singapore, and Thailand.

[26] ASEAN Secretariat, "Economic Achievement," n.d.

[27] ASEAN, *Declaration of ASEAN Concord II*, October 7, 2003, 6.

[28] Atje, "ASEAN Economic Community: In Search of a Coherent External Policy," 2007, chap. 10.

[29] For example, see Guerrero, "Regional Integration: The ASEAN Vision in 2020," January 2010, 54.

[30] ASEAN, "ASEAN Accelerates Integration of Priority Sectors," November 29, 2004.

[31] ASEAN, "Informal ASEAN Economic Ministers Meeting," January 19–20, 2004.

[32] ASEAN, "Joint Media Statement of the 41st ASEAN Economic Ministers' (AEM) Meeting," August 13–14, 2009.

[33] *Roadmap for an ASEAN Community 2009–2015*, 2009, 1; ASEAN Secretariat, "Fact Sheet: Initiative for ASEAN Integration and Narrowing the Development Gap," January 12, 2010. For more detailed information about the development gap and differences among the ASEAN member states, see, for example, Narjoko and Kartika, "Narrowing the Development Gap," 2009, 180–183; Severino, "The ASEAN Developmental Divide and the Initiative for ASEAN Integration," 2007, 35–44; and ADB, *Emerging Asian Regionalism*, 2008, 66–69, 75–84.

[34] See, for example, ASEAN, *Declaration of ASEAN Concord II,* October 7, 2003, 3; Cordenillo, "The Future of the ASEAN Free Trade Area and the Free Trade Areas Between ASEAN and Its Dialogue Partners," n.d.

[35] For more information on the status of ASEAN FTAs, see ADB, Free Trade Agreement Database for Asia, http://www.aric.adb.org/. ASEAN is involved with other regional projects as well, such as the Asia- Pacific Economic Cooperation (APEC) forum and the ASEAN+3 (with China, Japan, and Korea).

[36] See app. E for information on member state FTAs.
[37] ASEAN, "ASEAN Economic Community Blueprint," November 20, 2007, section III. Unlike the EU, which has strong institutions and a large regional bureaucracy, the ASEAN Secretariat is small and poorly funded (2010 budget: $14.5 million), limiting its ability to monitor compliance with agreements and commitments, collect and synthesize information, and conduct analytical work. Chia, "Trade and Investment Policies and Regional Economic Integration in East Asia," April 2010, 5; ASEAN Secretariat official, e-mail message to USITC staff, April 28, 2010.
[38] "AEC Scorecard," March 2010, 12–14. It is difficult to interpret the AEC Scorecard results because of the political sensitivity of the process.
[39] Malaysian Ministry of International Trade and Industry, "16th ASEAN Economic Ministers' Retreat," February 28, 2010.
[40] Pitsuwan, "Progress in ASEAN Economic Integration since the Adoption of the ASEAN Charter," June 29, 2009, 1.
[41] ASEAN Charter, article 24.
[42] ASEAN Secretariat, "Fact Sheet: ASEAN Dispute Settlement System," February 24, 2009.
[43] See, for example, Calvo-Pardo, Freund, and Ornelas, "The ASEAN Free Trade Agreement: Impact on Trade Flows and External Trade Barriers," June 2009, 3.
[44] Calvo-Pardo, Freund, and Ornelas, "The ASEAN Free Trade Agreement: Impact on Trade Flows and External Trade Barriers," June 2009, 10.
[45] Government official, interview by Commission staff, Singapore, March 1, 2010.
[46] Ibid.
[47] ASEAN, "ASEAN Integration in Services," August 2009.
[48] ASEAN, "ASEAN Integration in Services," August 2009, 10; Thangavelu, "Non-Tariff Barriers, Integration and Export Growth in ASEAN," January 2010, 15.
[49] ASEAN, "ASEAN Integration in Services," August 2009, 13–14. Also, for all general services information, see ASEAN Secretariat, "Fact Sheet: ASEAN Framework Agreement on Services," February 26, 2009.
[50] "AEC Scorecard," March 2010, 7.
[51] Capannelli, Lee, and Petri, "Developing Indicators for Regional Economic Integration and Cooperation," September 2009, 5; Plummer and Chia, "Assessing the Economics of the ASEAN Economic Community: Introduction," 2009, 8.
[52] Intraregional trade shares are upward biased because of double counting, which results from both Singapore's entrepot role and production fragmentation.
[53] See, for example, Capannelli, Lee, and Petri, "Developing Indicators for Regional Economic Integration and Cooperation," September 2009, 5; ESCAP, "ASEAN and Trade Integration," April 8, 2009, 9; and Guerrero, "Regional Integration: The ASEAN Vision in 2020," January 2010, 56. It also reflects the relatively small size of ASEAN's economies compared to the economies of the other groupings cited here. To compensate for the size bias, a trade intensity measure is also provided below.
[54] Plummer and Chia, "Assessing the Economics of the ASEAN Economic Community: Introduction," 2009, 3.
[55] It is interesting to note that for the broader East Asia region (ASEAN+6, which includes ASEAN and China, Japan, Korea, Australia, New Zealand, and India), there is a large and growing disparity between the share of intraregional imports and the share of intraregional exports, where such imports far outpace such exports, indicating a growing reliance on outside markets for the region's exports and a growing reliance on the region for its imports. See Chia, "Regional Trade Policy Cooperation and Architecture in East Asia," February 2010, 7.
[56] For more information about intraregional trade intensities, see chap. 1 of this chapter. Also, see ESCAP, "Trade Statistics in Policymaking," n.d., 52–53; Capannelli, Lee, and Petri, "Developing Indicators for Regional Economic Integration and Cooperation," September 2009, 6.
[57] Taking a longer viewpoint, ASEAN intraregional trade intensity was much higher in the past, reaching a peak in the 1960s. During that period, the economies were too small to trade much, and traded disproportionately with neighbors. The index declined erratically until the 1990s as these countries grew and increased trade with the world. In the mid-1990s, the index began to increase, possibly reflecting the implementation of AFTA and the growth of regional production networks. ADB, *Emerging Asian Regionalism*, 2008, 41–42; Capannelli, Lee, and Petri, "Developing Indicators for Regional Economic Integration and Cooperation," September 2009, 6–7.
[58] Capannelli, Lee, and Petri, "Developing Indicators for Regional Economic Integration and Cooperation," September 2009, 6.
[59] Plummer and Chia, "Assessing the Economics of the ASEAN Economic Community: Introduction," 2009, 9.
[60] Malaysian Institute of Economic Research, ADBI, and East-West Center, "Regional Market for Goods, Services, and Skilled Labor," 2009, 32.
[61] ASEAN, *Declaration of ASEAN Concord II,* October 7, 2003; Malaysian Institute of Economic Research, ADBI, and East-West Center, "Regional Market for Goods, Services, and Skilled Labor," 2009, 32; and Amin, Hamid, and Saad, "Economic Integration among ASEAN Countries: Evidence from Gravity Model," February 2009, 6.

[62] ADB, *Emerging Asian Regionalism*, 2008, 64.
[63] ADB, *Emerging Asian Regionalism*, 2008, 64. In addition to ASEAN, ADB includes Taiwan in this list.
[64] ADB, *Emerging Asian Regionalism*, 2008, 60. ADB notes that Singapore's productivity has caught up with the OECD average, while Malaysia and Thailand have made significant progress.
[65] ADB, *Emerging Asian Regionalism*, 2008, 60, 78; Petri, "Competitiveness and Leverage: Benefits from an ASEAN Economic Community," 2009, 223.
[66] Hiebert, written submission submitted to the USITC, March 10, 2010; Duek, "ASEAN Economic Integration: Evolving Patterns of Trade in AFTA," December 2008, 9–10; ADB, *Emerging Asian Regionalism*, 2008, 78; Soesastro, "Accelerating ASEAN Economic Integration: Moving Beyond AFTA," March 2005, 8; and Soesastro, "Implementing the ASEAN Economic Community (AEC) Blueprint," 2007, 8.
[67] See, for example, Hiebert, written testimony submitted to the USITC, March 10, 2010.
[68] Aldaba, Yap, and Petri, "Investment and Capital Flows: Implications of the AEC," 2009, 143.
[69] Aquino, written submission to the USITC, March 3, 2010, 7; ADB, *Emerging Asian Regionalism*, 2008, 58, 62, and 75.
[70] ADB, *Emerging Asian Regionalism*, 2008, 75; Soesastro, "Accelerating ASEAN Economic Integration: Moving Beyond AFTA," March 2005, 9.
[71] Petri, "Competitiveness and Leverage: Benefits from an ASEAN Economic Community," 2009, 216.
[72] For a list of such studies, see Petri, "Competitiveness and Leverage: Benefits from an ASEAN Economic Community," 2009, 217.
[73] Petri, "Competitiveness and Leverage: Benefits from an ASEAN Economic Community," 2009, 220.
[74] Duek, "ASEAN Economic Integration: Evolving Patterns of Trade in AFTA," December 2008, 9–10.
[75] Soesastro, "Accelerating ASEAN Economic Integration: Moving Beyond AFTA," March 2005, 9.
[76] ASEAN, "ASEAN Economic Community Blueprint," November 20, 2007, section II.A3.
[77] Under the AIA, items included in the Temporary Exclusion List (TEL) should be opened to ASEAN investors by 2010 and to all investors by 2020, for both market access and national treatment. Brunei, Indonesia, Malaysia, Burma, the Philippines, Singapore, and Thailand should have phased out their TEL by January 1, 2003, for the manufacturing sector. Cambodia, Laos, and Vietnam should have phased out their TEL by January 1, 2010. Items on the Sensitive List do not have a hard date for phaseout, but should be periodically reviewed by the AIA Council. ASEAN Secretariat Web site, "ASEAN Investment Area: An Update," http://www.aseansec.org/6480.htm (accessed April 22, 2010). The Temporary Exclusion and Sensitive Lists are posted on the ASEAN Web site http://www.aseansec.org/20633.htm (accessed April 22, 2010).
[78] ASEAN, "ASEAN Economic Community Blueprint," November 20, 2007, section II.A3. Under the ASEAN Framework Agreement on Services, member countries schedule commitments on services through four modes of delivery. Mode 3 is "commercial presence," i.e., a services provider establishing an affiliate in the host country, which in practice is foreign direct investment. This system of commitments follows the model set forth in the WTO General Agreement on Trade in Services (GATS). Like the WTO model, ASEAN services commitments follow a positive list format, meaning that countries are bound only to the particular industries and commitments included in their Schedules of Commitments.
[79] Under MFN treatment, a foreign investor is guaranteed to receive treatment equal to that of any other foreign investor. Under national treatment, a foreign investor is guaranteed the same treatment as a domestic investor would receive.
[80] Text of the Investment Guarantee Agreement, available on the ASEAN Secretariat Web site, http://www.aseansec.org/6464.htm (accessed April 23, 2010).
[81] ASEAN Secretariat, "Highlights of the ASEAN Comprehensive Investment Agreement (ACIA)," August 26, 2008.
[82] Performance requirements, such as local content targets or export quotas, are imposed as a condition for the establishment, acquisition, expansion, management, conduct, or operation of an investment. Office of the U.S. Trade Representative, "Bilateral Investment Treaties," n.d., http://www.ustr.gov/tradeagreements/bilateral-investment-treaties (accessed July 22, 2010).
[83] This holds "provided that this requirement does not materially impair the ability of the investor to exercise control over its investment."
[84] ASEAN Secretariat official, telephone interview with USITC staff, March 17, 2010.
[85] ASEAN Secretariat, *ASEAN Investment Report 2008*, 16; and text of the ACIA.
[86] Under a negative list system, all sectors except those specifically named as not included in each country's Schedule of Commitments are automatically covered in the agreement. By contrast, under a positive list framework, a country must positively include each service sector for which it agrees to coverage in its Schedule of Commitments attached to the agreement. In that case, sectors not included in the list are not covered by the agreement.
[87] ASEAN Secretariat official, telephone interview with USITC staff, March 17, 2010.
[88] Text of the ACIA, Articles 17 and 18.
[89] ASEAN Secretariat official, telephone interview with USITC staff, March 17, 2010.

[90] This recognition is included in the protocol to implement the seventh package of commitments under the AFAS, signed February 26, 2009.
[91] ASEAN, "ASEAN Economic Community Blueprint," November 20, 2007, section II.A2.
[92] ACIA, Article 3(5).
[93] This discussion of FDI trends reflects total FDI inflows into ASEAN. Data giving additional detail by country and industry are not available.
[94] ASEAN Secretariat, *ASEAN Investment Report 2009*, 10–11.
[95] Ibid., 11–12.
[96] United Nations Centre for Trade Facilitation and Electronic Business, cited in Asian Development Bank (ADB), *Designing and Implementing Trade Facilitation*, 2009, 3.
[97] There are no data for Burma.
[98] World Bank, Doing Business database (accessed April 20, 2010). For its time and cost estimates, Doing Business uses the hypothetical case of a 20-foot container of non-hazardous products that are not subject to special phytosanitary or environmental standards.
[99] Examples of ASEAN work on trade facilitation in the 1990s include the ASEAN Agreement on Customs (1997), the ASEAN Framework Agreement on the Facilitation of Goods in Transit (1998), and the Agreement on the Recognition of Commercial Vehicle Inspection Certificates for Goods Vehicles and Public Service Vehicles Issued by ASEAN Member Countries (1998).
[100] For evidence of this increased focus on trade facilitation, see, for example, the Recommendations of the High-Level Task Force on ASEAN Integration (2003), which call for numerous measures related to customs and standards.
[101] ASEAN Secretariat official, interview by USITC staff, Jakarta, Indonesia, March 1, 2010.
[102] ASEAN Economic Ministers, "Joint Media Statement," August 25–26, 2008.
[103] Some of the measures are particularly important for specific sectors examined in this study. Please see the profiles of ASEAN industries in chapters 3 through 8 for examples.
[104] ASEAN, "ASEAN Trade Facilitation Work Programme 2007–2015, Revised 08 July 2009," July 8, 2009.
[105] ASEAN Economic Ministers, "Joint Media Statement," September 28, 2005.
[106] ASEAN, "Agreement to Establish and Implement the ASEAN Single Window," December 9, 2005.
[107] The ASW Agreement named no specific deadline for completion of the ASW. However, another official document commits members to "accelerate the full implementation of the ASEAN Community's programme areas, measures and principles" with a view to realizing an "ASEAN Economic Community" by 2015. ASEAN, "Cebu Declaration on the Acceleration of the Establishment of an ASEAN Community by 2015," January 13, 2007.
[108] Le, "Briefing on ASEAN Single Window," March 9, 2010; government official, interview by USITC staff, Hanoi, Vietnam, March 8, 2010.
[109] Le, "Briefing on ASEAN Single Window," March 9, 2010.
[110] Benjelloun, "Technical Instructions," March 31, 2010, 5.
[111] Mealy, written testimony to the USITC, February 4, 2010, 3; industry representatives, interview by USITC staff, Singapore, March 4, 2010.
[112] Industry representative, interview by USITC staff, Hanoi, Vietnam, March 8, 2010.
[113] Industry representatives, interviews by USITC staff, Singapore, March 4, 2010.
[114] Industry representative, e-mail message to USITC staff, April 7, 2010.
[115] Hiebert, written testimony to the USITC, March 12, 2010, 5.
[116] Industry representative, interview by USITC staff, Singapore, March 4, 2010; Hiebert, written testimony to the USITC, March 12, 2010, 6.
[117] At the time of writing (April 2010), members were reviewing (but had not yet adopted) the expanded data model. The model spans 13 trade documents: the intra-ASEAN preferential certificate of origin (CEPT Form-D), the ACDD, air waybill, bill of lading, cargo manifest, commercial invoice, fishery certificate, non-preferential certificate of origin, packing list, phytosanitary certificate, purchase order, trade license, and veterinary certificate. Katsiak, "ASW Project Improves ASEAN Capacity," January 2, 2010.
[118] Le, "Briefing on ASEAN Single Window," March 9, 2010.
[119] ASEAN, "Memorandum of Understanding on the Implementation of the ASEAN Single Window Pilot Project," February 5, 2010; ASW Project, "ASW Legal Memorandum of Understanding Agreed," September 27, 2009.
[120] US-ASEAN Business Council Web site, http://www.usasean.org/newcalendar/index.php (accessed April 14, 2010).
[121] Hiebert, written testimony to the USITC, March 12, 2010, 5–6.
[122] Government officials, interview by USITC staff, Hanoi, Vietnam, March 9, 2010.
[123] ASEAN Secretariat official, interview by USITC staff, Jakarta, Indonesia, March 1, 2010; ASEAN Secretariat staff, e-mail to USITC staff, April 29, 2010. The ASEAN Charter stipulates that members make equal contributions to the annual budget; ASEAN, "ASEAN Leaders Sign ASEAN Charter," November 20, 2007.
[124] Laos Customs and Trade Facilitation Project Web site, http://webuPK=228424&Projectid=P101750 (accessed April 28, 2010).

[125] Vietnam Customs Modernization Project Web site, http://webmenuPK=228424&Projectid=P085071 (accessed April 28, 2010).
[126] USAID Philippines Web site, http://philippines.usaid.gov/eg_linc-eg.html (accessed April 28, 2010).
[127] USITC, *Logistic Services*, 2005, vii.
[128] Ibid., 1–3.
[129] De Souza et al., *An Investigation into the Measures*, 2007, 9–10. Toll Global Logistics is based in Singapore but is owned by an Australian parent company. Toll Group Web site, http://www.toll.com.sg/crp_brc.php (accessed April 22, 2010).
[130] ASEAN, "Roadmap for the Integration of Logistics Services," August 24, 2007. The ASEAN Economic Blueprint uses stronger language: it says that members will "remove substantially all restrictions" on trade in logistics services by 2013. ASEAN, "ASEAN Economic Community Blueprint," November 20, 2007, section II.A2.
[131] Hollweg and Wong, "Measuring Regulatory Restrictions in Logistics Services," May 2009, 27.
[132] USITC, *Logistic Services*, 2005, 3-1; De Souza et al., *An Investigation into the Measures*, 2007, i.
[133] USITC, *Logistic Services*, 2005, 3-15.
[134] Greenwald, "The ASEAN-China Free Trade Area," 2006, 206–08.
[135] Conference of Asia Pacific Express Carriers (CAPEC), "Express Delivery Services (EDS): ASEAN Regulatory Matrix and International Best Practices," May 2009.
[136] Industry representatives, interview by USITC staff, Singapore, March 4, 2010.
[137] Republic of Indonesia, Law Number 38 of 2009 regarding Postal Services, Article 4.
[138] Ibid., Article 12.
[139] ASEAN, "Roadmap for the Integration of Logistics Services," August 24, 2007; Law of the Republic of Indonesia 3 8/2009 Regarding Postal Services; industry representative, interview by USITC staff, Singapore, March 4, 2010.
[140] Prime Minister's Office of Malaysia, "Liberalisation of the Services Sector," April 22, 2009, 4–5.
[141] Industry representatives, interview by USITC staff, Singapore, March 4, 2010.
[142] WTO, "Working Party on the Accession of Viet Nam," October 27, 2006, 19 and 50.
[143] WTO, "Report of the Working Party," August 19, 2003, 9 and 22. Unlike Vietnam, Cambodia did not explicitly state that its commitments in courier services covered express delivery. Industry representatives consulted for this chapter did not cite concerns about this lack of specificity and made no mention of barriers to foreign investment in express delivery services in Cambodia.
[144] Hollweg and Wong, "Measuring Regulatory Restrictions in Logistics Services," May 2009, 21.
[145] Conference of Asia Pacific Express Carriers (CAPEC), "Express Delivery Services (EDS): ASEAN Regulatory Matrix and International Best Practices," May 2009.
[146] Acoca, *Empowering E-Consumers*, December 2009, 6.
[147] According to the OECD, "Broadband service is usually understood to be a connection providing highspeed Internet access, that is, a communication service that enables access to the Internet at data transmission rates above a specific threshold." Díaz-Pinés, *Indicators of Broadband Coverage,* December 10, 2009, 38.
[148] OECD, *OECD Information Technology Outlook 2008,* 2008, 239–40. This chapter presented evidence from several OECD countries showing that individuals in households with broadband access were more likely to buy goods and services online than individuals in households with non-broadband Internet access.
[149] Galexia, *Harmonisation of E-Commerce Legal Infrastructure*, April 2008; ASEAN Secretariat official, telephone interview by USITC staff, April 21, 2010.
[150] Vietnam E-Commerce and IT Agency (VECITA), *Vietnam E-commerce Report 2008*, 11.
[151] Government officials, interview by USITC staff, Hanoi, Vietnam, March 9, 2010.
[152] Industry representatives, interviews by USITC staff, Hanoi, Vietnam, March 8 and 9, 2010.
[153] Industry representative, interview by USITC staff, Hanoi, Vietnam, March 9, 2010.
[154] U.S. Embassy official, e-mail to USITC staff, April 27, 2010.
[155] Government officials, interview by USITC staff, Hanoi, Vietnam, March 9, 2010.
[156] ASEAN, "e-ASEAN Framework Agreement," November 24, 2000, Article 5.
[157] ASEAN, "ASEAN Economic Community Blueprint," November 20, 2007, section II.B6.
[158] ASEAN, "e-ASEAN Framework Agreement," November 24, 2000, Article 4.
[159] Galexia, *Harmonisation of E-Commerce Legal Infrastructure in ASEAN*, April 2008.
[160] ASEAN Secretariat official, telephone interview by USITC staff, April 21, 2010.
[161] Ibid.
[162] See app. F for a list of the Harmonized System numbers covered under this priority sector.
[163] The ASEAN Secretariat defines the scope of "electronics products" to include electronic data processing (EDP) equipment, electrical and electronic home appliances, medical and industrial equipment, telecommunication equipment, communications and radar, automotive electronics, instrumentation and controls, mechanical parts of electronic and electrical products, semiconductor devices (including inputs for the manufacture of semiconductor devices), and other machineries and equipment for the manufacture of semiconductors and printed circuit boards (PCBs).

[164] ASEAN, "Roadmap for Integration of Electronics Sector," November 29, 2004.
[165] ASEAN Secretariat representative, interview by USITC staff, Jakarta, Indonesia, March 1, 2010.
[166] Ibid.
[167] Parsons et al., "Measures Affecting the Integration of ASEAN's Priority Sectors (Phase 2)," June 2007,v.
[168] Ibid.
[169] Ibid.
[170] Mirza and Giroud, "Regional Integration and Benefits from Foreign Direct Investment in ASEAN Economies," 2004, 74.
[171] Industry representatives, interviews by USITC staff, Bangkok, Thailand; Binh Duong province, Vietnam; and Singapore, March 4–16, 2010.
[172] See app. F under "Electronics" for HS subheadings that apply to the computer components industry discussed in this chapter.
[173] WITS, Integrated Data Warehouse (accessed January 5, 2010). Exports based on partner country imports. The electronics product group is defined in ASEAN's "Roadmap for Integration of Electronics Sector," November 29, 2004.
[174] *The Straits Times*, "Intel to Close Philippines Plant," January 23, 2009.
[175] No specific information is available regarding the reasons for Brunei's lack of components production. However, it is likely that the country's small population restricts the availability of qualified labor. Some areas of concern in Brunei among electronics producers can be found in Parsons et al., "Measures Affecting the Integration of ASEAN's Priority Sectors (Phase 2)," June 2007, 37.
[176] Industry representative, interview by USITC staff, Bangkok, Thailand, March 16, 2010; industry representative, interview by USITC staff, Singapore, March 4, 2010.
[177] Industry representatives, interviews by USITC staff, Singapore, March 4–5, 2010. Singapore also specializes in producing the electronic components that are inputs into computer components, such as integrated circuits.
[178] Parsons et al., "Measures Affecting the Integration of ASEAN's Priority Sectors (Phase 2)," June 2007, 38; Rasiah, "Expansion and Slowdown in Southeast Asian Electronics Manufacturing," 2009, 124.
[179] Parsons, 38; Rasiah, 132.
[180] WITS, Integrated Data Warehouse (accessed January 5, 2010). Exports based on partner country imports.
[181] Intel Web site, http://www.intel.com/community (accessed April 29, 2010).
[182] Industry representative, interview by USITC staff, Ho Chi Minh City, Vietnam, March 11, 2010.
[183] WITS, Integrated Data Warehouse (accessed January 5, 2010). Exports based on partner country imports.
[184] Roadmap for Integration of Electronics Sector, ASEAN Framework Agreement for the Integration of Priority Sectors, 2004.
[185] Rasiah, "Expansion and Slowdown in Southeast Asian Electronics Manufacturing," 2009, 124.
[186] Ibid., 125.
[187] Ibid., 125.
[188] For the purposes of this section, the "cost" of labor refers not simply to workers' compensation but to the definition of "labor cost" as published by the Bureau of Labor Statistics (BLS), which factors in differences in worker productivity. BLS defines labor cost as total labor compensation divided by real output. For more information, see the BLS Web site, http://www.bls.gov/lpc/faqs.htm#P06 (accessed June 1, 2010).
[189] BLS, "Computer and Electronic Product Manufacturing." (accessed March 23, 2010).
[190] Multilateral Investment Guarantee Agency (MIGA), *Benchmarking FDI Competitiveness in Asia*, 2003,9.
[191] Industry representative, interview by USITC staff, Singapore, March 4, 2010.
[192] Industry representative, interview by USITC staff, Singapore, March 9, 2010.
[193] Industry representative, interview by USITC staff, Singapore, March 9, 2010; Mitarai, "Issues in the ASEAN Electric and Electronics Industry and Implications for Vietnam," February 2005, 16.
[194] Industry representative, interview by USITC staff, Bangkok, Thailand, March 16, 2010.
[195] Bradsher, "As Labor Costs Soar in China, Manufacturers Look to Vietnam," June 18, 2008.
[196] Martin, "For Jabil Circuit, Vietnam is the New China," June 8, 2008.
[197] Asian Productivity Organization, *APO Productivity Databook 2009*, 2009, 56.
[198] Irwin, *Free Trade under Fire*, 37.
[199] Industry representatives, interviews by USITC staff, Ho Chi Minh City, Vietnam, March 11, 2010.
[200] Multilateral Investment Guarantee Agency (MIGA), *Benchmarking FDI Competitiveness in Asia*, 2003,11.
[201] Industry representative, interview by USITC staff, Bangkok, Thailand, March 16, 2010; Mitarai, "Issues in the ASEAN Electric and Electronics Industry and Implications for Vietnam," February 2005, 7.
[202] Foster et al., *Technology and Organizational Factors in the Notebook Industry Supply Chain*, 2006, 14.
[203] Doner and Ritchie, "Economic Crisis and Technological Trajectories: Hard Disk Drive Production in Southeast Asia," 2001, 6.
[204] Ibid., 7.
[205] Industry representative, interview by USITC staff, Bangkok, Thailand, March 16, 2010.
[206] Industry representative, interview by USITC staff, Hanoi, Vietnam, March 9, 2010.

[207] Jabil Circuit's plant in Vietnam assembles printers for Hewlett-Packard. While printers are a computer accessory rather than a component, their experience with the local base of electronics suppliers is applicable to components firms.

[208] Martin, "For Jabil Circuit, Vietnam Is the New China," June 8, 2008.

[209] Industry representative, interview by USITC staff, Hanoi, Vietnam, March 9, 2010.

[210] Parsons et al., "An Investigation into the Measures Affecting the Integration of ASEAN's Priority Sectors (Phase 2)," June 2007, 36.

[211] Industry representative, interview by USITC staff, Ho Chi Minh City, Vietnam, March 10, 2010.

[212] Industry representatives, interviews by USITC staff, Ho Chi Minh City, Vietnam, March 10, 2010.

[213] Pecht and Tiku, "Bogus!" May 2006.

[214] ASEAN, "Roadmap for Integration of Electronics Sector," November 29, 2004.

[215] ASEAN Secretariat official, interview by USITC staff, Jakarta, Indonesia, March 2, 2010.

[216] Santiago, "Development of ASEAN Framework for Trade Negotiations: Electronics Industry," March 2007, 19.

[217] ASEAN Secretariat official, interview by USITC staff, Jakarta, Indonesia, March 2, 2010.

[218] Ibid.

[219] Santiago, "Development of ASEAN Framework for Trade Negotiations: Electronics Industry," March 2007, 25.

[220] Fukase and Martin, "A Quantitative Evaluation of Vietnam's Accession to the ASEAN Free Trade Area," November 1999, 47–48.

[221] After China, the EU, and the U.S. WITS, Integrated Data Warehouse (accessed January 5, 2010). Exports based on partner country imports.

[222] Montevirgen, "Industry Surveys: Semiconductors," November 2009, 21.

[223] A portion of the decrease between the 2004–06 period and the 2007–08 period could be due to 2007 changes to the Harmonized System classification of some computer components that may not be fully reflected in the data. The overall trend in the data, however, is consistent whether or not the items for which the classification changed are included in the analysis.

[224] IBISWorld, "Global Computer Hardware Manufacturing" (accessed June 2009).

[225] For example, Dell operates a computer assembly facility in Penang, Malaysia.

[226] WTO, Information Technology Agreement: Schedules of Concessions (accessed April 5, 2010).

[227] The ASEAN-6, of which all except Brunei are signatories to the ITA, do not apply duties on these items. Vietnam became a member of the WTO and a signatory to the ITA in 2007. Under its schedule of concessions (WTO, "Information Technology Agreement: Schedules of Concessions, Vietnam," 2007), it has agreed to eliminate all duties on computer components on an MFN basis by 2012. Cambodia and Burma are members of the WTO but are not signatories to the ITA, and both apply duties to imports of these products from non-ASEAN countries. Laos is not a WTO member and also applies duties to imports of these products.

[228] Parsons et al., "Measures Affecting the Integration of ASEAN's Priority Sectors (Phase 2)," June 2007,

[229] USTR, "Indonesia," 2009, 249.

[230] IBISWorld, "Computer Body Manufacturing in China" (accessed June 2009).

[231] Industry representative, interview by USITC staff, Singapore, March 4, 2010.

[232] BLS, "Computer and Electronic Product Manufacturing" (accessed March 23, 2010).

[233] Rasiah, "Expansion and Slowdown in Southeast Asian Electronics Manufacturing," 2009, 125; industry representative, interview by USITC staff, Bangkok, Thailand, March 16, 2010.

[234] Rasiah, "Expansion and Slowdown in Southeast Asian Electronics Manufacturing," 2009, 126.

[235] Ernst, "Global Production Networks in East Asia's Electronics Industry and Upgrading Prospects in Malaysia," 2004, 94.

[236] ISuppli, "Seagate Maintains Top Ranking in Hard Disk Drive Market in Q4," March 22, 2010.

[237] Montevirgen, "Industry Surveys: Semiconductors," November 2009.

[238] Corbett and Umezaki, "Table 6: FDI Policy," 2008.

[239] Industry representative, interview by USITC staff, Ho Chi Minh City, Vietnam, March 11, 2010.

[240] Singapore competes mainly with countries outside of ASEAN, such as Hong Kong, for regional headquarters investment, according to Brown and Tucker, "Business in Asia: A High-Flying Rivalry," April 25, 2010 (accessed April 27, 2010).

[241] Industry representatives, interviews by USITC staff, Singapore; Ho Chi Minh City, Vietnam; and Bangkok, Thailand, March 4–16, 2010.

[242] Industry representatives, interviews by USITC staff, Singapore, March 4–5, 2010.

[243] "Pioneer" status carries different definitions, requirements, and benefits in each country. For example, according to Malaysia's Ministry of Finance Web site, pioneer status for high-technology companies carries a five-year tax exemption. Meanwhile, in Vietnam, pioneer status may confer tax exemptions for as long as 50 years.

[244] Industry representative, interview by USITC staff, Singapore, March 4, 2010; Stoler and Phan, "Vietnam: Intel and the Electronics Sector," 2009, 184.

[245] Brown and Tucker, "Business in Asia: A High-Flying Rivalry," April 25, 2010.

[246] All data in this paragraph are from the Zephyr database (accessed May 3, 2010). Among intra-ASEAN investments listing Singapore as the source country, only a small number were made by Singapore-based

subsidiaries of large multinational companies (headquartered in a third country). The majority of these investments were made by firms indigenous to Singapore.

[247] Industry representative, interview by USITC staff, Ho Chi Minh City, Vietnam, March 11, 2010.

[248] Stoler and Phan, "Vietnam: Intel and the Electronics Sector," 2009, 184.

[249] Ibid.

[250] Industry representative, interview by USITC staff, Ho Chi Minh City, Vietnam, March 11, 2010.

[251] World Bank Governance Indicators database (accessed April 22, 2010).

[252] Industry representative, interview by USITC staff, Ho Chi Minh City, Vietnam, March 11, 2010.

[253] Industry representatives, interviews by USITC staff, Ho Chi Minh City, Vietnam, March 10–11, 2010.

[254] Industry representative, interview by USITC staff, Ho Chi Minh City, Vietnam, March 11, 2010.

[255] Industry representative, interview by USITC staff, Bangkok, Thailand, March 16, 2010.

[256] Industry representatives, interviews by USITC staff, Bangkok, Thailand, Ho Chi Minh City, Vietnam, and Singapore, March 4–16, 2010.

[257] Industry representative, interview by USITC staff, Singapore, March 4, 2010.

[258] Industry officials, interviews by USITC staff, Singapore, March 4, 2010 and Bangkok, Thailand, March 16, 2010.

[259] Industry representative, interview by USITC staff, Binh Duong province, Vietnam, March 11, 2010.

[260] Industry representatives, interviews by USITC staff, Singapore and Ho Chi Minh City, Vietnam, March 4–11, 2010.

[261] Ernst, "Global Production Networks in East Asia's Electronics Industry and Upgrading Prospects in Malaysia," 2004, 137.

[262] Intel Web site, "RosettaNet" (accessed April 26, 2010).

[263] Ernst, "Global Production Networks in East Asia's Electronics Industry and Upgrading Prospects in Malaysia," 2004, 138.

[264] Intel Web site, "RosettaNet" (accessed April 26, 2010).

[265] RosettaNet Web site, http://www.rosettanet.org/ (accessed May 7, 2010).

[266] Ernst, "Global Production Networks in East Asia's Electronics Industry and Upgrading Prospects in Malaysia," in *Global Production Networking and Technological Change in East Asia*, ed. Yusuf, Altaf, and Nabeshima (Washington, D.C.: World Bank and Oxford University Press, 2004), 138.

[267] Santiago, "Development of ASEAN Framework for Trade Negotiations: Electronics Industry" (paper for the United States Agency for International Development, March 2007), 22.

[268] See app. F for a list of the Harmonized System numbers covered under this priority sector.

[269] These products are those listed in the "Roadmap for Integration of Textiles and Apparel Products Sector."

[270] External tariffs on imports of these products are maintained by the ASEAN countries.

[271] Industry representative, e-mail message to USITC staff, April 30, 2010.

[272] Industry representative, interview by USITC staff, Kuala Lumpur, Malaysia, March 10, 2010.

[273] Industry representative, interview by USITC staff, Kuala Lumpur, Malaysia, March 10, 2010. Note that virtually no information was available on Burma's textile and apparel industry. Also, according to industry sources, Brunei's textile and apparel industry has virtually disappeared since the ATC quotas were removed in 2005. Industry representative, telephone interview by USITC staff, January 28, 2010.

[274] Note that after the ATC quotas expired, the EU and the United States maintained safeguard quotas on selected textile and apparel imports from China, as permitted under the China WTO accession agreement. However, these safeguards were not as comprehensive as the quotas that were in place under the ATC. The safeguard quotas expired for the EU at the end of 2007 and for the United States at the end of 2008.

[275] Industry representative, interview by USITC staff, Kuala Lumpur, Malaysia, March 10, 2010.

[276] For the purposes of this discussion, "cost of labor" means the cost to make one unit of a good. Differences in actual wage rates may be moderated or intensified by differences in productivity.

[277] As of spring 2010, Vietnam had the fastest-growing textile and apparel sector in the ASEAN region.

[278] Industry representative, interview by USITC staff, Kuala Lumpur, Malaysia, March 10, 2010.

[279] For example, the government of the Philippines is seeking preferential duty-free treatment in the U.S. market for its apparel exports that are made with U.S. fabric. Galvez, "RP Eyes Duty-Free Deal with US to Boost Textile Industry," April 20, 2010.

[280] Malaysian apparel factories, for example, tend to be small and produce higher-quality brand-name apparel in smaller quantities. Industry representative, interview by USITC staff, Kuala Lumpur, Malaysia. March 10, 2010.

[281] Industry representatives, interviews by USITC staff, Jakarta, Indonesia, March 3, 2010, and Ho Chi Minh City, March 10, 2010.

[282] Industry representative, telephone interview by USITC staff, January 28, 2010. As explained below, another key driver of integrated production of apparel and other textile finished products is the ASEAN- Japan FTA.

[283] See app. F under "Cotton Woven Apparel" for HS subheadings that apply to the cotton woven apparel industry discussed in this chapter. The fabrics considered part of the cotton woven apparel supply chain in this profile

consist of cotton woven fabrics containing 85 percent or more by weight of cotton. These fabrics are classified in HS headings 5208 and 5209.

[284] WITS Integrated Data Warehouse (accessed January 5, 2010).

[285] The United States and the EU established quotas on imports of woven cotton apparel (and other textile and apparel articles) from China after the quotas under the ATC expired, as permitted under the China WTO accession agreement. However, these safeguards were not as comprehensive as the quotas that were in place under the ATC. The safeguard quotas expired for the EU at the end of 2007 and for the United States at the end of 2008.

[286] For example, the duty rate for men's and boys' cotton trouser and shorts imported under U.S. Harmonized Schedule subheading 6203 .42.40 dropped from 90 percent ad valorem to 16.6 percent ad valorem.

[287] U.S. Department of State, "Background Note on Vietnam," October 2009.

[288] Industry representative, interview by USITC staff, Bangkok, Thailand, March 15, 2010.

[289] Industry representatives, e-mail messages to USITC staff, March 11, 2010, March 16, 2010; March 22, 2010, March 26, 2010; industry representatives, telephone interviews by USITC staff, March 22 and March 24, 2010.

[290] Industry representative, interview by USITC staff, Bangkok, Thailand, March 17, 2010.

[291] Government official, interview by USITC staff, Jakarta, Indonesia, March 1, 2010.

[292] Ibid., March 2, 2010.

[293] The ACE project's initiative to develop "fast fashion" production.

[294] Industry representative, interview by USITC staff, Kuala Lumpur, Malaysia, March 10, 2010; and government official, interview by USITC staff, Jakarta, Indonesia, March 2, 2010.

[295] Industry representative, e-mail message to USITC staff, April 30, 2010.

[296] Virtual vertical factories (VVFs) are apparel factories and fabric weaving mills that would partner to create finished apparel. Industry representative, interview by USITC staff, Bangkok, Thailand, March 15, 2010.

[297] Industry representative, interview by USITC staff, Bangkok, Thailand, March 15, 2010; industry representative, e-mail to USITC staff, April 20, 2010.

[298] Likhikiatikul, "Global Fashion Brands Sign 23 MOU with ASEAN Suppliers." June 15, 2010.

[299] Fee and Shannon, "Free Trade Agreements: Japan's Free Trade Network," March 16, 2010.

[300] Industry representative, meeting with USITC staff, Jakarta, Indonesia, March 3, 2010.

[301] Industry representative, interview by USITC staff, Bangkok, Thailand, March 15, 2010.

[302] For example, see Blake, "Blaming China: Indonesian Garment Makers Say Free Trade Pact Leaves Them on Brink of Collapse," April 26, 2010.

[303] Industry representatives, telephone interviews by Commission staff, March 22 and March 24, 2010.

[304] For example, China's exports of cotton woven fabrics to the ASEAN countries nearly doubled from 2004 to 2008 (table 4.2).

[305] VITAS, "Garments Top Nation's Export Lists," January 12, 2010; VITAS, "Vietnam Receives Just a Small Slice of the Garment Industry Pie," January 26, 2010.

[306] VITAS, "Vietnam Receives Just a Small Slice of the Garment Industry Pie," January 26, 2010.

[307] Industry representative, interview by USITC staff, Ho Chi Minh City, Vietnam, March 10, 2010.

[308] Unless otherwise noted, information in this paragraph is based on the following sources: Industry representatives, e-mail messages to USITC staff, March 11, March 16, March 22, and March 26, 2010; industry representatives, telephone interviews by USITC staff, March 22 and March 24, 2010.

[309] Joe Ayling, "Analysis: Emerging Exporters Eye China Market Share," April 16, 2010.

[310] Industry representative, meeting by USITC staff, Jakarta, Indonesia, March 3, 2010.

[311] Industry representative, interview by USITC staff, Kuala Lumpur, Malaysia, March 10, 2010.

[312] Industry representative, meeting by USITC staff, Jakarta, Indonesia, March 3, 2010.

[313] Ibid.

[314] Joe Ayling, "Analysis: Emerging Exporters Eye China Market Share," April 16, 2010; Joe Ayling, "Speaking with Style: Janet Fox, JC Penney," April 27, 2010.

[315] Industry representative, interview by Commission staff, Kuala Lumpur, Malaysia, March 10, 2010.

[316] BDLINK Cambodia and ACE Project Team, "Diagnostics Report, Textile and Apparel Supply Chain Corridors," July 2009, 25–26.

[317] BDLINK Cambodia and ACE Project Team, "Diagnostics Report, Textile and Apparel Supply Chain Corridors," July 2009, 15–16.

[318] Industry representative, interview by Commission staff, Bangkok, Thailand, March 17, 2009.

[319] BDLINK Cambodia and ACE Project Team, "Diagnostics Report, Textile and Apparel Supply Chain Corridors," July 2009, 15–16.

[320] ACE Project Team, "Textile and Apparel Supply Chain Corridor Diagnostics, SWOT Analysis," May 2009, 23–24.

[321] Textile Outlook International, "World Textile and Apparel Trade and Production Trends," September– October 2008, 26.

[322] Ibid., 27.

[323] Ibid., 29.
[324] Galvez, "RP Eyes Duty-Free Deal with US to Boost Textile Industry," April 20, 2010. The "Save Our Industries Act" (Senate Bill 3170, 111th Congress) would provide duty preferences for U.S. imports of certain apparel manufacturing in the Philippines using U.S. yarns and fabrics.
[325] Industry representative, interview by USITC staff, Kuala Lumpur, Malaysia, March 10, 2010.
[326] Textile Outlook International, "World Textile and Apparel Trade and Production Trends," September– October 2008, 2 1–22.
[327] Industry representatives, interviews by USITC staff, Jakarta, Indonesia, March 3, 2010, and Ho Chi Minh City, Vietnam, March 10, 2010.
[328] Industry representative, interview by USITC staff, Kuala Lumpur, Malaysia, March 10, 2010.
[329] Greenwood, "WFP Laos-Logistics Overview." May 7, 2010.
[330] Industry representative, telephone interview by USITC staff, January 28, 2010.
[331] Industry representative, interview by USITC staff, Bangkok, Thailand, March 15, 2010.
[332] Industry representative, interview by USITC staff, Ho Chi Minh City, Vietnam, March 10, 2010.
[333] Industry representative, meeting by USITC staff, Jakarta, Indonesia, March 3, 2010.
[334] Ibid.
[335] BDLINK Cambodia and ACE Project Team, "Diagnostics Report, Textile and Apparel Supply Chain Corridors," July 2009, 27.
[336] Textile Outlook, "World Textile Trade and Production Trends," September/October 2008, 33.
[337] Ibid.
[338] For the purposes of this discussion, "cost of labor" means the cost to make one unit of a good. Differences in actual wage rates may be moderated or intensified by differences in productivity.
[339] Industry representative, interview by USITC staff, Ho Chi Minh City, Vietnam, March 10, 2010.
[340] Industry representative, telephone interview by USITC staff, March 24, 2010.
[341] Industry representative, interview by USITC staff, Jakarta, Indonesia, March 3, 2010.
[342] Ibid.
[343] Industry representative, interview by USITC staff, Bangkok, Thailand, March 17, 2010.
[344] Industry representative, interview by USITC staff, Ho Chi Minh City, Vietnam, March 10, 2010.
[345] Industry representative, e-mail message to USITC staff, March 16, 2010; industry representatives, telephone interviews by USITC staff, March 22 and 24, 2010.
[346] Government official, interview by USITC staff, Jakarta, Indonesia, March 15, 2010.
[347] Industry representative, interview by USITC staff, Ho Chi Minh City, Vietnam, March 10, 2010. Time periods reported above may differ from the World Bank "Trading Across Borders" data cited in chapter 2 because of practices used in the World Bank's methodology.
[348] Government official, interview by USITC staff, Jakarta, Indonesia, March 15, 2010.
[349] Industry representative, interview by USITC staff, Kuala Lumpur, Malaysia, March 10, 2010.
[350] See app. F for a list of the Harmonized System numbers covered under this priority sector.
[351] Industry representative, interview by USITC staff, Bangkok, March 15, 2010.
[352] WITS, Integrated Data Warehouse (accessed March 29, 2010).
[353] Industry representative, telephone interview by USITC staff, Singapore, January 27, 2010.
[354] Industry representative, telephone interview by USITC staff, Jakarta, March 2, 2010.
[355] Ibid.
[356] See app. F under "Wood-based Products" for HS subheadings that apply to the hardwood plywood and flooring industry discussed in this chapter.
[357] Resosudarmo and Yusuf, "Is the Log Export Ban Effective?" June, 2006, 1.
[358] Although unprocessed wood products (timber, roundwood, posts, etc.) are not included within the trade liberalization scope of the ASEAN Sectoral Integration Protocol for Wood-based Products, the Roadmap addresses illegal logging and related perceptions about the region's wood products.
[359] Some of the programs and processes to address legal trade and forest sustainability are discussed in the regional integration subsection. See also table 5.3.
[360] Estimated based on FAO production data and unit export values, and value of global wood furniture imports from ASEAN members based on Global Trade Information Service (GTIS) data, accessed March 15, 2010.
[361] Calculated using Food and Agriculture Organization (FAO) plywood export value as a percent of derived value of plywood shipments (production times unit export value).
[362] See table 2.4 for total exports of wood-based products.
[363] FAO, FAOSTAT database,.
[364] See table 5.3.
[365] Government representative, interview by USITC staff, March 2, 2010.
[366] Industry representative, telephone interview by USITC staff, March 2, 2010.
[367] Industry representative, interview by USITC staff, Kuala Lumpur, March 12, 2010.

[368] Government representatives, interviews by USITC staff, Hanoi, March 8, 2010; FAO, FAOSTAT database, accessed January 7, 2010; Philippines Forest Management Bureau, 2007 Forestry Statistics, accessed March 17, 2010.

[369] Philippines Forest Management Bureau, 2007 Forestry Statistics, accessed March 17, 2010.

[370] Based on GTIS trade data (accessed March 15, 2010) and confirmed by industry representative, interviewed by USITC staff, Washington, DC, March 30, 2010.

[371] Industry representatives, interviews by USITC staff, Jakarta, March 2, 2010 and Kuala Lumpur, March12, 2010.

[372] Industry representative, interview by USITC staff, Kuala Lumpur, March 12, 2010. Engineered wood flooring is typically made from a plywood core with a "wear" layer of the desired species adhered to its surface.

[373] Industry representative, interview by USITC staff, Kuala Lumpur, March 12, 2010.

[374] Tachibana, "Forest-Related Industries and Timber Exports," 1999, 70–71.

[375] ITTO, *Tropical Timber Market Report*, April 30, 2010, 19; log export price data are limited to ASEAN countries that permit log exports.

[376] Industry representative, interview by USITC staff, March 12, 2010; EC-FAO Partnership Programme, December, 2002, 176.

[377] Industry representative, interview by USITC staff, Hanoi, March 9, 2010. Vietnam lacks large diameter logs and thus has only a small plywood industry.

[378] Industry representative, interview by USITC staff, Kuala Lumpur, March 12, 2010.

[379] Ibid.

[380] Enforcement of harvesting regulations and restrictions is believed by many to be ineffective. Allegations of illegal logging are widespread.

[381] Simula, "Pros and Cons of Procurement," April, 2010, 15.

[382] Industry representative, interview by USITC staff, February 16, 2010.

[383] Industry representative, interview by USITC staff, Kuala Lumpur, March 11, 2010.

[384] The Lacey Act (16 U.S.C. 3371 et seq.) was first enacted in 1900 to protect wildlife. The most recent amendments to the Lacey Act were included in the Food, Conservation, and Energy Act of 2008 and expanded its provisions to cover a broad range of plants and plant products (Section 8204, Prevention of Illegal Logging Practices). The Lacey Act amendments exempt common cultivars and food crops from their provisions. They also exempt packing materials and live plants from the import declaration requirement.

[385] Industry representative, interview by USITC staff, Kuala Lumpur, March 11, 2010.

[386] Industry representative, interview by USITC staff, February 16, 2010.

[387] Based on extrapolations from FAOSTAT production (accessed January 7, 2010) and GTIS trade data (accessed March 15, 2010).

[388] Market share estimates are based on the GTIS database (accessed March 15, 2010).

[389] Industry representative, telephone interview by USITC staff, February 11, 2010.

[390] Oliver, "Monitoring the Competitiveness of Tropical Timber," 65.

[391] ASEAN official, interview by USITC staff, March 1, 2010. ASEAN work on wood products standards is pursued through the ASEAN Expert Group on Research and Development in Forest Products.

[392] Industry representative, interview by USITC staff, Kuala Lumpur, March 12, 2010.

[393] The California Air Resources Board (CARB) rule requires producers of composite wood products and products made with them that are sold in California to be certified by a third-party testing agency accredited by CARB. See CARB Web site: http://www.arb.ca.gov/toxics/compwood/tpc/listofmills.htm (accessed March 20, 2010).

[394] Public Law 111-199, 124 Stat. 1359.

[395] Seneca Creek Associates, "Illegal Logging and Global Wood Markets," October, 2004. See also Chatham House Web site for reports and news articles: www.illegal-logging.info (accessed various dates).

[396] FAO, "Global Forest Resources Assessment," 2005, table 4, 197. The FAO Forest Resources Assessment showed the deforestation rate for the region as a whole in the 1990s averaged 1.2 percent annually and continued at 0.7 percent per annum during the first half of the past decade.

[397] Roadmap, 9–14.

[398] ASEAN Forest Clearing House Mechanism Web site. http://www.aseanforestchm.org/issue_pages/sfm_ implementation/ forest_and_timber_certification (accessed March 8, 2010).

[399] ASEAN Senior Officials on Forestry (ASOF) signed the Vientiane Action Programme (VAP) in 2008; an ASEAN Wildlife Law Enforcement Network (ASEAN-WEN) was created in 2005; an agreement on transboundary haze pollution was signed in 2006; and the ASEAN Declaration on Environmental Sustainability was signed in 2007.

[400] Boungnakeo, "Forest Law Enforcement and Governance in Lao PDR," 15.

[401] ASEAN, "ASEAN Economic Community Blueprint," November 20, 2007, section II.A7.

[402] ASEAN Forest Clearing House Mechanism Web site.

[403] PEFC is an internationally recognized certification body that endorses national certification schemes that conform to its core standards.

[404] See FSC Web site for information about global FSC certificates, including types of certificates and global distribution: http://www.fsc.org/facts-figures.html (accessed February 16, 2010); see also the LEI web site: http://www.lei.or.id/.
[405] Ozinga, EU Forest Watch February 2010.
[406] USTR, press release, September 2, 2009.
[407] Simula, IITO, 30.
[408] The trade data are not an exact match. Some of the HTS 6-digit categories that include flooring are broad groups that contain moldings, panels, and other wood products, so these figures overstate the actual trade in flooring.
[409] Industry representative, telephone interview by USITC staff, Jakarta, March 2, 2010.
[410] Industry representative, interview by USITC staff, Washington, DC, March 23, 2010. Bamboo is technically a grass species but is marketed with hardwood when utilized for plywood and flooring products.
[411] WITS, Integrated Data Warehouse (accessed January 5, 2010).
[412] Ibid.
[413] Industry representative, telephone interview by USITC staff, February 12, 2010.
[414] ASEAN China Free Trade Area. Annex 2, App. 1, 99.
[415] Industry representatives, interview by USITC staff, Kuala Lumpur, March 12, 2010.
[416] Industry representative, interview by USITC staff, Bangkok, March 15, 2010. China is currently a major purchaser of Malaysian logs (from Sarawak and Sabah).
[417] *Jakarta Post*, "Thousands Rally Against Free Trade Treaty," January 21, 2010.
[418] Industry representative, interview by USITC staff, Kuala Lumpur, March 11, 2010.
[419] Industry representative, telephone interview by USITC staff, April 22, 2010.
[420] *Saigon Times*, "Sumitomo Forestry to Build Plywood Factory," January 13, 2010.
[421] Industry representative, interview by USITC staff, Washington, DC, March 23, 2010.
[422] The ACIA replaces the Framework Agreement on the ASEAN Investment Area (AIA) signed in 1998, but has not yet entered into force. See chap. 2 for additional information on the ACIA and AIA.
[423] U.S. Department of State, "Investment Climate Statements" for ASEAN countries; Economic Research Institute for ASEAN and East Asia (ERIA), "Deepening East Asian Economic Integration," Chap. 5, table 6.
[424] Malaysia Timber Industry Board, "Credit Opportunities in Forest Plantations," n.d.
[425] Industry representative, telephone interview by USITC staff, Singapore, January 27, 2010.
[426] Industry representative, interview by USITC staff, Kuala Lumpur, March 11, 2010.
[427] Ibid., March 12, 2010.
[428] See discussion on trade facilitation in chap. 2.
[429] ASEAN official, e-mail message, April 21, 2010. Not all ASEAN members are reporting trade and tariff data at the same level of detail for external as well as intra-ASEAN trade, nor are all members using the most recent 2007 ASEAN Harmonized Tariff Nomenclature (AHTN).
[430] Industry representative, interview by USITC staff, Bangkok, March 15, 2010.
[431] Nathan Associates, "Toward a Roadmap of Integration," March 2007, 33.
[432] The Cross-Border Transit Agreement is part of a Greater Mekong Subregion (GMS) Program supported by the Asian Development Bank. It was crafted in consultation with ASEAN and the United Nations Economic and Social Commission for Asia and the Pacific.
[433] Government official, interview by USITC staff, Hanoi, March 9, 2010.
[434] Malaysia Timber Industry Board Web site. Cess rates range from less than US$1.00 per m^3 to as much as US$80.00 per m^3 for certain species of timber. MDF products (that could be used to manufacture laminate wood flooring) have no cess rate.
[435] See Vietnam Ministry of Natural Resources and Environment web site and Global Trade Alert.
[436] NGO representative, interview by USITC staff, Hanoi, March 8, 2010.
[437] Industry representative, interview by USITC staff, Kuala Lumpur, March 12, 2010.
[438] See app. F for a list of the Harmonized System numbers covered under this priority sector. Healthcare services do not have Harmonized System numbers, and thus are not included in this list.
[439] The healthcare priority sector, as designated by the ASEAN Secretariat, is categorized into five subsectors: healthcare services; cosmetics; medical devices and equipment; pharmaceuticals; and traditional medicines and health supplements. The priority sector product basket (excluding healthcare services) comprises 245 goods, almost half of which are pharmaceutical products.
[440] The use of the term "healthcare goods" is understood to refer only to those healthcare goods covered by the Roadmap, which are only a portion of all healthcare goods traded.
[441] Intra-ASEAN trade volumes are estimates based on available import trade data and may be undervalued. Brunei, Burma, Indonesia, and Laos do not report trade data. Cambodia did not report trade data for 2005–08, and the Philippines did not report trade data for 2004–06.
[442] ASEAN Secretariat, "Appendix 1," n.d.; Gross and Minot, "ASEAN Medical Device Regulatory Integration," January 1, 2009.

443 The Agreement on the Harmonized Cosmetic Regulatory Scheme was signed in 2005 and is in varying stages of implementation among member countries. Badan Standardisasi Nasional, "Indonesia Will Soon Ratify ASEAN Cosmetic Directive," September 9, 2009.

444 Lee-Brago, "ASEAN Agrees to Further Promote Integration of Traditional Medicine," January 7, 2010. In the medical devices, pharmaceuticals, and traditional medicines subsectors, working groups for each respective sector have been formed and continue to meet in efforts to coordinate industry regulations and requirements. Seerangam, "ASEAN Pharmaceutical Harmonization," November 17–19, 2008.

445 CEPT rates for healthcare goods in the priority basket were successfully eliminated in the ASEAN-6 (Brunei, Indonesia, Malaysia, the Philippines, Singapore, and Thailand) and are a maximum of 5 percent in the ASEAN-4 (Burma, Cambodia, Laos, and Vietnam). Agreement on the Common Effective Preferential Tariff (CEPT) Scheme for the ASEAN Free Trade Area, Consolidated Package of Tariff Reductions for 2008, http://www.aseansec.org/19802.htm (accessed June 11, 2010); Royal Malaysia Customs Department, HS-Explorer (accessed April 28, 2010); World Trade Organization (WTO), Tariff Analysis Online (accessed April 28, 2010).

446 SingaporeMedicine, "Growing Biomedical Manufacturing Industry," n.d. (accessed June 11, 2010); Singapore Economic Development Board, "Medical Technology: Industry Background," n.d. (accessed June 11, 2010); Malaysian Industrial Development Authority, "Industries in Malaysia: Medical Devices Industry," n.d. (accessed June 11, 2010).

447 Jain, "Malaysia Poised for Leap of Growth–II," May 11, 2010. Medical technologies range from advanced medical electronics, such as MRI scanners, to implantable devices, such as cardiac stents, and single-use technologies, such as gloves and needles. Eucomed, "Medical Technology Brief," May 2007.

448 WITS, Integrated Data Warehouse (accessed March 29, 2010).

449 The ASEAN Secretariat does not report data on foreign investment for the healthcare sector. Instead, information supporting this investment discussion was gathered from reports of individual deals, anecdotal information, and discussions with industry representatives. Bureau van Dijk, Zephyr database (accessed April 27, 2010).

450 Malaysian Industrial Development Authority, "Incentives for the Medical Devices Industry."

451 Gross and Minot, "The Roar of Vietnam's Device Market," May 2009.

452 Pacific Bridge Medical, "Growing Foreign Investment in Vietnam's Pharmaceutical Industry," February 4, 2008.

453 Data used in this section, where available, are largely from national statistical agencies, as the ASEAN Secretariat does not report statistics on healthcare services. Data for 2007 were the most recent available. *Hospitals.sg*, "Singapore's Medical Tourism Figures Revealed by Health Minister," January 28, 2009.

454 Healthcare services exports refer to consumption abroad, or medical treatment provided to foreign patients by healthcare facilities, often referred to as medical travel. Medical travel is a primary focus of the ASEAN integration efforts and of national governments.

455 In the following discussion, healthcare services refer specifically to the services categorized under the United Nation's Central Product Classification (CPC) code Human Health Services (931), which is consistent with the commitments in healthcare services made in the ASEAN Framework Agreement on Services (AFAS). Services industries cannot be categorized or defined using tariff numbers, but the CPC system provides a parallel classification system. CPC codes do not categorize industries, per se, but rather the output of an industry. To the degree possible, products or services provided by a single industry are grouped under one CPC: for example, Human Health Services (931) encompass most of the services provided by hospitals and other healthcare facilities. Human health services encompass the following subclasses: inpatient services (9311); medical and dental services (9312), which include outpatient medical treatments; and other human health services (9319), which include laboratory services. UN, Statistics Division, "CPC Ver. 2," December 31, 2008.

456 Trade in healthcare services can be divided into four modes, or categories, of supply. Cross-border trade (mode 1) parallels trade in goods, wherein suppliers and consumers are located in different countries. Cross-border trade in healthcare services usually involves pathology or radiology services (specimens or films can be sent across borders), or telemedicine (the provision of services using information technology). Consumption abroad (mode 2) occurs when the consumer, or patient, travels to the country of the supplier. Commercial presence (mode 3) is when the supplier, or provider, establishes a presence through a juridical person in the consumer's market; this is closely linked to foreign investment. Movement of service providers (mode 4) describes the movement of an individual healthcare professional to the country of the patient. See below for a further discussion of trade in healthcare services. WTO, "Definition of Services Trade and Modes of Supply," n.d.; Arunanondchai and Fink, "Trade in Health Services in the ASEAN Region," March 2007, 11.

457 Generally, "public" healthcare facilities are defined as facilities which belong to the state; conversely, "private" facilities are those owned by non-government individuals or groups. Giusit, Criel, and De Behune, "Viewpoint," 1997, 193.

458 The focus of this analysis is the private healthcare services sector, as publicly provided services are a public good and generally do not compete with private services.

459 Governments frequently participate in the provision of healthcare services to address economic and public health concerns, often serving as the primary healthcare provider in lower-income economies. However, as countries

experience economic development, the private healthcare system develops and treats public health issues. Additionally, the population's preference for healthcare services shifts, due to the increased incidence of noncommunicable diseases as well as rising household expenditure on health. As a result, healthcare systems frequently grow over time to include a mixture of public and private providers as governments seek private sector participation to further develop healthcare infrastructure and meet growing demand. Arunanondchai and Fink, "Trade in Health Services in the ASEAN Region," 2007, 2, 8; Hanson and Berman, "Private Health Care Provision in Developing Countries," n.d., 10; Global Health Council, "Private- Public Sector," n.d.

[460] Ministry of Health Singapore, "Healthcare Facilities," n.d.; Ministry of Health Singapore, *Health Facts 2008*, 2008.

[461] *International Medical Travel Journal Online*, "Indochina," November 20, 2008.

[462] In Burma, private hospitals were prohibited prior to 2007, and in Laos, the government granted the first licenses for private hospitals in 2009. Irrawaddy, "Burma Licenses Private Hospitals, Clinics," October 8, 2009; Phouthonesy, "Govt Grants First Private Hospital Licenses," September 22, 2009.

[463] Imports of healthcare services refer to the movements of individuals across borders to receive medical treatment and care outside their home market.

[464] As mentioned above, healthcare services can be traded via four modes; however, each of these four modes is captured by discrete and incomparable trade data, where such data exist. For the purposes of this chapter, exports of healthcare services will refer to consumption abroad (medical treatment provided to foreign patients by healthcare facilities, often referred to as medical travel). Medical travel is a primary focus of the ASEAN integration efforts and of national governments. Trade via mode 3, or commercial establishment, is closely related to investment, and so will be addressed in the investment discussion of this chapter. Data for the other two modes of trade in healthcare services do not exist, and so, where necessary, will be discussed anecdotally. For example, data on cross-border trade in healthcare services in the ASEAN region, if captured, falls under trade in information services. Similarly, statistics on the volume or value of services provided by foreign medical personnel to patients in their (patient's) home country (mode 4) do not exist; instead, trade through movement of professionals is captured through data on the number of foreign professionals operating in a country. This data will be presented where relevant to the discussion.

[465] *International Medical Travel Journal*, "Malaysia," September 8, 2009; Arunanondchai and Fink, "Trade in Health Services in the ASEAN Region," March 2007, 13; Chee, "Medical Tourism in Malaysia," January 2007, 10, 13, 18.

[466] There are some exceptions to this, such as travel by U.S. citizens to Thailand or India specifically for healthcare services. However, the majority of healthcare exports are provided to foreign patients from neighboring countries, and industry representatives indicate that only a minority of patients are outside a 6–7 hour flight radius from the healthcare facility. Industry representative, interview with USITC staff, Singapore, March 8, 2010.

[467] Industry representatives, interview with USITC staff, Singapore, March 8 and 9, 2010; eTurboNews, "Medical Tourism Health Travel Magazine Launched in Thailand," August 27, 2009.

[468] There is no international body that governs the private healthcare industry, and due to the combination of public and private provision, there is no specific indicator for competitiveness across the global market.

[469] The Joint Commission International (JCI) is the international division of Joint Commission Resources, the U.S. hospital accreditation body. JCI accreditation is considered an indicator of quality, as accreditation indicates a facility has met baseline safety and quality standards for international care. Other standards and accreditations used throughout the hospital industry include those set by the International Organization for Standards (ISO) 9000. Joint Commission International Website, "About Joint Commission International"; Timmons, "The Value of Accreditation," October 17, 2007.

[470] Industry representative, interview with USITC staff, Bangkok, Thailand, March 16, 2010.

[471] Industry representative, interview by USITC staff, Bangkok, Thailand, March 16, 2010; Chee, "Medical Tourism in Malaysia," January 2007, 9–10.

[472] The ASEAN Framework Agreement on Services (AFAS) is the regional agreement on liberalization in services trade, which is intended to bind regional commitments beyond commitments made under the General Agreement on Trade in Services. As of 2009, seven rounds of commitments had been negotiated and implemented. ASEAN Secretariat official, e-mail to USITC staff, April 25, 2010; ASEAN Secretariat Web site, "ASEAN Framework Agreement on Services."

[473] ASEAN Secretariat, "Appendix 1," n.d.

[474] Indonesia, Malaysia, and Vietnam did not make commitments on hospital services. Industry representatives, interviews by USITC staff, Singapore, March 7 and 8, 2010; industry representative, interview by USITC staff, Bangkok, Thailand, March 17, 2010.

[475] Industry representative, interview with USITC staff, Singapore, March 7, 2010; industry representative, interview with USITC staff, Bangkok, Thailand, March 17, 2010.

[476] ASEAN economies exhibited strong growth in gross domestic product (GDP) per capita, particularly from 2005 through 2007. ASEAN Secretariat, *ASEAN Yearbook of Statistics 2008,* 38.

[477] Arunanondchai and Fink, "Trade in Health Services in the ASEAN Region," 2007, 8; Hanson and Berman, "Private Health Care Provision in Developing Countries," n.d., 10.

[478] See the discussion on investment below for further information on regional expansion of healthcare providers. KPJ Healthcare Berhad Web site, "Hospital Network"; World Eye Reports, "Healthcare Tourism on the Rise," November 14–15, 2010.

[479] Industry representative, interview by USITC staff, March 7, 2010; *Xinhua News Agency*, "Myanmar, Malaysian Airlines Increase Flight Capacity," October 6, 2005; *Airline Industry Information*, "Tiger Airways Increases Capacity on Singapore-Kuala Lumpur Route," September 29, 2008; ASEAN Secretariat, "Eleventh ASEAN Transport Ministers Meeting," November 17, 2005.

[480] ASEAN Secretariat, "Appendix 1," n.d.

[481] Arunondchai and Fink, "Trade in Health Services in the ASEAN Region," 2007, 2.

[482] Industry representative, interview by USITC staff, March 9, 2010. Acquisition of an existing firm, or a controlling interest, is reportedly the easiest way for healthcare firms to enter a market, as it requires less investment in infrastructure and leverages local knowledge of consumption patterns and culture. Industry representative, interview by USITC staff, Singapore, March 9, 2010.

[483] ASEAN Framework Agreement on Services, 2007.

[484] A memorandum of understanding (MOU) is a binding document that expresses mutual accord on an issue between two or more parties. "Memorandum of Understanding," Business Dictionary.com, n.d. (accessed June 27, 2010).

[485] Ministry of Health, Singapore, "MOU between Singapore and Brunei Ministries of Health to Promote Trainings," December 6, 2004; Chee, "Medical Tourism in Malaysia," 2007, 20.

[486] Noor, "Thailand Looks to Set Up Private Hospital Here," March 30, 2010.

[487] Neither ASEAN nor the individual governments of these countries collect comprehensive statistics or comparable statistics on the volume of medical travelers on a regular basis, so statistics are estimates. Data on the value of healthcare services exports are not widely available or consistent. Instead, data on volumes of foreign patients are presented. These data are not specific to the private healthcare sector; however, based on anecdotal evidence, it is assumed that the majority of services are provided by private facilities. Industry representative, interview with USITC staff, Singapore, March 8, 2010; *Hospitals.sg*, "Singapore's Medical Tourism Figures Revealed by Health Minister," January 28, 2009.

[488] *International Medical Tourism Journal*, "Thailand: Health and Medical Tourism Update," April 15, 2010; industry representative, interview with USITC staff, Bangkok, Thailand, March 16, 2010; UN, ESCAP, "Medical Travel in Asia and the Pacific," 2009, 16.

[489] *International Medical Tourism Journal*, "Thailand: Health and Medical Tourism Update," April 15, 2010.

[490] UN, ESCAP, "Medical Travel in Asia and the Pacific," 2009, 12; Chee, "Medical Tourism in Malaysia," January 2007, 19; industry representatives, interview by USITC staff, Bangkok, Thailand, March 16 and 17, 2010.

[491] Chee, "Medical Tourism in Malaysia," January 2007, 20; eTurboNews, "Health Travel Magazine Launched in Thailand," August 27, 2009.

[492] Industry representative, interview by USITC staff, Singapore, March 8, 2010; Malaysian-German Chamber of Commerce and Industry, "Market Watch 2009—the Healthcare Sector," 7.

[493] Socio-Economic and Environmental Research Institute, "Medical Tourism," April 2009, 13.

[494] Davis and Erixon, "The Health of Nations," 2008, 10.

[495] Industry representative, interview by USITC staff, Bangkok, Thailand, March 16, 2010.

[496] Industry representative, interview by USITC staff, Singapore, March 9, 2010.

[497] Industry representative, interview by USITC staff, Singapore, March 8, 2010; Chee, "Medical Tourism in Malaysia," 2007, 20.

[498] Industry representative, interview by USITC staff, Singapore, March 9, 2010.

[499] Ibid.

[500] Singaporeans account for 10 percent of foreign patients treated in Malaysia. Industry representative, interview by USITC staff, Singapore, March 8, 2010; *International Medical Travel Journal*, "Malaysia Promoting Health Tourism from Brunei and Singapore," March 25, 2010; Malaysian-German Chamber of Commerce and Industry, "Market Watch 2009," 6.

[501] Chee, "Medical Tourism in Malaysia," 2007, 18.

[502] Ibid., 25.

[503] ASEAN Secretariat, "Appendix 1," n.d.

[504] ASEAN Secretariat official, e-mail to USITC staff, April 25, 2010.

[505] Industry representative, interview by USITC staff, Kuala Lumpur, March 11, 2010.

[506] Industry representatives, interviews by USITC staff, Singapore, March 7 and 8, 2010. ASEAN countries that do report statistics on foreign investment, such as Singapore, often do not report data specific to healthcare services, but instead aggregate investment in health care services with other industries. For example, Singapore reports healthcare services in the "other" category, which also includes agriculture and related services; fishing, operation of fishing hatcheries, and fish farms; extraction of crude petroleum and natural gas; other mining and quarrying; electricity and gas supply; collection, purification, and distribution of water;

sewage and refuse disposal, sanitation; recycling; education; social and community activities; arts, entertainment, and recreation; repair and maintenance of vehicles, office equipment, and personal and household goods; other services activities; public administration and defense; and domestic work activities. As a result, the investment discussion below is based on data for specific transactions, and anecdotal information. Singapore Department of Statistics, "Singapore's Investment Abroad," April 2010, 32.

[507] Foreign Direct Investment (FDI) is only relevant to private healthcare markets. Smith, "Foreign Direct Investment and Trade in Health Services," 2004, 2318.

[508] World Health Organization (WHO), Regional Office for the Western Pacific, "Cambodia," 2009; industry representative, interview with USITC staff, Singapore, March 9, 2010.

[509] Bangkok Hospital Web site, "Overseas Network Affiliates."

[510] Bureau van Dijk, Zephyr database, April 27, 2010.

[511] Ibid.

[512] Industry representative, interview by USITC staff, Singapore, March 9, 2010.

[513] Currently, out of the 10 ASEAN member countries, at least 6 governments have announced national initiatives promoting their healthcare services to foreign patients.

[514] For example, healthcare providers in Thailand report seeing a drop in patient volumes from Bangladesh as a result of the opening of advanced hospitals in Bangladesh. Industry representative, interview by USITC staff, Bangkok, Thailand, March 17, 2010.

[515] For example, ParkwayHealth (Singapore) has a majority interest or more in hospitals in Singapore, Malaysia, Indonesia, and Brunei; KPJ Healthcare (Malaysia) holds controlling interests in facilities in Malaysia, Indonesia, Bangladesh, and Saudi Arabia.

[516] ParkwayHealth Patient Assistance Center Web site.

[517] Industry representative, interview by USITC staff, Singapore, March 9, 2010.

[518] Industry representative, interview by USITC staff, Bangkok, Thailand, March 16, 2010.

[519] Industry representative, interview by USITC staff, Singapore, March 9, 2010; Chee, "Medical Tourism in Malaysia," January 2007, 23.

[520] Industry representative, interview by USITC staff, Bangkok, Thailand, March 17, 2010.

[521] Ibid.

[522] Industry representative, interview by USITC staff, Singapore, industry representatives, interview by USITC staff, Bangkok, Thailand, March 17 and 18, 2010.

[523] In 2003, the number of Internet subscribers/users per 1,000 people in ASEAN was 58.7; in 2007, it was 114.9. Telemedicine is another application of e-commerce seen with increasing frequency in the global healthcare market. However, literature reviews and interviews with regional officials indicated that currently, telemedicine in ASEAN is predominantly used by public providers to outsource services such as radiology and to increase domestic access (Malaysia's Telehealth initiative is an example). As such, telemedicine applications do not contribute to trade volumes and do not increase the export competitiveness of healthcare services. ASEAN Secretariat, *ASEAN Statistical Yearbook 2008*.

[524] Industry representative, interview by USITC staff, Singapore, March 9, 2010.

[525] UN, ESCAP, "e-Health in Asia and the Pacific," 2009, 14.

[526] Tele-education is defined as "the use of information and communication technologies to provide Distance Education." International Telecommunications Union and Inter-Americana Telecommunication Commission Organization of American States, *Tele-Education in the Americas*, December 2001, 9.

[527] Industry representative, interview by USITC staff, Singapore, March 8, 2010.

[528] UN, ESCAP, "e-Health in Asia and the Pacific," 2009, 10.

[529] Industry representative, interview by USITC staff, Singapore, March 8, 2010.

[530] Industry representative, interview by USITC staff, Singapore, March 9, 2010.

[531] ASEAN Secretariat official, e-mail to USITC staff, April 25, 2010.

[532] See app. F for a list of the Harmonized System numbers covered under this priority sector.

[533] The automotive priority sector includes a wide variety of automotive and transportation-related products, including tires, engines, air conditioning systems, self-propelled trucks, bearings, gaskets, batteries, electrical equipment, truck tractors, golf carts and similar vehicles, special-purpose vehicles (e.g., ambulances, hearses), passenger cars, trucks, motor vehicle parts, tanks and other armored vehicles, mopeds, motorcycles and their parts, and trailers and semi-trailers.

[534] The AICO is an ASEAN industrial cooperation scheme designed to increase industrial production, achieve closer integration, increase investment, increase intra-ASEAN trade, improve economies of scale, enhance the technology base, create internationally competitive industries, increase private sector participation, and increase industrial complementation. ASEAN, ASEAN Industrial Cooperation Scheme (accessed May 6, 2010). The program encourages joint manufacturing activities between ASEAN-based companies, with a minimum of two companies in two different countries receiving preferential rates of duty on goods exchanged between the signatory countries, subject to a government-approved AICO arrangement. For more information on the AICO system, see box 7.2.

[535] Indonesia has the lead on this priority sector Roadmap.

[536] Government official, interview by USITC staff, Jakarta, March 1, 2010.

[537] For the purposes of this chapter, the ASEAN motor vehicle parts industry includes electronic components (e.g., lighting and signaling equipment, wiring harnesses) and parts and accessories (e.g., gearboxes, wheels, brake systems and components). See app. F under "Automotive" for HS subheadings that apply to the motor vehicle parts industry discussed in this chapter.

[538] According to one group, the most useful ASEAN initiative has been the establishment of the ASEAN Free Trade Area (AFTA), which has particularly helped the auto industry. Association official, interview by USITC staff, Singapore, March 4, 2010. See ch. 2 for more information on AFTA.

[539] The AICO system was not the first effort by the ASEAN members to encourage industrial integration. The Brand-to-Brand Complementation scheme for the automotive industry was a precursor to AICO, with the objective being industrial complementation through improved economies of scale for all suppliers and increased intra-ASEAN trade.

[540] Industry official, interview by USITC staff, Bangkok, March 16, 2010.

[541] The China-ASEAN FTA was frequently cited by industry representatives as a motivating factor for cooperation within the region. China is one of the world's leading motor vehicle manufacturers, and its parts industry is increasingly sophisticated. Paired with its low manufacturing costs, China is seen by some as a major threat to the regional auto parts industry, although others see China as a major export opportunity. However, China and the four leading ASEAN auto-producing countries placed their automotive industries on the "sensitive" list within the FTA, providing continued protection of their respective industries for an extended yet undefined timeframe. Chrysler, "New ASEAN-China Trade Pact," February 11, 2010.

[542] Just-auto.com, "ASEAN Automotive Market Review," March 2007, 2.

[543] ECORYS, "Trade Sustainability Impact Assessment," April 3, 2009, 125.

[544] JAMA, "ASEAN and the Auto Supporting Industry – Part 2 of 2," March 2010.

[545] Although production data for motor vehicle parts are not available for the region, these four countries accounted for over 95 percent of regional auto production of nearly 2.0 million units in 2009. OICA, "World Motor Vehicle Production."

[546] OEM parts makers are commonly referenced by their position in the industry "tier." Tier One producers are generally large multinationals that supply components, systems, and modules directly to automakers. In addition to manufacturing, these firms may undertake supply chain management, inventory control, systems integration, foreign investment, and extensive design and R&D. Tier Two and Tier Three suppliers, which number in the tens of thousands, are generally smaller in size and product/function scope and are often less likely to have the financial resources and customer base to support significant foreign investment. Tier Two suppliers generally provide parts and materials for finished components/assemblies to Tier One producers, whereas Tier Three suppliers often provide raw materials or parts to a wide variety of industries, including the motor vehicle sector.

[547] WTO, Thailand Trade Policy Review (Revision), February 6, 2008, 115.

[548] WTO, Malaysia Trade Policy Review (Revision), March 9, 2006, 89.

[549] Industry official, interview by USITC staff, Kuala Lumpur, March 11, 2010.

[550] USAID/SENADA, "Automotive Component Value Chain Overview," August 2007, 8–9.

[551] Pratiwi, "Automotive Industry," n.d.

[552] Aldaba, "Assessing the Competitiveness of the Philippine Auto Parts Industry," November 2007, 18.

[553] Findlay, "An Investigation into the Measures," April 2007, 3.

[554] Lau, "Distinguishing Fiction from Reality," 2006, 461.

[555] Industry officials, interviews by USITC staff, Bangkok, March 15, 2010 and Kuala Lumpur, March 12, 2010.

[556] Industry official, interview by USITC staff, Singapore, March 9, 2010.

[557] USAID/SENADA, "Automotive Component Value Chain Overview," August 2007, 18.

[558] Auto Industry, "Quality Cost Delivery," Auto Industry [UK] Web site (accessed June 1, 2010).

[559] The ISO/TS 16949:2009 standard, for example, is specific to the automotive industry and notes the particular requirements for the application of ISO 9001:2008 for automotive production and relevant service part organizations. ISO, "New Edition of ISO/TS 16949 Quality Specification for Automotive Industry Supply Chain," July 2, 2009.

[560] Ford, for example, requires a commitment from its suppliers to bring leading-edge technological innovations. Ford Motor Company, "Ford, Key Suppliers Roll Out Innovative Business Model," n.d.

[561] Industry official, interview by USITC staff, Bangkok, March 15, 2010.

[562] Industry official, interview by USITC staff, Bangkok, March 16, 2010.

[563] Industry official, interview by USITC staff, Kuala Lumpur, March 11, 2010.

[564] Ibid.

[565] *The Nation*, "Major firms see ASEAN as second investment base," October 27, 2009.

[566] Industry official, interview by USITC staff, Singapore, March 9, 2010.

[567] U.S. Chamber of Commerce, written submission to the USITC, March 12, 2010, 2.

[568] Industry official, interview by USITC staff, Singapore, March 9, 2010.

[569] Ibid.

[570] Association official, interview by USITC staff, Singapore, March 4, 2010; Borhan, "Indonesia to Boost Automotive Industry," International Enterprise Singapore, April 16, 2010.
[571] Industry official, interview by USITC staff, Kuala Lumpur, March 12, 2010.
[572] The UNECE 52 standards come out of the World Forum for Harmonization of Vehicle Regulations (Working Party 29). WP.29 is a permanent working party of the United Nations that functions as a global forum allowing open discussions on motor vehicle regulations among UN member countries and any regional economic integration organizations set up by country members of the United Nations. Malaysia and Thailand have already acceded to the 1958 agreement. World Forum for Harmonization of Vehicle Regulations (WP.29) Web site, http://www.unece.org/trans/main/welcwp29.htm?expandable=99.
[573] U.S. Chamber of Commerce, written submission to the USITC, March 12, 2010, 12.
[574] Mutual recognition agreements would allow companies to test their products once at a designated testing center, after which the products would be certified as meeting or exceeding the mutually agreed upon standards and would not require in-country testing. Malaysia is reportedly drafting the MRA as the first step in technical harmonization. Government official, interview by USITC staff, Jakarta, March 1, 2010.
[575] One automaker feels that the ASEAN market is weak and not yet ready for self-certification, a process that allows a firm to self-certify that its products meet all necessary standards and requirements stipulated by the foreign market. Industry official, interview by USITC staff, Bangkok, March 15, 2010.
[576] Industry official, interview by USITC staff, Singapore, March 9, 2010.
[577] Industry officials, interviews by USITC staff, Singapore, March 9, 2010, and Bangkok, March 15, 2010.
[578] Industry official, interview by USITC staff, Bangkok, March 15, 2010.
[579] Industry official, interview by USITC staff, Singapore, March 9, 2010.
[580] Government official, interview by USITC staff, Jakarta, March 1, 2010.
[581] Industry official, interview by USITC staff, Kuala Lumpur, March 12, 2010.
[582] Industry official, interview by USITC staff, Bangkok, March 17, 2010.
[583] Austria, *The Pattern of Intra-ASEAN Trade*, August 2004, 9.
[584] Industry official, interview by USITC staff, Singapore, March 9, 2010.
[585] Industry official, interview by USITC staff, Bangkok, March 15, 2010.
[586] USITC hearing transcript, February 3, 2010, 27.
[587] WTO, Malaysia Trade Policy Review, December 14, 2009, 55.
[588] Industry official, interview by USITC staff, Bangkok, March 15, 2010.
[589] MITI, Review of National Automotive Policy, October 28, 2009.
[590] Industry official, interview by USITC staff, Kuala Lumpur, March 12, 2010.
[591] Manufacturers of transmission systems, brake systems, airbag systems, and steering systems are eligible for better fiscal incentives (e.g., Pioneer Status of 100 percent fiscal deduction for 10 years or Investment Tax Allowance of 100 percent for five years). MITI, Review of National Automotive Policy, October 28, 2009.
[592] Pioneer Status of 100 percent for 10 years or Investment Tax Allowance of 100 percent for 5 years for manufacture of selected critical components supporting hybrid and electric vehicles (e.g., electric motors, electric batteries, battery management systems, inverters, electric air conditioning, and air compressors). MITI, Review of National Automotive Policy, October 28, 2009.
[593] Malaysia's indigenous population, which benefit from many preferential government programs.
[594] Industry official, interview by USITC staff, Singapore, March 9, 2010.
[595] Industry official, interview by USITC staff, Kuala Lumpur, March 11, 2010.
[596] Industry official, interview by USITC staff, Bangkok, March 15, 2010.
[597] The excise tax rate ranges between 65 percent to 125 percent, depending on the vehicle type. USTR, *National Trade Estimate Report*, 2008, 381.
[598] Industry official, interview by USITC staff, Bangkok, March 15, 2010
[599] ECORYS, "Trade Sustainability Impact Assessment," April 3, 2009, 125. The excise tax on pickups is 3 percent, whereas those for passenger cars with engines over 2,000 cc range from 30 percent to 50 percent. Thailand Board of Investment Web site, http://www.boi.go.th/english/how/tax_rates_and_double_taxation (accessed April 16, 2010).
[600] ECORYS, "Trade Sustainability Impact Assessment," April 3, 2009, 122.
[601] Nag, Banerjee, and Chatterjee, "Changing Features of the Automobile Industry in Asia," n.d., 1.
[602] USAID/ADVANCE, *Evaluation of Proposed Target Sectors,* June 2008, 26 and 30.
[603] The CEPT scheme is discussed in more detail in chapter 2 of this chapter.
[604] ECORYS, "Trade Sustainability Impact Assessment," April 3, 2009, 125; USITC hearing transcript, February 3, 2010, 45.
[605] Industry official, interview by USITC staff, Bangkok, March 15, 2010.
[606] Ibid.
[607] U.S. Chamber of Commerce, written submission to the USITC, March 12, 2010, 12.
[608] AutomotiveWorld.com, "Malaysia: Govt wants more parts companies to be Tier 1," April 22, 2010.
[609] Embassy of the United States, Jakarta, "Indonesia's Automotive Industry," February 12, 2009.

[610] The Roadmap identifies staggered timelines for compliance with the measure to eliminate investment restrictions for sensitive industry sectors: 2010 for the ASEAN-6; 2013 for Vietnam; and 2015 for the remaining ASEAN countries.
[611] U.S. Chamber of Commerce, written submission to the USITC, March 12, 2010, 4.
[612] Industry official, interview by USITC staff, Singapore, March 9, 2010.
[613] *ASIAtalk*, "Thailand: Auto Parts," September 2008.
[614] *The Nation*, "Major firms see ASEAN as second investment base," October 27, 2009.
[615] Data collected by USITC staff from FDIMarkets database (accessed January 10, 2010).
[616] Japan's net outflow of investment focused on Thailand and Indonesia, which accounted for 63 percent and 21 percent, respectively, of the total. Malaysia, the Philippines, and Vietnam were secondary FDI recipients. Ministry of Finance, "Japan's FDI Flows."
[617] Data collected by USITC staff from FDIMarkets database (accessed January 10, 2010).
[618] Zephyr is a database of mergers and acquisitions, initial public offerings, and venture capital deals. Zephyr database (accessed January 10, 2010).
[619] *ASIAtalk*, "Thai Auto-Electronics Spark Investment," August 2008.
[620] Data collected by USITC staff from FDIMarkets database (accessed January 10, 2010).
[621] *The Japan Automotive Digest*, "Malaysia Opens Auto Industry," November 2, 2009, 4.
[622] Industry official, interview by USITC staff, Kuala Lumpur, March 11, 2010.
[623] CIA, *The World Factbook: Indonesia*, April 21, 2010.
[624] The vehicle ownership rate in Indonesia is estimated at 2.8 units per 100 people, compared to 4 units per 100 people in Thailand and 28 units per 100 people in Malaysia. EIU, *Indonesia: Automotive Report*, September 21, 2009.
[625] Invest in China database (accessed January 20, 2010).
[626] Chrysler, "New ASEAN-China Trade Pact," February 11, 2010.
[627] Industry official, interview by USITC staff, Singapore, March 9, 2010.
[628] Industry official, interview by USITC staff, Bangkok, March 15, 2010.
[629] Ibid., March 16, 2010.
[630] Ibid., March 15, 2010.
[631] Industry consultant, interview by USITC staff, Bangkok, March 17, 2010, The American Chamber of Commerce in Thailand, "Customs Meeting," April 24, 2009.
[632] Industry official, interview by USITC staff, Bangkok, March 17, 2010.
[633] Industry official, interview by USITC staff, Kuala Lumpur, March 12, 2010.
[634] Industry official, interview by USITC staff, Bangkok, March 17, 2010.
[635] Industry official, interview by USITC staff, Singapore, March 9, 2010.
[636] Industry official, interview by USITC staff, Kuala Lumpur, March 12, 2010.
[637] Industry consultant, interview by USITC staff, Bangkok, March 17, 2010.
[638] Industry official, interview by USITC staff, Hanoi, March 8, 2010.
[639] Palm oil covers crude and refined palm oil but not palm oil that has been chemically modified, such as by hydrogenation. See app. F under "Agro-based products" for HS subheadings that apply to the palm oil industry discussed in this chapter.
[640] Roadmap for Integration of Agro-based Products Sector, ASEAN, November 29, 2004.
[641] A number of important ASEAN agricultural products industries were not included in the Roadmap, as reflected in the 90 percent of ASEAN agricultural trade not covered during 2004–08. Trade in animal products such as beef and poultry and trade in rice are some notable omissions. It is likely that these industries were omitted for food security reasons, especially because rice in particular is a staple of the diet for many ASEAN countries and has historical importance for a number of countries' attempts to feed their people within their own borders. Furthermore, trade in rice was subject to restrictions or outright bans during the height of the commodity price increases that took place worldwide during 2007–08, whereas palm oil is by comparison freely traded despite global price fluctuations, although ASEAN countries also place a premium on maintaining targeted domestic supplies of palm oil. These actions suggest that the beef, poultry, and rice industries in ASEAN countries may be hindered in any efforts to expand beyond their national borders, even if an ASEAN focus on these industries had been present.
[642] The Codex Alimentarius Commission was created in 1963 by the Food and Agricultural Organization of the United Nations and the World Health Organization to develop international food standards, guidelines, and codes of practice.
[643] Smallholders account for less than one-third of the planted area in Indonesia and Malaysia. Biodiversity and Agricultural Commodities Program (BACP), "Market Transformation Strategy for Palm Oil," May 1, 2008, 15-16.
[644] American Palm Oil Council (APOC), "Palm Oil: Gift to Nature, Gift to Life," n.d.
[645] Basiron, "Palm Oil Production through Sustainable Plantations," 2007, 290.
[646] Government official, telephone interview by USITC staff, April 16, 2010; Southern Group, "Manufacturing: Palm Oil Mill"; Agriprods.com, "Anglo Eastern Plantations PLC."

[647] USITC, hearing transcript, February 3, 2010, 35–36 (testimony of Rosidah Radzian, Embassy of Malaysia).

[648] A small portion of southern Vietnam is located between 9 and 10 degrees north latitude; Indonesian sources indicate that the Vietnamese are trying to grow oil palm commercially despite the less-than-favorable conditions. Palm production in the Philippines is also reportedly hindered by soil conditions. Industry representatives, interview by USITC staff, Jakarta, Indonesia, March 2, 2010.

[649] USITC, hearing transcript, February 3, 2010, 42 (testimony of Rosidah Radzian, Embassy of Malaysia).

[650] One industry representative estimated that 80 percent of oil palm plantation land in Indonesia and Malaysia is owned by large companies. Industry representative, interview by USITC staff, Bangkok, Thailand, March 17, 2010.

[651] Industry representatives, interview by USITC staff, Kuala Lumpur, Malaysia, March 12, 2010.

[652] Basiron, "Palm Oil Production through Sustainable Plantations," 290.

[653] One analysis led to the conclusion that it is unlikely that any other palm oil–producing country will challenge the supplier roles of Indonesia and Malaysia. BACP, "Market Transformation Strategy for Palm Oil," 16–17.

[654] USITC, hearing transcript, February 3, 2010, 6–7 (testimony of Rosidah Radzian, Embassy of Malaysia); Basiron, "Palm Oil Production through Sustainable Plantations," 290; Law, "Can Palm Oil Help Indonesia's Poor?"

[655] Industry representatives, interview by USITC staff, Jakarta, Indonesia, March 2, 2010; industry representative, interview by USITC staff, Singapore, March 5, 2010.

[656] For example, in the United States, soybean oil and canola oil dominate the U.S. market because soybeans are grown throughout the United States, providing nearby, ready markets, and Canada produces large amounts of rapeseed, from which canola oil is produced, and has enlarged its share of the U.S. market by capitalizing on canola oil's health benefits and ready supply.

[657] See industry representatives, interview by USITC staff, Kuala Lumpur, Malaysia, March 12, 2010.

[658] WITS, Integrated Data Warehouse (accessed January 5, 2010).

[659] Than, "Mapping the Contours of Human Security Challenges in Myanmar," 172–202; Oil Crops in Myanmar Project, "Oil Crops in Myanmar"; USDA, FAS, Production, Supply and Distribution database (accessed May 4, 2010).

[660] Export competitiveness factors include growing demand and increasing productivity. For a more detailed discussion of export competitiveness, see chap. 1.

[661] Industry representatives, interview by USITC staff, Kuala Lumpur, Malaysia, March 12, 2010.

[662] See Sime Darby Plantation, "South East Asia," 2008.

[663] For a similar examination of the importance of non-ASEAN markets, see Sally, "Regional Economic Integration in Asia," 2010.

[664] APOC, "Palm Oil: Gift to Nature, Gift to Life," n.d.; industry representatives, interview by USITC staff, Jakarta, Indonesia, March 2, 2010.

[665] Industry representative, interview by USITC staff, Singapore, March 5, 2010.

[666] Industry representatives, interview by USITC staff, Jakarta, Indonesia, March 2, 2010.

[667] World Bank, "Indonesia Economic Quarterly: Building Momentum," March 2010, 15.

[668] Basiron, "Palm Oil Production through Sustainable Plantations," 2007, 290; Valentine, "Reframing Global Warming," n.d., 21; USITC, hearing transcript, February 3, 2010, 37–38 (testimony of Rosidah Radzian, Embassy of Malaysia); industry representatives, interview by USITC staff, Jakarta, Indonesia, March 2, 2010.

[669] USTR, "2010 National Trade Estimate Report on Foreign Trade Barriers," 2010, 247.

[670] Thoenes, "Biofuels and Commodity Markets–Palm Oil Focus," 2006, 2.

[671] As one example, Vietnam consumed 485,000 MT of palm oil in MY2009/10, but had no domestic production. See table 8.1 and box 8.1.

[672] USITC, hearing transcript, February 3, 2010, 34 (testimony of Rosidah Radzian, Embassy of Malaysia).

[673] Sime Darby Plantation, "South East Asia," 2008.

[674] Ibid.

[675] See Chandran, "Commodity Sector Overview," March 2010, 161.

[676] Indonesian domestic palm oil consumption for cooking was relatively steady in the last few years and represented about 20 percent of the total supply.

[677] To encourage palm oil consumer price stability, the Indonesian government launched the Minyak Kita program (literally translated as "Our Oil") in March 2009. Large Indonesian crude palm oil producers can provide cheaper oil as part of their initiative to help low-income households. Bromokusumo, "Indonesia: Oilseeds and Products Annual," May 5, 2009, 3; Bromokusumo and Slette, "Indonesia: Oilseeds and Products Annual," March 18, 2010, 12. It is unclear whether this program is an Indonesian government mandate or a government initiative, although the government and the industry maintain a close relationship because of palm oil's importance in the Indonesian agricultural sector and to the economy overall.

[678] See Chandran, "Commodity Sector Overview," March 2010, 161.

[679] USITC, hearing transcript, February 3, 2010, 13, 15 (testimony of Mohd Salleh Kassim, American Palm Oil Council); Sime Darby Plantation, "Overview," 2008; IOI Group, "Agronomy Support," 2009. See Basiron, "Palm Oil Production through Sustainable Plantations," 2007, 290.

680 USITC, hearing transcript, February 3, 2010, 36, 52 (testimony of Rosidah Radzian, Embassy of Malaysia).

681 Industry representatives, interview by USITC staff, Kuala Lumpur, Malaysia, March 12, 2010.

682 Ibid.

683 In May 2010, six Malaysian and Indonesian palm oil producer associations reportedly created the Indonesia-Malaysia Palm Oil Group to address concerns over the sustainability of palm oil production. International Centre for Trade and Sustainable Development, "Malaysia Sees Possible WTO Case against EU Palm Oil Limits," May 19, 2010.

684 Roundtable on Sustainable Palm Oil (RSPO), "Who Is RSPO?" 2009; Malaysian Palm Oil Promotion Council, "Business With Palm Oil," n.d., 15.

685 RSPO homepage, http://rspo.org/ (accessed April 26, 2010).

686 A Thai palm oil industry representative confirmed this impression for Thai palm oil production as well. Industry representative, interview by USITC staff, Bangkok, Thailand, March 17, 2010. However, some Indonesian producers reportedly believe that satisfying the national requirements of their purchasers is more important than RSPO certification. Industry representatives, interview by USITC staff, Jakarta, Indonesia, March 2, 2010.

687 USITC, hearing transcript, February 3, 2010, 36–38, 51–52 (testimony of Rosidah Radzian, Embassy of Malaysia); industry representatives, interview by USITC staff, Kuala Lumpur, Malaysia, March 12, 2010.

688 RSPO, "RSPO Certified Mills & CSPO Production," 2009. Estimates of RSPO-certified palm oil amounts in 2009 range from 1.4 million to 2 million tons. Industry representative, interview by USITC staff, Singapore, March 5, 2010; USITC, hearing transcript, February 3, 2010, 18 (testimony of Mohd Salleh Kassim, American Palm Oil Council). A separate account estimated that 3 percent of global palm oil production is "sustainable." BBC News, "Orangutan Survival and the Shopping Trolley," February 22, 2010. In 2010, Malaysia is targeting half of its production (9 million tons) to be RSPO certified. USITC, Hearing transcript, February 3, 2010, 13 (testimony of Mohd Salleh Kassim, American Palm Oil Council).

689 A Thai palm oil industry representative estimated this premium only to be 5 percent currently. Industry representative, interview by USITC staff, Bangkok, Thailand, March 17, 2010. Palm oil consumers, however, indicated at an RSPO meeting in November 2006 that they do not anticipate sustainable palm oil to cost significantly more or any more than nonsustainable palm oil. BACP, "Market Transformation Strategy for Palm Oil," May 1, 2008, 19.

690 Cadbury announced in August 2009 that it would stop using palm oil in its chocolate bars because of consumer environmental concerns. Chan, "World Bank Arm Halts Financing to Palm Oil Industry," September 10, 2009. Cargill is reportedly reconsidering its supply arrangements with an Indonesian company over environmental concerns. Creagh, "Cargill Considers Dropping Sinar Mas Over Land Clearing Claims," March 25, 2010.

691 Notably, both Indonesian producers, one of which is the world's second-largest palm oil producer, were members of the RSPO, a fact that has contributed to allegations that RSPO standards and procedures for monitoring and certification have not been effective. Mongabay.com, "BBC Documentary Leads Unilever to Blacklist Indonesian Palm Oil Company," Febriaru 24. 2010.

692 Hickman, "Online Protest Drives Nestlé to Environmentally Friendly Palm Oil," May 19, 2010.

693 The reportedly largest food processor, manufacturer, and trader in China is the only RSPO member there. Similarly, India has four or five RSPO members, and Pakistan has none, leading some Indonesian industry representatives to believe that there is no demand for RSPO palm oil in these countries. Industry representatives, interview by USITC staff, Jakarta, Indonesia, March 2, 2010; BACP, "Market Transformation Strategy for Palm Oil," May 1, 2008, 13.

694 Industry representative, interview by USITC staff, Singapore, March 5, 2010.

695 Industry representative, interview by USITC staff, Washington, D.C., February, 18, 2010.

696 USITC, hearing transcript, February 3, 2010, 36–38, 51–52 (testimony of Rosidah Radzian, Embassy of Malaysia); industry representatives, interview by USITC staff, Kuala Lumpur, Malaysia, March 12, 2010.

697 Bromokusumo and Slette, "Indonesia: Oilseeds and Products Annual," March 18, 2010, 14.

698 Industry representatives, interview by USITC staff, Jakarta, Indonesia, March 2, 2010. Most Malaysian palm oil investment overseas is not directed at plantations, but rather at processing and manufacturing facilities. Industry representatives, interview by USITC staff, Kuala Lumpur, Malaysia, March 12, 2010.

There are examples of investments outside the dominant Malaysia-to-Indonesia pattern. A Thai palm oil industry representative indicated that Malaysian business interests currently are considering investments in the Thai palm oil industry. Industry representative, interview by USITC staff, Bangkok, Thailand, March 17, 2010. Some Indonesian industry representatives stated that Malaysian investment is flowing into African and South American palm oil ventures instead of additional Indonesian projects. These representatives also indicated that some Indonesian producers are exploring the possibility of investing abroad. Industry representatives, interview by USITC staff, Jakarta, Indonesia, March 2, 2010.

699 USITC, hearing transcript, February 3, 2010, 37–38 (testimony of Rosidah Radzian, Embassy of Malaysia); industry representatives, interview by USITC staff, Jakarta, Indonesia, March 2, 2010; Sime Darby Plantation, "South East Asia," 2008; industry representatives, interview by USITC staff, Kuala Lumpur, Malaysia, March 12, 2010.

[700] Industry representatives, interview by USITC staff, Kuala Lumpur, Malaysia, March 12, 2010.

[701] USITC, hearing transcript, February 3, 2010, 13, 34–35 (testimony of Rosidah Radzian, Embassy of Malaysia); Malaysian Palm Oil Council, "Oil Palm: Tree of Life," 2006, 10; industry representatives, interview by USITC staff, Kuala Lumpur, Malaysia, March 12, 2010; BACP, "Market Transformation Strategy for Palm Oil," May 1, 2008, 13. The cross-border investment figures in the table above do not reflect Indonesian domestic investment on palm oil projects, which would be expected to be much greater in monetary terms than its investment throughout ASEAN.

Investment facilitation in certain ASEAN countries' palm oil industries can occur at a very local level, as evidenced by 2009 requests made to the Small & Medium Enterprises Development Cluster of the East ASEAN Growth Area (EAGA), a subregional economic cooperation initiative, by a Philippines economic development council. Brunei Indonesia Malaysia Philippines East ASEAN Growth Area (BIMP-EAGA), "Report of the 9th Small & Medium Enterprises Development (SMED) Cluster Meeting," 2009, 7. See BIMP-EAGA, http://www.bimpbc.org.

[702] Industry representatives, interview by USITC staff, Jakarta, Indonesia, March 2, 2010; industry representative, interview by USITC staff, Singapore, March 5, 2010.

[703] Industry representative, interview by USITC staff, Washington, D.C., February 18, 2010.

[704] One industry representative indicated that the Malaysian government recently opened a new sector for foreign investment but capped foreign ownership at 49 percent. Industry representative, interview by USITC staff, Singapore, March 5, 2010.

[705] Industry representative, interview by USITC staff, Washington, D.C., February 18, 2010; USITC, hearing transcript, February 3, 2010, 47 (testimony of Murray Hiebert, U.S. Chamber of Commerce).

[706] According to some Malaysian industry representatives, all trading is done electronically. Industry representatives, interview by USITC staff, Kuala Lumpur, Malaysia, March 12, 2010.

[707] *The Star Online*, "El Nino May Lift CPO Futures Prices," March 10, 2010.

[708] One industry representative in Thailand offered the observation that transportation of crude palm oil between the mills and the refineries, for which tanker trucks are used, is not an issue. Industry representative, interview by USITC staff, Bangkok, Thailand, March 17, 2010. Trade facilitation and logistics services are discussed in chap. 2 as they relate to the ASEAN economy as a whole.

[709] This appendix reflects only the principal points made by the particular party. The views summarized are those of the submitting parties and not the Commission. Commission staff did not undertake to confirm the accuracy of, or otherwise correct, the information described. For the full text of hearing testimony and written submissions, see entries associated with investigation no. 332-511 at the Commission's Electronic Docket Information System (https://edis.usitc.gov/edis3-internal/app).

[710] Thomas G. Aquino, Philippine Senior Undersecretary of Trade and Industry on behalf of the Republic of the Philippines, written submission to the USITC, March 3, 2010.

[711] USITC, hearing transcript, February 3, 2010, 13–20 (testimony of Mohd Salleh Kassim, APOC); APOC Web site, http://www.americanpalmoil.com/ (accessed March 29, 2010).

[712] USITC, hearing transcript, February 3, 2010, 4–13 (testimony of Rosidah Radzian, MPOB); MPOB Web site, http://www.mpob.gov.my/html/about/pdf/cb_complete.pdf (accessed March 29, 2010).

[713] NMPF and USDEC, written submission to the USITC, March 10, 2010.

[714] US-ASEAN Business Council, written submission to the USITC, February 4, 2010.

[715] Unless otherwise specified, the source for this position of interested party is U.S. Chamber of Commerce, written submission to the USITC, March 10, 2010.

[716] USITC hearing transcript, February 3, 2010, 26.

[717] Ibid., 27

CHAPTER SOURCES

The following chapters have been previously published:

Chapter 1 – This is an edited, excerpted and augmented edition of a United States Congressional Research Service publication, Report Order Code R40933, dated November 16, 2009.

Chapter 2 – This is an edited, excerpted and augmented edition of a United States Congressional Research Service publication, Report Order Code R40583, dated October 26, 2009.

Chapter 3 – This is an edited, excerpted and augmented edition of a United States International Trade Comission publication 4176, dated August 2010.

INDEX

A

B

C

D

E

F

G

H

I

J

K

L

M

N

O

P

Q

R

S

T

U

V

W

Y